——Principles of——
Three-Dimensional Imaging
in Confocal Microscopes

T0332955

Min Gu

Dept. of Applied Physics
Victoria University of Technology
Victoria, Australia

World Scientific
Singapore • New Jersey • London • Hong Kong

Published by

World Scientific Publishing Co Pte Ltd

P O Box 128, Farrer Road, Singapore 912805

USA office: Suite 1B, 1060 Main Street, River Edge, NJ 07661

UK office: 57 Shelton Street, Covent Garden, London WC2H 9HE

Library of Congress Cataloging-in-Publication Data
Gu, Min.
    Principles of three dimensional imaging in confocal microscopes / Min Gu.
       p.   cm.
    Includes bibliographical references and index.
    ISBN 9810225504
    1. Confocal microscopy.   2. Three-dimensional display systems.
    I. Title.
    QH224.G8   1996
    681'.413--dc20   95-48425
                                                        CIP

British Library Cataloguing-in-Publication Data
A catalogue record for this book is available from the British Library.

This book is printed on acid-free paper.

Printed in Singapore by Uto-Print

To my wife, Yunshan Hu, and my son, Henry.

Contents

Contents

PREFACE

While confocal scanning microscopy has become a useful tool in various practical fields because of its ability of three-dimensional (3-D) imaging, the research work in one of the groups in the Department of Physical Optics within the School of Physics, the University of Sydney, Australia, has been focused on the understanding of its 3-D imaging properties and various practical effects such as aberration, size of pinhole, pupil functions of lenses, novel optical arrangements, and etc. on 3-D imaging performance. Of most importance, over the last five years, the concept of the 3-D transfer function has been developed to cope with these factors. According to 3-D transfer functions, resolving power in 3-D confocal imaging can be defined in a unified way, different optical arrangements can be compared with an insight into their inter-relationship, and images of thick objects can be modelled or calculated in terms of the Fourier transform, which makes the analysis easy. In addition, with the help of the 3-D transfer functions, the image quality can be improved in post digital processing. For example, an inverse electronic filter can be designed in terms of the transfer function for the system used in the imaging process, so that weakly-imaged spatial frequencies in the object may be enhanced.

Many of the results on the 3-D transfer functions have been published in various journals. While they are important, it may be difficult for readers to apply them to particular systems they work on. This is because the symbols and the ways used to deal with various problems may be different from case to case. It is therefore necessary to write a book which integrates the obtained results together in a systematic way. It is this motivation that has driven me to complete this book. Efforts have been made to give the inter-relation between different methods. Some of the sections in the book have been thought to third- and fourth-year undergraduates and graduates in the course of modern optics in the School of Physics, the University of Sydney. I have attempted to make the text self-contained and easy to understand.

The book is aimed for researchers already involved in or wishing to enter the field of confocal optical microscopy. Many figures in the book can be used for scientists to develop practical confocal microscopes or to understand the performance of their systems. For this reason, the book may be considered to be a handbook. The book should be also useful for scientists who are interested in general optical microscopic imaging: the method of the 3-D transfer function can be developed in the other fields.

I would like to thank many people who offered me various helps in the course of completing this book. Prof. Colin J. R. Sheppard was one of the first persons who introduced confocal microscopy into my research fields, while I worked in the University of Sydney. I am very grateful for his long-time encouragement, valuable collaborations, and numerous stimulating discussions. My particular thanks go to Prof. P. C. Cheng, State University of New York (USA), who encouraged me to complete this book. I am thankful to Drs. T. Wilson, Oxford University (UK), and A. Kriete, University of Giessen (Germany), for their helpful discussions on many occasions. Thanks are also given to my students, K. Brain, X. S. Gan, D. Jackson, T. Siu, T. Tannous, E. Yap, and H. Zhou, who made direct and indirect contributions to the subject. The excellent support from the administrative assistant, Ms. N. McIntoch, at the University of Sydney is gratefully acknowledged. I would like to acknowledge the important support from the Australian Research Council (ARC) in various aspects over the last five years. Since September of 1995, I have moved to Melbourne of Australia for an academic position in the Department of Applied Physics, Victoria University of Technology (VUT). Prof. D. Booth and staff at VUT are very supportive, so that I can finally complete this book.

Finally, I would like to deeply thank my lovely wife, Yunshan Hu, for her patience, understanding, encouragement and support. Without her, it would not be possible for me to complete the book.

Melbourne, 1995.

Min Gu

Chapter 1

INTRODUCTION

This introductory chapter is devoted to providing an overview of three-dimensional (3-D) imaging properties in confocal scanning microscopy. Section 1.1 is a summary of the main advantages of confocal microscopy and recent practical and theoretical developments in this area. In Section 1.2, the principles of 3-D confocal imaging are described in terms of the degree of coherence. The previous studies on 3-D conventional optical microscopy are briefly summarized in Section 1.3. In Section 1.4, the concept of the 3-D transfer function is introduced. Section 1.5 gives an overview of how the book is orginized to discuss the 3-D imaging principles in various confocal microscopes in terms of the 3-D transfer function.

1. 1 Confocal Microscopy

Optical microscopy is a century-old subject of viewing the microscopic world. The non-destructive nature of optical microscopy is important in diverse fields such as biological studies and material sciences. The main drawback of conventional optical microscopy is its limited resolution: the resolution is of an order of wavelength of the light used for illumination. Another limitation associated with conventional microscopy is its finite depth of focus, which makes it difficult to achieve an image of the object with depth structures. It is therefore impossible to obtain a true 3-D image of a thick object in practice.

The concept of confocal microscopy was first proposed by Minsky[1.1, 1.2] in 50's when he attempted to reduce the amount of the multiply scattered light from the sample under inspection in a conventional optical microscope. In this novel imaging system, a sample is illuminated by a small spot of light and the signal from the illuminated spot is collected by a small detector which is ideally a point detector. By moving the sample, a

1

map of information from the different parts in the sample can be recorded, which gives rise to the image of the sample. Unfortunately although subsequent breakthroughs in the limit of imaging resolution were reported by a number of researchers under different conditions in the 60's,[1.3-1.5] it did not attract much attention as expected. It was until the 70's that the first detailed investigation[1.6] of the imaging performance in confocal microscopy indicated that transverse resolution in confocal microscopy is 1.4 times as large as that in a conventional microscope. The improvement in resolution is obtained at the expense of the reduction of the field of view. This drawback can be easily compensated for by using the scanning mechanism. It was soon understood that confocal scanning microscopy has a strong depth discrimination property: the out-of-focus signal is detected much less strongly than the in-focus information whereas in a conventional system it is only blurred.[1.7] Another advantage is that the noise associated with flare and unwanted scattered light is significantly reduced because a small pinhole mask is employed in front of the detector.[1.8-1.10]

Among these advantages, the depth discrimination property, which is also called the optical sectioning property[1.8] in this book, is of particular importance because it is this feature that allows one to record an image of the particular transverse section in a

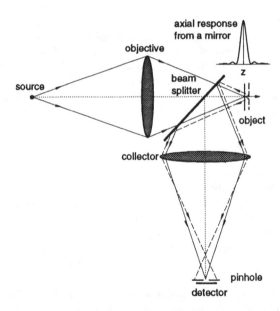

Fig. 1.1.1   Schematic diagram of a reflection-mode confocal scanning microscope. The strength of the optical sectioning property is demonstrated by considering a mirror scanned in the axial direction. The narrower the axial response the higher the axial resolution in 3-D imaging.

thick object and to form a 3-D image accordingly.

Fig. 1.1.1 demonstrates the principle of optical sectioning in a reflection-mode confocal scanning imaging system. Imagine that the object is a perfect reflector (e. g. a mirror) scanned in the axial direction of the objective lens. When it is located in the focal plane of the objective (solid lines in Fig. 1.1.1), the reflected beam is focused exactly onto a point detector which collects a large amount of the incident energy. However, if the reflector moves away from the focal plane (dotted lines in Fig. 1.1.1), the signal is focused at a position either before or behind the point detector, depending on the scanning direction. Thus, the detector collects only a small amount of the incident energy. Therefore the signal strength detected when the reflector is at the defocus position is weaker than that at the in-focus position, as shown in Fig. 1.1.1. The larger the defocus distance the weaker the signal strength. This property implies that the location of the reflector can be determined and the image of the reflector is efficiently recorded only when it is in the focal plane. For a real thick object, by recording a series of sections of the image at different depths of the object, a complete image of the thick object can be built up in a form of a stack of sections. In other words, a 3-D image of the thick object can be recorded without necessarily slicing the object. It is obvious that the broadness of the axial response shown in Fig. 1.1.1 determines axial resolution of 3-D imaging because the narrower the axial response the weaker the cross-talk between two adjacent sections of the image. It is this 3-D imaging ability that confocal scanning microscopy has been applied to a wide range of fields including biology, bio-medicine, industry inspection, metrology and so on.[1.8-1.14]

Confocal microscopy can be operated in different modes. Confocal bright-field microscopy can be achieved in transmission and reflection modes.[1.8] The 3-D imaging performance in these two systems are completely different although they behave identically in the case of two-dimensional (2-D) imaging. A confocal image of a biological samples labelled with fluorescent materials can also be recorded by measuring the fluorescence radiation from the sample.[1.8, 1.15] This method is termed confocal fluorescence microscopy. If the incident light is intense enough, two-photon fluorescence emission can be stimulated in the sample, so that confocal two-photon fluorescence microscopy was proposed[1.16] and has been recently demonstrated.[1.17]

Another recent developments in confocal microscopy is the introduction of optical fibres and fibre-optical components to replace the conventional small source and detector.[1.18-1.22] As a result, a fibre-optical confocal scanning microscope (FOCSM) can be built in a compact form and the system is less affected by vibrations caused by lasers and other electronic devices as they can be arranged remotely from the microscope. The most important feature of the FOCSM is that bright-field imaging is purely coherent,[1.20, 1.21] which allows one to perform quantitative microscopy.

Recent developments of ultrashort pulsed lasers provide an excellent opportunity for confocal microscopy to add the forth dimension, i. e., the time-dimension, into imaging[1.23, 1.24] This technique allows one to measure time-dependent 3-D images. Consequently, the information regarding the dynamics of the sample can be derived and imaging of an object hidden in highly-scattering media becomes possible. Confocal microscopy under ultrashort pulse illumination is currently an active research field.

In order to obtain further improvement in resolution, in particular, in the axial direction, 4Pi confocal microscopy has been proposed recently.[1.25, 1.26] In this configuration, a sample is illuminated by two beams in the opposite directions and the signal is collected from both sides. As a result, the numerical aperture of the microscope objective is effectively increased. Accordingly, axial resolution of confocal imaging can be improved appreciably.

The theory concerning 2-D confocal imaging has been well documented in a number of books and review articles.[1.8-1.10] Over the past five years, theoretical investigations have been focused on the understanding of 3-D imaging properties in confocal microscopy to promote the application of the technique. The concept of 3-D transfer functions has been introduced for a complete description of various 3-D confocal imaging modes.[1.27, 1.28] With the 3-D transfer functions, images of any arbitrary object can be derived. The 3-D transfer functions are thus applicable to describing not only 3-D imaging but also 2-D and one-dimensional (1-D) imaging.[1.27, 1.28]

1. 2 Coherence of Confocal Microscopy

Although all confocal microscopes commonly use a small mask in front of the detector, the imaging performance is quite different from each other. In particular, the degree of coherence in the imaging processes is dependent on samples, size of the detector, and optical components. As a result, a confocal microscope can be operated in coherent, or incoherent, or partially-coherent imaging modes, depending on these conditions.

For bright-field imaging, a confocal microscope is purely coherent if a point detector and a point source are used.[1.8] Introducing a finite-sized incoherent detector, which is usually a case in practice, results in a partially-coherent imaging system.[1.29-1.31] When a sample is labelled with fluorescent materials, the coherent nature of the illumination light is completely destroyed by the sample, so that confocal fluorescence microscopy is incoherent.[1.15] This coherence property is independent of the size of the detector.

Although the tip of an optical fibre has finite size, bright-field imaging in a FOCSM is purely coherent regardless of the size of the fibre spot.[1.20, 1.21] This nature contrasts with that for a confocal bright-field system consisting of a finite-sized detector and is ensured because a fibre is a waveguide which responds to both the amplitude and the phase of the light impinged on the fibre. However, imaging becomes incoherent again when one performs fluorescence imaging in a FOCSM.[1.32, 1.33] It is therefore impossible to operate a FOCSM in a partially-coherent mode.

The coherence nature becomes complicate in confocal bright-field imaging under ultrashort pulse illumination because the response time of the detector affects the degree of imaging coherence.[1.24] Imaging with a detector of a long response time is usually partially-coherent even when a point detector is used. The purely-coherent property can be achieved only when the response time of the detector becomes impulse. When a fluorescent sample is employed imaging becomes incoherent as expected.[1.34]

To describe 3-D image formation in confocal microscopes, one can use a 3-D point spread function (PSF) which is the image of an ideal point object. However this approach does not give a complete description of confocal imaging. As an example, in a confocal bright-field microscope with a finite-sized circular detector and circular lenses (Chapter 4), the corresponding 3-D PSF is given by[1.31]

$$I(v,u) = |h_1(v,u)|^2 \left[|h_2(v,u)|^2 \otimes_3 D(v) \right], \tag{1.2.1}$$

where $h_1(v, u)$ and $h_2(v, u)$ are the 3-D amplitude point spread functions for objective and collection lenses in object space (Fig. 1.1.1). $\otimes_3$ denotes the 3-D convolution operation in object space and $D(v)$ is the intensity sensitivity of the detector. v and u are the normalized optical coordinates of the scan point. It can be shown (Chapter 5) that Eq. (1.2.1) also holds for confocal single-photon fluorescence microscopy with a finite-sized circular detector. As a result, the description of the 3-D PSF does not give the difference of imaging properties between these two systems. An alternative description for optical imaging is the use of the concept of the transfer function.[1.27, 1.28, 1.31-1.38] With this approach, the imaging performance of an optical system is investigated in Fourier space, which is an analogy to the frequency-response analysis in electricity. In the former case, the Fourier space is usually a 3-D space while the latter is a 1-D space. Throughout this book, the transfer-function description is adopted to discuss various principles in confocal scanning microscopy.

1. 3 Previous Studies on Three-Dimensional Optical Microscopy

Before we introduce the concept of the transfer function, let us briefly review the previous studies on three-dimensional optical microscopy. Although the ability of 3-D imaging in a conventional microscope is rather poor because the image becomes blurred when the object is in a defocus plane, there have been a few people who have investigated the 3-D imaging performance in a conventional system. McCutchen[1.35] and Wolf[1.36] initiated the early studies of the 3-D imaging theory by considering the 3-D image formation in a coherent microscopic system. The concept of the 3-D aperture, which has been termed the 3-D coherent transfer function (CTF) later on, was introduced. Frieden [1.37] and Mertz[1.38] developed the 3-D imaging theory in an incoherent system in which the 3-D imaging performance was investigated by the 3-D optical transfer function

(OTF). Recently the effect of primary spherical aberration on the 3-D OTF for conventional incoherent imaging has also been studied.[1.39] For a partially-coherent conventional microscope, 3-D imaging was investigated by Streibl[1.40], and Sheppard and Mao.[1.41] Both the analyses were based on the first Born approximation and the 3-D transmission cross-coefficient (TCC) was introduced for the description of partially-coherent conventional imaging.

1. 4 Concept of Transfer Functions

Let us start to consider the concept of the transfer function in one dimension. An object can be represented by a transmission function $o(x)$, which can be resolved into a series of periodic components. For a periodic function $o(x)$, there are only a series of discrete sinusoidal components which are of appropriate amplitude and phase. In Fig. 1.4.1, it is shown how a square wave grating (i. e., an object), consisting of alternative bright and dark regions, is resolved into a series including a constant term, a first-order harmonic, a third-order harmonic, and so on. For this particular case, symmetry dictates that there is no second-order harmonic, or any even orders. The sum of the first three terms is also shown in Fig. 1.4.1. The more the terms in summation, the closer the result to the original object.

The spatial periods of the components are denoted by L which is shown for the first harmonic. The spatial frequency, m, of the component is defined as $1/L$. The higher the spatial frequency the finer the grating. An optical system usually behaves as a low-pass filter, that is it transmits the low spatial frequency information efficiently, but does not transmit high spatial frequencies representing very fine detail.

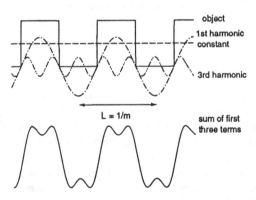

Fig. 1.4.1   A square wave object is resolved into sinusoidal components.

To understand this property further, let us consider the diffraction of a particular periodic component (i. e., a periodic grating) of spatial frequency m (Fig. 1.4.2). If a plane wave is normally incident on the grating, the diffracted light has three components: one is the transmitted beam in the direction of the incident beam and the other two beams propagate at angles of $\pm\theta$. The magnitude of the angle θ is determined by the spatial frequency of the grating m and the incident wavelength λ:

$$\theta = m\lambda. \tag{1.4.1}$$

It is clear that the larger the spatial frequency the larger the angle θ. An imaging lens has a maximum angle, α, of convergence of a ray. When θ increases up to α, the corresponding spatial frequency is

$$m_0 = \alpha/\lambda, \tag{1.4.2}$$

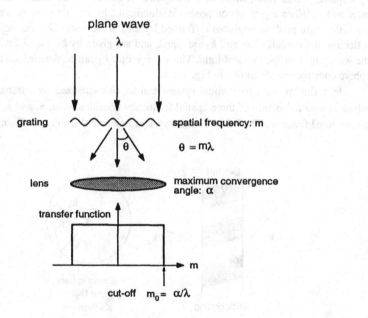

Fig. 1.4.2   A plane wave of wavelength λ is diffracted by a sinusoidal grating of spatial frequency m. The lens collects only the diffracted light of an angle θ less than the maximum convergence angle α.

giving rise to the maximum spatial frequency transmitted in the imaging process. Any spatial frequency higher than this maximum is then cut off. In other words, the fine structures of the object corresponding to higher spatial frequencies cannot be imaged efficiently.

The efficiency with which the different spatial frequencies are transmitted is called the transfer function. In Fig. 1.4.2, the transfer function is a constant when $m \leq m_0$. In an optical system, the spatial frequency at which the transfer function drops to zero is called the cut-off frequency. In general the bigger the range of spatial frequencies transmitted, the better the image. It is also required that the transfer function is a smoothly varying function, as otherwise ringing occurs, which is observed as fringing of the image.

The spatial frequency approach can be used also with non-periodic transmission functions, but in this case we need an integral rather than a sum over the periodic components: this is a Fourier integral, or a Fourier transform.

The above concept can be generalized to including variations in two or three dimensions. A thick object (Fig. 1.4.3) can be considered to consist of periodic grating components of appropriate amplitude and phase. Each component can be characterized by a spatial period $1/|m|$, where m is the spatial frequency vector with three components, m, n, and s. When a grating component is illuminated by a plane wave of wave vector k_1, the diffraction process produces diffracted light of wave vector k_2. For transmission or reflection the moduli of k_1 and k_2 are equal, and are given by $k_1 = k_2 = 2\pi/\lambda$, where λ is the wavelength of the incident light. The wave vector k_2 can be determined by the Ewald sphere construction[1.42] shown in Fig. 1.4.3.

In a similar way, the optical system can be characterized by a transfer function, which is now a function of three spatial frequency coordinates m, n, and s, i. e., a three-dimensional function. The 3-D CTF for a system in which the object is illuminated with a

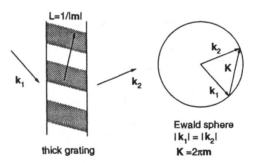

Fig. 1.4.3   A thick object consisting of a series of sinusoidal grating components can diffract the incident light, wave vector k_1, into light with wave vector k_2.

plane wave, wave vector k_1, is given by a cap of the Ewald sphere, as any optical system has a finite numerical aperture, so that only part of the scattered light within the aperture can be transmitted. For an illumination of non plane waves, one can resolve it into an angular spectrum of plane waves. As a result, a similar 3-D coherent transfer function can be calculated for general microscope systems.

The above discussion is based on a coherent imaging process. A practical imaging system is more complicate. One must distinguish between the OTF, which is applicable for incoherent systems including, for example, fluorescence microscopes, and the CTF, which is used for coherent systems such as a reflection-mode confocal microscope, and the TCC, which is applied to partially-coherent systems. For the OTF, the object transmission function represents the *intensity* variations in the object, whereas the CTF and the TCC are used with an *amplitude* transmission function.

The 3-D transfer function is a useful concept for describing image formation in confocal scanning microscopy. The description of imaging using 3-D transfer functions is a generalized method which describes completely imaging of *any* object, if scattering is weak enough that the first Born approximation holds. For example, one can derive from the 3-D transfer functions the corresponding 2-D transfer function for in-focus imaging and the 1-D transfer function for on-axis imaging. In addition, by investigating the 3-D transfer functions, one can judge the performance of the different confocal imaging systems, gain an understanding of the inter-relationship between different imaging processes, and obtain assistance in post image processing.

1. 5   Overview

The aim of this book is to provide a systematic introduction of the concept of the 3-D transfer functions in various confocal microscopes, to describe the methods for the derivation of different 3-D transfer functions, and to explain how to understand the principles of 3-D confocal imaging in terms of these functions. It is assumed that readers have gained the basic knowledge of Fourier optics,[1.43] Fourier transformation,[1.44] and confocal scanning microscopy.[1.45, 1.46]

When a beam of light is incident upon an object in an optical imaging system, the interaction of light with the object is a complicate process. Usually, multiple scattering of light is caused in the object. Throughout this book, it is assumed that the first Born approximation is valid, i. e., multiple scattering and depletion of the incident beam are neglected. The imaging theory for the case when the first Born approximation does not hold can be found from a few references.[1.42, 1.47]

The second assumption used throughout the book is the scaler approximation. This treatment is equivalent to ignoring the effect of polarization of light. In fact, the methods described in this book can be generalized to a vectorial form according to the early vectorial theory of light diffraction.[1.48-1.50] But the generalization is not within the scope of this book.

Another approximation used in most of the chapters is the paraxial approximation. This approach implies that the numerical aperture of the microscope objective is not large. In general, the results derived under the paraxial approximation hold for a lens of a numerical aperture less than $1/\sqrt{2}$.[1.10] This assumption is removed in Chapter 9, where the 3-D transfer functions are developed for high-aperture objectives.

The chapters of this book are orginized to minimize the need of cross-referencing. Chapter 2 gives the background materials and the mathematical tool used in understanding the subsequent chapters. Chapters 3-5 consider imaging performance in three basic confocal scanning microscopes which behave as coherent, partially-coherent, and incoherent systems, respectively. Chapters 6-9 discuss four new confocal imaging methods developed recently. Chapter 10 is a conclusion of the whole book. The following brief outline of each chapter provides an overview of the text.

Chapter 2 provides the necessary theoretical methods and the mathematical framework, which are useful for reading the remaining chapters. The properties of the 3-D diffraction of light by a single thin lens are first summarized. A 3-D space-invariant PSF for a single thin lens is then developed. As a result, the 3-D CTF and the 3-D OTF for coherent and incoherent imaging with a single lens are derived, respectively.

In Chapter 3, 3-D image formation in a confocal bright-field microscope with a point detector and a point source is described. Because imaging is purely coherent in this special case, the 3-D CTF is introduced to understand imaging performance. The influence of primary spherical aberration on the 3-D CTF and on imaging resolution is considered. The effect of the detector offset, which sometimes happens in practice when a detector is not perfectly aligned, is discussed in terms of the 3-D CTF. Another effect, i. e., the effect of the pupil function, on imaging is also discussed.

Chapter 4 considers a confocal bright-field microscope consisting of a finite-sized detector including circular and slit masks. Since such a system behaves as a partially-coherent microscope, it is necessary to introduce the 3-D TCC to describe imaging properties. As an application of the 3-D TCC, the effects of a finite-sized detector on axial and transverse resolution are carefully examined. When scattering of light in an object is weak, a special form of the TCC, termed the weak-object transfer function (WOTF), can be introduced for the description of imaging.

Chapter 5 is devoted to the discussion of 3-D confocal single-photon fluorescence imaging. Due to the incoherent imaging nature, the 3-D OTF is introduced. The effects of the detector size and of the pupil functions are in detail considered in terms of the 3-D OTF. Signal level as a function of the detector size and the optimization of axial resolution by the combination of annular pupils and a finite-sized detector are described. The effect of the finite-sized source is briefly discussed.

Chapter 6 examines the 3-D imaging property of a confocal two-photon fluorescence microscope by introducing the 3-D OTF. The influence of a finite-sized detector on the 3-D OTF and on the signal level is considered. A comparison of the 3-D imaging performance such as resolution in two-photon and single-photon confocal fluorescence microscopy is given.

Chapter 7 gives a comprehensive description of the imaging performance of a fibre-optical confocal scanning microscope (FOCSM). The 3-D CTF is introduced for bright-field imaging. Resolution in axial and transverse directions, and signal level are investigated in terms of the 3-D CTF. Fibre-optical confocal interferometry based on the purely-coherent nature in the FOCSM is also described. It is then shown that the 3-D image properties including resolution for a fluorescent object can be studied in terms of the 3-D OTF.

3-D imaging properties in a confocal microscope under ultrashort pulse illumination are, in Chapter 8, described for bright-field and fluorescence imaging. After the 3-D PSF for a single lens is generalized to including the effect of a pulsed beam, two bright-field imaging modes: time-resolved and time-averaged imaging, are first discussed using the 3-D CTF and the 3-D TCC, respectively. Two-photon and single-photon fluorescence imaging methods are then investigated.

In practice, a microscope objective has a high numerical aperture for high resolution. It is necessary to develop the 3-D transfer functions without the paraxial approximation. Chapter 9 is the description of the 3-D transfer functions for confocal microscopy with high-aperture objectives. One of the effects associated with the high-aperture objective, i. e., the effect of apodization, is incorporated into the transfer functions. Transfer functions for 4Pi confocal microscopy, which uses two objectives of high aperture for high axial resolution, are developed.

As a conclusion of the book, Chapter 10 points out the significance of the 3-D transfer functions in various cases and summarizes the main features of 3-D imaging in confocal microscopy.

References

1.1. M. Minsky, U. S. patent 3013467, *Microscopy Apparatus*, Dec. 19, 1961 (Filed Nov. 7, 1957).
1.2. M. Minsky, *Scanning*, **10** (1988) 128.
1.3. W. Lukosz and M. Marchand, *Optica Acta*, **10** (1963) 241.
1.4. M. D. Egger and M. Petran, *Science*, **157** (1967) 305.
1.5. M. Petran, M. Hadravsky, M. D. Egger, and R. Calambos, *J. Opt. Soc. Am.*, **58** (1968) 661.
1.6. C. J. R. Sheppard and A. Choudhury, *Optica Acta*, **24** (1977) 1051.
1.7. C. J. R. Sheppard and T. Wilson, *Opt. Lett*, **3** (1978) 115.
1.8. C. J. R. Sheppard, Scanning optical microscopy, in *Advances in Optical and Electron Microscopy*, Vol. 10, eds. R. Barer and V. E. Cosslett (Academic, London, 1987), p. 1-98.
1.9. T. Wilson, *Confocal Microscopy* (Academic, Loddon, 1990).

1.10.   T. Wilson and C. J. R. Sheppard, *Theory and Practice of Scanning Optical Microscopy* (Academic, London, 1984).

1.11.   J. B. Pawley, *Handbook of Biological Confocal Microscopy* (Plenum, New York, 1994).

1.12.   A. Kriete, *Visualization in Biomedical Microscopies* (VCH, Weinheim, 1992).

1.13.   J. K. Stevens, L. R. Mills, and J. E. Trogadis, *Three-Dimensional Confocal Microscopy* (Academic, London, 1994).

1.14.   P. C. Cheng, *Computer-Assisted Multidimensional Microscopies* (Springer, New York, 1993).

1.15.   A. F. Slomba, D. F. Wasserman, G. I. Gaufman, and J. F. Nester, *J. Assoc. Adv. Med. Instrum.*, **6** (1972) 230.

1.16.   C. J. R. Sheppard, R. Kampler, J. Gannaway, and D Walsh, *IEEE J. Quantum Electron.*, **QE-13** (1977) 100D.

1.17.   W. Denk, J. H. Strickler, and W. W. Webb, *Science*, **248** (1990)151.

1.18.   J. Benshchop and G. von Rosmalen, *Applied Optics*, **30** (1991) 1179.

1.19.   T. Dabbs and M. Glass, *Applied Optics*, **31** (1992) 705.

1.20.   M. Gu, C. J. R. Sheppard, and X. Gan, *J. Opt. Soc. Am. A*, **8** (1991) 1755.

1.21.   S. Kimura and T. Wilson, *Applied Optics*, **30** (1991)2143.

1.22.   K. Ghiggino, M. R. Harris and P. G. Spizzirri, *Rev. Sci. Instrum.*, **63** (1992) 2999.

1.23.   M. Kempe and W. Rudolph, *J. Opt. Soc. Am. A*, **10** (1993) 240.

1.24.   M. Gu and C. J. R. Sheppard, *J. Modern Optics*, **42** (1995) 747.

1.25.   S. Hell, *European Patent Application* 91121368.4 (1991/1992).

1.26.   S. Hell and E. H. K. Stelzer, *J. Opt. Soc. Am. A*, **9** (1992) 2159.

1.27.   C. J. R. Sheppard and M. Gu, *J. Microscopy*, **165** (1992) 377.

1.28.   C. J. R. Sheppard and M. Gu, 3-D transfer functions in confocal scanning microscopy, in *Visualization in Biomedical Microscopies*, ed. A. Kriete (VCH, Weinheim, 1992), p. 251-285.

1.29.   C. J. R. Sheppard and T. Wilson, *Optica Acta*, **25** (1978) 315.

1.30.   C. J. R. Sheppard and X. Q. Mao, *J. Modern Optics*, **35** (1988) 1169.

1.31.   M. Gu. and Sheppard, *J. Modern Optics*, **41** (1994), 1701.

1.32.   X. Gan, M. Gu, and C. J. R. Sheppard, *J. Modern Optics*, **39** (1992) 825.

1.33.   M. Gu and C. J. R. Sheppard, *J. Opt. Soc. Am. A*, **9** (1992) 1991.

1.34.   M. Gu, T. Tannous, and C. J. R. Sheppard, *Opt. Commun.*, **117** (1995) 406.

1.35.   C. W. McCuthen, *J. Opt. Soc. Am.*, **54** (1964) 240.

1.36.   E. Wolf, *Opt. Commun.*, **1** (1969) 153.

1.37.   B. R. Frieden, *J. Opt. Soc. Am.*, **57** (1967) 56.

1.38.   L. Mertz, *Transformation in Optics* (Wiley, New York, 1965).

1.39.   S. Wang and B. R. Frieden, *Applied Optics*, **18** (1990) 2424.

1.40.   N. Streibl, *J. Opt. Soc. Am. A*, **2** (1985) 121.

1.41.   C. J. R. Sheppard and X. Q. Mao, *J. Opt. Soc. Am. A*, **6** (1989) 1260.

1.42.   C. J. R. Sheppard, *Eur. J. Cell. Biol.*, **48**, suppl. 25 (1989) 29.

1.43. J. W. Goodman, *Introduction to Fourier Optics* (McGraw-Hill, New York, 1968).

1.44. B. R. Bracewell, *The Fourier Transform and Its Applications* (McGraw-Hill, New York, 1965).

1.45. S. G. Anderson, *Laser Focus World*, Feb., (1994) 83.

1.46. G. S. Kino, and T. J. Corle, *Physics Today*, Sept., (1989) 55.

1.47. C. J. R. Sheppard and J. T. Sheridan, *Proc SPIE*, **1139** (1989) 32.

1.48. M Born and E. Wolf, *Principles of Optics* (Pergamon, New York, 1980).

1.49. B. Richards and E. Wolf, *Proc. R. Soc. London*, **A253** (1959) 349.

1.50. B. Richards and E. Wolf, *Proc. R. Soc. London*, **A253** (1959) 358.

Chapter 2

THREE-DIMENSIONAL FOURIER OPTICS
OF A THIN LENS

A confocal scanning microscope usually includes two optical lenses, an objective lens and a collector lens, which are responsible for light illumination and signal collection, respectively (see Chapter 3). For an understanding of the three-dimensional (3-D) imaging principles in confocal scanning microscopy, it is necessary to study the 3-D imaging properties of a single lens. The objective of this chapter is to describe 3-D Fourier optics of a single thin lens which is a basic and key element in a confocal microscope. We start with, in Section 2.1, the derivation of the transmittance of a single thin lens and its 3-D diffraction properties. The 3-D coherent imaging properties of a thin lens are described in Section 2.2. In Section 2.3, a 3-D space-invariant form of the point spread function (PSF) for a thin lens is derived, which is of importance in understanding the 3-D imaging principles in confocal microscopy. Based on this 3-D PSF, we develop the 3-D coherent transfer function (CTF) for 3-D coherent imaging of a single lens in Sections 2.4. In Section 2.5, the 3-D incoherent imaging features of a single lens are analysed in terms of the 3-D optical transfer function (OTF). 3-D image properties associated with a conventional microscope are briefly discussed in Section 2.6.

2. 1 Diffraction of a Thin Lens

In this section, the transmittance of a single thin lens is derived and the 3-D diffraction properties of the lens are discussed.[2.1, 2.2] The results presented in this section provide a basis of understanding 3-D imaging performance in conventional imaging systems as well as in confocal imaging systems which will be described in next chapters. They are also useful for other 3-D optical imaging systems.

2. 1. 1   Transmittance of a Thin Lens

When a light wave of wavelength λ (λ is the wavelength in vacuum) passes through an optical lens, the light field impinging on it experiences two effects: i) the phase change of the field and ii) the amplitude change of the field. To study these effects, we should consider the lens to be a diffraction screen. So the transmittance of a lens may be a complex function $t(x, y)$, i. e.,

$$t(x,y) = P(x,y)\exp[-i\phi(x,y)],\qquad\qquad(2.1.1)$$

where $P(x, y)$ and $\phi(x, y)$ are responsible for the amplitude and phase changes of the incident light, respectively. Consider the lens to be optically thin and have the uniform refractive index $\tilde{n}$.[2.1] Assume that the light fields in the planes immediately before and behind the lens are $U_1(x_1, y_1)$ and $U_2(x_2, y_2)$ as shown in Fig. 2.1.1. For a thin lens, the displacement of the beam caused by the refraction of the lens can be neglected, i. e.,

$$\begin{cases} x_1 = x_2 = x \\ \\ y_1 = y_2 = y. \end{cases}\qquad\qquad(2.1.2)$$

The front and back surfaces of the lens can be considered to be spherical surfaces with radii of curvature, R_1 and $-R_2$, where the negative sign represents that the two surfaces

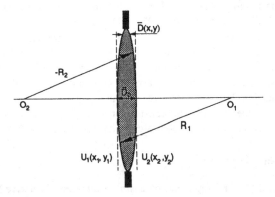

Fig. 2.1.1   A single thin lens can be considered to have two spherical surfaces with radii of curvature, R_1 and $-R_2$, subtended at O_1 and O_2, respectively. $\overline{D}_0$ is the thickness of the lens on the axis.

face the opposite directions. If the geometric thickness of the lens at the center is $\overline{D}_0$, the geometric thickness at any point on the lens, $\overline{D}(x, y)$, can be derived according to the conditions given in Fig. 2.1.1. For an imaging system, light waves usually propagate along the direction close to the optical axis. Therefore the paraxial approximation can be made and accordingly one can obtain[2.1]

$$\overline{D}(x,y) = \overline{D}_0 - \frac{x^2 + y^2}{2}\left(\frac{1}{R_1} - \frac{1}{R_2}\right). \tag{2.1.3}$$

Therefore the phase delay caused by the lens can be expressed as

$$\phi(x,y) = k\tilde{n}\overline{D}(x,y) + k[\overline{D}_0 - \overline{D}(x,y)], \tag{2.1.4}$$

where $k = 2\pi/\lambda$ is the wave number of the incident light. Suppose that the amplitude transmittance of the lens is $P(x, y)$. The field immediately behind the lens is

$$U_2(x,y) = U_1(x,y)P(x,y)\exp[-i\phi(x,y)], \tag{2.1.5}$$

where $P(x, y)$ is sometimes called the pupil function of the lens. By using Eqs. (2.2.1)-(2.2.4), Eq. (2.1.5) can be expressed as

$$U_2(x,y) = U_1(x,y)P(x,y)\exp(-ik\tilde{n}\overline{D}_0)\exp\left[ik(\tilde{n}-1)\frac{x^2+y^2}{2}\left(\frac{1}{R_1} - \frac{1}{R_2}\right)\right]. \tag{2.1.6}$$

Let

$$\frac{1}{f} = (\tilde{n}-1)\left(\frac{1}{R_1} - \frac{1}{R_2}\right), \tag{2.1.7}$$

which is called the focal length of the lens in geometric optics.[2.1] Thus the transmittance of the lens is given by

$$t(x,y) = \frac{U_2(x,y)}{U_1(x,y)}, \tag{2.1.8}$$

or

$$t(x,y) = P(x,y)\exp(-ik\tilde{n}\overline{D}_0)\exp\left[\frac{ik(x^2+y^2)}{2f}\right]. \tag{2.1.9}$$

The first factor $\exp(-ik\bar{n}\,\overline{D}_0)$ represents a constant phase term, so that it can be neglected. Finally, the complex transmittance of a thin lens is

$$t(x,y) = P(x,y)\exp\left[\frac{ik(x^2+y^2)}{2f}\right].$$

$$(2.1.10)$$

The quadratic phase factor in the lens plane represents that the lens causes either a convergent wave if f is positive or a divergent wave if f is negative. The former lens is called the positive lens while the latter is the negative lens.

If a plane wave passes through a positive lens, the wave converges to a point at a distance f behind the lens. This point is called the focus of the lens in geometric optics. However, the distribution of the light field near the focal region is not an ideal point because of the diffraction of light.[2.1, 2.2] The exact distribution of the field in the focal region can be investigated by using the diffraction formula.[2.1, 2.2]

2.1.2   Fresnel Approximation

From Maxwell's equations,[2.1, 2.2] a vectorial form of the wave equation for the propagation of light can be obtained. When the polarization properties of light are neglected, a scaler form of the wave equation can be derived and is sometimes called the Helmholtz equation. The Kirchhoff diffraction formula is a solution to the Helmholtz equation when the Kirchhoff boundary conditions[2.1, 2.2] are applied, and can be expressed as

$$U_2(x_2,y_2) = \frac{i}{\lambda}\int\int_{-\infty}^{\infty} U_1(x_1,y_1)\frac{\exp(-ikr)}{r}\cos(n,r)dx_1dy_1,$$

$$(2.1.11)$$

where $U_1(x_1, y_1)$ is the light field at a point Q on the diffraction plane, i. e., the x_1 - y_1

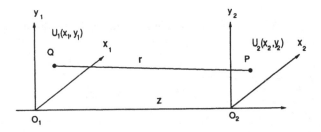

Fig. 2.1.2   Coordinates in the diffraction plane, i. e., the x_1 - y_1 plane, and in the observation plane, i. e., the x_2 - y_2 plane.

plane, and $U_2(x_2, y_2)$ is the light field at a point P on the observation plane, i. e., the $x_2 - y_2$ plane, both of which are shown in Fig. 2.1.2. The distance between the two planes is z and the distance between the points Q and P is r. The factor

$$\frac{\exp(-ikr)}{r}$$

in Eq. (2.1.11) represents a spherical wavelet originating from the point Q and observed at the point P. The factor $\cos(n,r)$ denotes the cosine of the angle between the wave vector (n) of $U_1(x_1, y_1)$ and the observation direction (r). The condition for Eq. (2.1.11) to hold is $\lambda \ll r$.[2.1, 2.2]

According to the coordinates shown in Fig. 2.1.2, one can express the distance r as

$$r^2 = z^2 + (x_2 - x_1)^2 + (y_2 - y_1)^2 = z^2 \left[1 + \frac{(x_2 - x_1)^2 + (y_2 - y_1)^2}{z^2} \right]. \qquad (2.1.12)$$

If $(x_2 - x_1)^2 + (y_2 - y_1)^2 \ll z^2$, which corresponds to a situation that the observation plane is far away from the diffraction plane in comparison with the distance between the observation point and the optical axis, we can approximately assume

$$r \approx z \left[1 + \frac{(x_2 - x_1)^2 + (y_2 - y_1)^2}{2z^2} \right]. \qquad (2.1.13)$$

This condition is called the Fresnel approximation. If the observation point is close to the optical axis z, we can approximately have $\cos(n,r) \approx 1$, which is a result of the paraxial approximation, and the distance r in the denominator of Eq. (2.1.11) can be replaced approximately by z. Under these conditions, Eq. (2.1.11) can be reduced to

$$U_2(x_2, y_2) = \frac{i \exp(-ikz)}{\lambda z} \iint_{-\infty}^{\infty} U_1(x_1, y_1) \exp\left\{ -\frac{ik}{2z} \left[(x_2 - x_1)^2 + (y_2 - y_1)^2 \right] \right\} dx_1 dy_1. \qquad (2.1.14)$$

This is the formula for calculating the Fresnel diffraction pattern, when the observation plane is not far away from the diffraction screen, which is usually the case in an optical imaging system. The pre-factors $\exp(-ikz)$ and $i/\lambda z$ are important in confocal microscopy, in particular, under ultrashort pulsed beam illumination (see Chapter 8). One of the properties associated with Eq. (2.1.14) is the quadratic phase variation on the observation plane. Because of the nonlinear phase variation, the calculation of the Fresnel diffraction pattern becomes complicate. However, if the observation screen is located remotely from the diffraction screen, one can derive the Fraunhofer diffraction pattern,[2.1, 2.2] which is simply the Fourier transform of the incident field $U_1(x_1, y_1)$.

2. 1. 3   Diffraction Pattern in the Focal Plane

To find the light distribution in the focal plane of a thin lens, we should consider the lens to be a diffraction screen. Once the light field immediately after the lens is known, the field in the focal plane can be derived from Eq. (2.1.14). Assume that the lens is illuminated by a uniform plane wave, so that the field immediately before the lens is $U_1(x_1, y_1) = U_0$. The field immediately behind the lens is, if the transmittance of the lens is given by Eq. (2.1.10),

$$U_2(x,y) = U_0 P(x,y) \exp\left[\frac{ik}{2f}(x^2 + y^2)\right].$$

(2.1.15)

Using the Fresnel diffraction formula (2.1.14), we can express the light field on a screen placed at the focus, i. e., at $z = f$, as

$$U_3(x_3, y_3) = \frac{iU_0}{\lambda f} \int\int_{-\infty}^{\infty} P(x,y) \exp\left[\frac{ik}{2f}(x^2 + y^2)\right] \exp\left[-\frac{ik}{2f}(x_3^2 + y_3^2)\right]$$

(2.1.16)

$$\exp\left[-\frac{ik}{2f}(x^2 + y^2)\right] \exp\left[\frac{ik}{f}(x_3 x + y_3 y)\right] dx dy.$$

Obviously, the quadratic phase caused by the lens can be cancelled by the quadratic phase resulting from the Fresnel diffraction, so that

$$U_3(x_3, y_3) = \frac{iU_0}{\lambda f} \exp(-ikf) \exp\left[-\frac{ik}{2f}(x_3^2 + y_3^2)\right]$$

(2.1.17)

$$\int\int_{-\infty}^{\infty} P(x,y) \exp\left[\frac{ik}{f}(x_3 x + y_3 y)\right] dx dy.$$

The integration in Eq. (2.1.17) is the two-dimensional (2-D) Fourier transform of the pupil function $P(x, y)$ at spatial frequencies of $m = x_3/(f\lambda)$ and $n = y_3/(f\lambda)$, according to the definition of the 2-D Fourier transform in Appendix 1, and therefore gives a Fraunhofer diffraction pattern of $P(x, y)$. It should be pointed out that although $U_3(x_3, y_3)$ takes a form of the Fraunhofer diffraction of the pupil function, the diffraction process caused by a lens is the Fresnel diffraction.

Assume that the lens is circularly symmetric. Its pupil function is thus dependent only on the radial coordinate, i. e., $P(x, y) = P(r)$, where $r = (x^2 + y^2)^{1/2}$. Using the

Hankel transform (Appendix 1) and neglecting U_0, we have

$$U_3(r_3) = \frac{i}{\lambda f}\exp(-ikf)\exp\left(-\frac{i\pi r_3^2}{\lambda f}\right)\int_0^\infty P(r)J_0\left(\frac{2\pi rr_3}{\lambda f}\right)2\pi r\,dr, \qquad (2.1.18)$$

where J_0 is a Bessel function of the first kind of order zero[2.1] and $r_3 = (x_3^2 + y_3^2)^{1/2}$. If $P(r)$ is a circular uniform aperture with radius a, its pupil function becomes

$$P(r) = \begin{cases} 1 & , \quad r \leq a, \\ \\ 0 & , \quad otherwise. \end{cases} \qquad (2.1.19)$$

Eq. (2.1.18) accordingly reduces to

$$U_3(r_3) = \frac{i\pi a^2}{\lambda f}\exp(-ikf)\exp\left(-\frac{i\pi r_3^2}{\lambda f}\right)\left[\frac{2J_1\left(\frac{2\pi r_3 a}{\lambda f}\right)}{\left(\frac{2\pi r_3 a}{\lambda f}\right)}\right]. \qquad (2.1.20)$$

Here J_1 is a Bessel function of the first kind of order unity.[2.1] Three important parameters can be introduced in order to simplify Eq. (2.1.20).

i)   Numerical aperture of the lens, $N.A.$:

$$N.A. = \sin\alpha \approx \frac{a}{f}. \qquad (2.1.21)$$

ii)   Radial (transverse) optical coordinate v:

$$v = \frac{2\pi}{\lambda}\frac{a}{f}r_3 \approx \frac{2\pi}{\lambda}r_3\sin\alpha. \qquad (2.1.22)$$

iii)   Fresnel number N:

$$N = \frac{a^2}{\lambda f}. \qquad (2.1.23)$$

Using Eqs. (2.1.21) - (2.1.23), we can rewrite Eqs. (2.1.18) and (2.1.20) as

$$U_3(v) = 2\pi iN\exp(-ikf)\exp\left(-\frac{iv^2}{4N\pi}\right)\int_0^1 P(\rho)J_0(v\rho)\rho\,d\rho \qquad (2.1.24)$$

and

$$U_3(v) = i\pi N \exp(-ikf) \exp\left(-\frac{iv^2}{4N\pi}\right)\left[\frac{2J_1(v)}{v}\right],$$
(2.1.25)

respectively, where $\rho = r/a$ is the normalized radial coordinate over the lens aperture and $P(\rho)$ is the pupil function with the normalized radius of unity and is given by

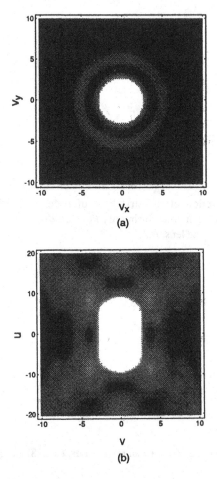

(a)

(b)

Fig. 2.1.3   Diffraction pattern near the focal region of a single circular lens: (a) the intensity distribution in the focal plane; (b) the intensity distribution in the meridional plane of the optical axis.

$$P(\rho) = \begin{cases} 1 & , \quad \rho \le 1, \\ \\ 0 & , \quad otherwise, \end{cases}$$

(2.1.26)

for a circular uniform pupil. The intensity in the focal plane is the modulus squared of Eq. (2.1.25):

$$I(v) = (\pi N)^2 \left[\frac{2J_1(v)}{v} \right]^2,$$

(2.1.27)

which is called the Airy pattern, as shown in Fig. (2.1.3a). Approximate 80% of the incident energy is concentrated within the central bright spot.

2. 1. 4 Diffraction Pattern in a Defocus Plane

Eq. (2.1.25) gives the diffraction pattern of a thin lens in the plane placed exactly at the focus. If the observation plane is placed at a defocus position, what is the field distribution on the observation plane? Let us consider the defocus distance to be Δz, so that the distance between the observation plane and the lens is $z = f + \Delta z$. For a uniform plane wave U_0 incident upon the lens, the field immediately behind the lens $U_2(x_2, y_2)$ is the same as Eq. (2.1.15). The field $U_3(x_3, y_3)$ on the observation plane at $z = f + \Delta z$ is given, in terms of the Fresnel diffraction formula (2.1.14), by

$$U_3(x_3, y_3) = \frac{iU_0}{\lambda z} \exp(-ikz) \int \int_{-\infty}^{\infty} P(x, y) \exp\left[\frac{ik}{2f}(x^2 + y^2) \right]$$

(2.1.28)

$$\exp\left\{ -\frac{ik}{2z} \left[(x_3 - x)^2 + (y_3 - y)^2 \right] \right\} dx dy.$$

For a circularly symmetric lens, the Hankel transform (Appendix 1) can be used in Eq. (2.1.28) and we have

$$U_3(r_3) = \frac{i}{\lambda z} \exp(-ikz) \exp\left(-\frac{i\pi r_3^2}{\lambda z} \right) \int_0^{\infty} P(r) \exp\left[\frac{ikr^2}{2} \left(\frac{1}{f} - \frac{1}{z} \right) \right] J_0\left(\frac{2\pi r r_3}{\lambda z} \right) 2\pi r dr,$$

(2.1.29)

where U_0 has been assumed to be unity without losing generality. If we let

$$P(r, z) = P(r) \exp\left[\frac{ikr^2}{2} \left(\frac{1}{f} - \frac{1}{z} \right) \right],$$

(2.1.30)

which is called the defocused pupil function for the lens, then Eq. (2.1.29) becomes

$$U_3(r_3) = \frac{i}{\lambda z}\exp(-ikz)\exp\left(-\frac{i\pi r_3^2}{\lambda z}\right)\int_0^\infty P(r,z)J_0\left(\frac{2\pi r r_3}{\lambda z}\right)2\pi r dr.$$ (2.1.31)

Eq. (2.1.31) is a 2-D Hankel transform of $P(r, z)$. In other words, $U_3(r_3)$ in the defocus plane is given by a 2-D Fourier transform of the defocused pupil function. For a circular lens of radius a, two optical coordinates can be introduced.

i)   Radial (transverse) optical coordinate v:

$$v = \frac{2\pi}{\lambda}\frac{a}{z}r_3 \approx \frac{2\pi}{\lambda}\frac{a}{f}r_3 \approx \frac{2\pi}{\lambda}r_3\sin\alpha,$$ (2.1.32)

ii)   Axial optical coordinate u:

$$u = \frac{2\pi}{\lambda}a^2\left(\frac{1}{f}-\frac{1}{z}\right) = \frac{2\pi}{\lambda}\Delta z\frac{a^2}{f^2}.$$ (2.1.33)

Using v and u in Eq. (2.1.31) and expressing U_3 as an explicit function of the defocus distance u gives

$$U_3(v,u) = 2\pi iN\exp(-ikf)\exp\left(-\frac{iv^2}{4N\pi}\right)\int_0^1 P(\rho)\exp\left(\frac{iu\rho^2}{2}\right)J_0(v\rho)\rho d\rho,$$ (2.1.34)

where $\rho = r/a$. For a lens of a circular uniform aperture, Eq. (2.1.34) reduces to

$$U_3(v,u) = 2\pi iN\exp(-ikf)\exp\left(-\frac{iv^2}{4N\pi}\right)\int_0^1 \exp\left(\frac{iu\rho^2}{2}\right)J_0(v\rho)\rho d\rho.$$ (2.1.35)

This expression gives the 3-D distribution of the diffraction pattern near the region of the focal plane. When $u = 0$, i. e., when the observation plane is at the focus, the in-focus intensity becomes

$$I(v) = |U_3(v,u=0)|^2 = (\pi N)^2\left[\frac{2J(v)}{v}\right]^2,$$ (2.1.36)

which is, as expected, the same as Eq. (2.1.27). When $v = 0$, the intensity along the axial direction is

$$I(u) = \left| U_3(v = 0, u) \right|^2 = (\pi N)^2 \left[\frac{\sin(u/4)}{u/4} \right]^2.$$

(2.1.37)

In general, $U_3(v, u)$ can be expressed by Lommel functions or evaluated by numerical integration.[2.2] The intensity distribution $I(v, u)$ in an axial plane including the optical axis is shown in Fig. 2.1.3b. A contour plot corresponding to Fig. 2.1.3b can be found in standard textbooks.[2.2]

2. 2   Three-Dimensional Coherent Image Formation

We now consider optical imaging by a single lens. A thin object with an amplitude transmittance of $o(x_1, y_1)$ is placed in a plane at a distance d_1 before the lens and the image of the object is observed in a plane at distance d_2 after the lens (Fig. 2.2.1). These two planes are called the object and image planes, respectively. The corresponding x_1-y_1-z_1 and x_3-y_3-z_3 spaces are called the object and image spaces for a thin lens, respectively. Assume that the pupil function of the lens is $P(x_2, y_2)$. If a uniform plane wave ($U_0 = 1$) illuminates the object, the field immediately after the object is $U_1(x_1, y_1) = o(x_1, y_1)$. The field before the lens, $U_2(x_2, y_2)$, can be calculated by using the Fresnel diffraction formula (see Eq. (2.1.14)):

$$U_2(x_2, y_2) = \frac{i \exp(-ikd_1)}{\lambda d_1} \int\int_{-\infty}^{\infty} o(x_1, y_1) \exp\left\{ -\frac{ik}{2d_1} \left[(x_2 - x_1)^2 + (y_2 - y_1)^2 \right] \right\} dx_1 dy_1.$$

(2.2.1)

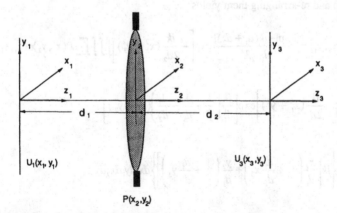

Fig. 2.2.1   Geometry of an imaging system with a single lens. A thin object is placed in the x_1 - y_1 plane and its image is observed in the x_3 - y_3 plane. The lens has a pupil function $P(x_2, y_2)$. The x_2 - y_2 plane is called the lens plane.

The transmittance of the lens is given by Eq. (2.1.10) and thus the field immediately after the lens is

$$U_2(x_2,y_2) = \frac{i\exp(-ikd_1)}{d_1\lambda} P(x_2,y_2)\exp\left[\frac{ik}{2f}(x_2^2+y_2^2)\right]$$

(2.2.2)

$$\iint_{-\infty}^{\infty} o(x_1,y_1)\exp\left\{-\frac{ik}{2d_1}[(x_2-x_1)^2+(y_2-y_1)^2]\right\}dx_1dy_1.$$

Finally, using the Fresnel diffraction formula (2.1.14) again we can find the field in the object plane:

$$U_3(x_3,y_3) = \frac{\exp[-ik(d_1+d_2)]}{d_1d_2\lambda^2}\iiint\int_{-\infty}^{\infty} P(x_2,y_2)o(x_1,y_1)\exp\left[\frac{ik}{2f}(x_2^2+y_2^2)\right]$$

$$\exp\left\{-\frac{ik}{2d_1}[(x_2-x_1)^2+(y_2-y_1)^2]\right\}\exp\left\{-\frac{ik}{2d_2}[(x_3-x_2)^2+(y_3-y_2)^2]\right\}$$

$$dx_1dy_1dx_2dy_2.$$

(2.2.3)

Here a pre-negative sign has been omitted. Expanding the two quadratic phase factors in Eq. (2.2.3) and re-arranging them yields

$$U_3(x_3,y_3) = \frac{\exp[-ik(d_1+d_2)]}{d_1d_2\lambda^2}\exp\left[-\frac{ik}{2d_2}(x_3^2+y_3^2)\right]\iiint\int_{-\infty}^{\infty} P(x_2,y_2)o(x_1,y_1)$$

$$\exp\left[-\frac{ik}{2d_1}(x_1^2+y_1^2)\right]\exp\left[\frac{ik}{2}\left(\frac{1}{f}-\frac{1}{d_1}-\frac{1}{d_2}\right)(x_2^2+y_2^2)\right]$$

$$\exp\left\{ik\left[\frac{x_2}{d_1}\left(x_1+\frac{d_1}{d_2}x_3\right)+\frac{y_2}{d_1}\left(y_1+\frac{d_1}{d_2}y_3\right)\right]\right\}dx_1dy_1dx_2dy_2.$$

(2.2.4)

This expression includes three quadratic phase terms in the object, lens, and image planes, respectively.

Let us first consider that the lens law[2.1] is satisfied, i. e.,

$$\frac{1}{f} = \frac{1}{d_1} + \frac{1}{d_2}.$$ (2.2.5)

In this case, the quadratic phase in the lens plane disappears. We can introduce a de-magnification factor of the lens, defined as

$$M = d_1/d_2,$$ (2.2.6)

so that Eq. (2.2.4) becomes

$$U_3(x_3, y_3) = \frac{M \exp[-ikd_1(1+1/M)]}{d_1^2 \lambda^2} \exp\left[-\frac{ikM}{2d_1}(x_3^2 + y_3^2)\right] \iiiint_{-\infty}^{\infty} P(x_2, y_2) o(x_1, y_1)$$

$$\exp\left[-\frac{ik}{2d_1}(x_1^2 + y_1^2)\right] \exp\left\{\frac{ik}{d_1}[x_2(x_1 + Mx_3) + y_2(y_1 + My_3)]\right\} dx_1 dy_1 dx_2 dy_2.$$

(2.2.7)

In order to evaluate $U_3(x_3, y_3)$, we first perform the integration with respect to x_2 and y_2 and define a function $h(x, y)$:

$$h(x, y) = \iint_{-\infty}^{\infty} P(x_2, y_2) \exp\left[\frac{ik}{d_1}(x_2 x + y_2 y)\right] dx_2 dy_2,$$ (2.2.8)

which is the Fourier transform of the pupil function $P(x_2, y_2)$. Then Eq. (2.2.7) becomes

$$U_3(x_3, y_3) = \frac{M \exp[-ikd_1(1+1/M)]}{d_1^2 \lambda^2} \exp\left[-\frac{ikM}{2d_1}(x_3^2 + y_3^2)\right]$$

(2.2.9)

$$\iint_{-\infty}^{\infty} o(x_1, y_1) \exp\left[-\frac{ik}{2d_1}(x_1^2 + y_1^2)\right] h(x_1 + Mx_3, y_1 + My_3) dx_1 dy_1.$$

To understand the significance of the function $h(x, y)$, we consider the object to be a single point, i. e.,

$$o(x_1, y_1) = \delta(x_1)\delta(y_1).$$

The image of a single point is thus

$$U_3(x_3, y_3) = \frac{M \exp[-ikd_1(1+1/M)]}{d_1^2 \lambda^2} \exp\left[-\frac{ikM}{2d_1}(x_3^2 + y_3^2)\right] h(Mx_3, My_3).$$ (2.2.10)

It is now understood that the function $h(x, y)$ defined in Eq. (2.2.8) is related to the image of a single point object. Therefore we call $h(x, y)$ the 2-D point spread function (PSF), or the 2-D amplitude point spread function (APSF) as it gives the complex amplitude of the light field. It is also sometimes called the impulse response of the optical system.[2.1] From the definition of $h(x, y)$, it is seen that $h(x, y)$ is the 2-D Fourier transform of the pupil function $P(x, y)$ for the imaging lens.

For a good imaging system, we hope that $h(x, y)$ is close to the point object. This means that $h(x, y)$ falls off quickly. As a result, we can assume that $x_1 + Mx_3$ in Eq. (2.2.9) is small , i. e.,

$$x_1 \approx -Mx_3,$$ (2.2.11)

so that Eq. (2.2.9) can be rewritten as

$$U_3(x_3, y_3) = \frac{M \exp[-ikd_1(1+1/M)]}{d_1^2 \lambda^2} \exp\left[-\frac{ikM}{2d_1}(x_3^2 + y_3^2)(1+M)\right]$$ (2.2.12)

$$\int\int_{-\infty}^{\infty} o(x_1, y_1) h(x_1 + Mx_3, y_1 + My_3) dx_1 dy_1.$$

Here the quadratic phase term in the object plane disappears. It can be recognized that Eq. (2.2.12) is a convolution relation, so that the image field is given by the 2-D convolution of the transmittance of the object with the 2-D amplitude point spread function for the imaging lens. In the limiting case when the size of lens is quite large, the 2-D amplitude point spread function becomes

$$h(x, y) = \delta(x)\delta(y),$$

and the image is thus

$$U_3(x_3, y_3) = \frac{M \exp[-ikd_1(1+1/M)]}{d_1^2 \lambda^2} \exp\left[-\frac{ikM}{2d_1}(x_3^2 + y_3^2)(1+M)\right] o(-Mx_3, -My_3).$$ (2.2.13)

Therefore we conclude that the image predicated by geometric optics is a magnified and inverted replica of the object in the image plane.

In terms of Eq. (2.2.12), the intensity of the image is

$$I_3(x_3, y_3) = \left(\frac{M}{d_1^2 \lambda^2}\right)^2 \left|\iint_{-\infty}^{\infty} o(x_1, y_1) h(x_1 + Mx_3, y_1 + My_3) dx_1 dy_1\right|^2.$$ (2.2.14)

We now turn to the case when the positions of the object and image planes do not satisfy the lens law (see Eq. (2.2.5)), i. e.,

$$\frac{1}{d_1} + \frac{1}{d_2} - \frac{1}{f} \neq 0.$$

Assume that

$$\frac{1}{d_1} + \frac{1}{d_2} - \frac{1}{f} = \frac{1}{d_0},$$ (2.2.15)

so that an effective pupil function $P_{eff}(x_2, y_2)$ can be introduced:

$$P_{eff}(x_2, y_2) = P(x_2, y_2) \exp\left[-\frac{ik}{2d_0}(x_2^2 + y_2^2)\right],$$ (2.2.16)

which is also called the defocused pupil function for the imaging system. So Eq. (2.2.4) can be rewritten as

$$U_3(x_3, y_3) = \frac{M \exp[-ikd_1(1 + 1/M)]}{d_1^2 \lambda^2} \exp\left[-\frac{ikM}{2d_1}(x_3^2 + y_3^2)(1 + M)\right]$$ (2.2.17)

$$\iint_{-\infty}^{\infty} o(x_1, y_1) h'(x_1 + Mx_3, y_1 + My_3) dx_1 dy_1,$$

where

$$h'(x, y) = \iint_{-\infty}^{\infty} P_{eff}(x_2, y_2) \exp\left[\frac{ik}{d_1}(x_2 x + y_2 y)\right] dx_2 dy_2.$$ (2.2.18)

Eq. (2.2.18) is called the defocused amplitude point spread function. It is clear that the image is again given by the convolution of the transmittance of the object with the defocused amplitude point spread function. $h'(x, y)$ is also called the 3-D amplitude point spread function as it describes the image of a single point object in a 3-D space (x, y and defocus distance). This property will be, in detail, discussed in the next section. The difference $1/d_0$ should be small in order that Eq. (2.2.11) is satisfied. Otherwise, $h'(x, y)$ becomes broad, i. e., $h'(x, y)$ falls off slowly.

The intensity of the image in the focal region is therefore

$$I_3(x_3, y_3) = \left(\frac{M}{d_1^2 \lambda^2}\right)^2 \left|\iint_{-\infty}^{\infty} o(x_1, y_1) h'(x_1 + Mx_3, y_1 + My_3) dx_1 dy_1\right|^2. \tag{2.2.19}$$

For a circular lens, the following optical coordinates can be introduced:

$$\begin{cases} v = \dfrac{2\pi}{\lambda} r \dfrac{a}{d_1} \approx \dfrac{2\pi}{\lambda} r \sin\alpha_o, \\[3mm] u = \dfrac{2\pi}{\lambda} a^2 (\dfrac{1}{d_1} + \dfrac{1}{d_2} - \dfrac{1}{f}), \end{cases} \tag{2.2.20}$$

where $\sin\alpha_o$ is the numerical aperture of the lens in object space. Eq. (2.2.18) is thus reduced to

$$h'(v,u) = \int_0^1 P(\rho) \exp(-iu\rho^2/2) J_0(v\rho) 2\pi\rho d\rho, \tag{2.2.21}$$

which has the same distribution as Eq. (2.1.35) except the pre-phase factor. When $u = 0$, i. e., when the lens law is satisfied, Eq. (2.2.21) gives rise to the 2-D amplitude point spread function:

$$h'(v, u = 0) = \pi \left[\frac{2J_0(v)}{v}\right]. \tag{2.2.22}$$

2. 3   Three-Dimensional Space-Invariant Point Spread Function

As has been mentioned in the last section, $h'(x, y)$ is the 3-D amplitude point spread function. This conclusion is based on the fact that it is the function of $1/d_0$ and is 3-D space-invariant. Therefore, it can be used to describe 3-D imaging of an object with finite thickness. To understand this property further,[2.3] we express $h'(x, y)$ explicitly as

$$h'(x,y) = \iint_{-\infty}^{\infty} P(x_2, y_2) \exp\left[\frac{ik}{2}\left(\frac{1}{f} - \frac{1}{d_1} - \frac{1}{d_2}\right)(x_2^2 + y_2^2)\right]$$

$$\exp\left[\frac{ik}{d_1}(x_2 x + y_2 y)\right] dx_2 dy_2, \tag{2.3.1}$$

where Eq. (2.2.15) has been employed. The difference $1/d_0$ in Eq. (2.2.15) can be caused by the displacement of either the object plane or the image plane from the position at which the lens law is satisfied. We can express d_1 and d_2, according to the coordinates described in Fig. 2.2.1, as

$$\begin{cases} d_1 = d_{10} - z_1, \\ \\ d_2 = d_{20} + z_3, \end{cases}$$

(2.3.2)

where d_{10} and d_{20} are the distances satisfying the lens law:

$$\frac{1}{d_{10}} + \frac{1}{d_{20}} = \frac{1}{f}.$$

(2.3.3)

Coordinates z_1 and z_3 are the axial coordinates of the object and image planes in the coordinates systems shown in Fig. 2.2.1. If the situation that the lens law is satisfied is called the focusing condition of the imaging system, z_1 and z_3 can be called the defocus distances of the object and image planes, respectively.

In a practical microscope, the minimum distance d_{10} is approximately 3 mm. If a sample is defocused by 100 μm, which can be considered to be the thickness in the case of a thick object, $z_1 / d_{10} \approx 3 \times 10^{-2}$. The same justification applies to the ratio of z_3 / d_{20}. Therefore we can have the following approximate relations:

$$\begin{cases} \dfrac{1}{d_1} = \dfrac{1}{d_{10} - z_1} \approx \dfrac{1}{d_{10}}(1 + \dfrac{z_1}{d_{10}}), \\ \\ \dfrac{1}{d_2} = \dfrac{1}{d_{20} + z_3} \approx \dfrac{1}{d_{20}}(1 - \dfrac{z_3}{d_{20}}). \end{cases}$$

(2.3.4)

Accordingly, we have

$$\frac{1}{d_1} + \frac{1}{d_2} - \frac{1}{f} = \frac{1}{d_{10}^2}(z_1 - M^2 z_3),$$

(2.3.5)

where $M = d_{10}/d_{20}$. Substituting Eq. (2.3.5) into Eq. (2.3.1) and replacing d_1 in the linear phase factor approximately by d_{10}, one can obtain

$$h'(x,y) = \iint_{-\infty}^{\infty} P(x_2, y_2) \exp\left[-\frac{ik}{2}\left(\frac{1}{d_{10}}\right)^2 (z_1 - M^2 z_3)(x_2^2 + y_2^2)\right]$$

$$\exp\left[\frac{ik}{d_{10}}(x_2 x + y_2 y)\right] dx_2 dy_2. \tag{2.3.6}$$

It is noted that Eq. (2.3.6) has been expressed as a space-invariant form in the axial direction. This feature is important for 3-D imaging of a thick object. Substituting Eq. (2.3.6) into Eq. (2.2.17) yields the image of a thin object:

$$U_3(x_3, y_3) = \frac{M \exp[-ik(d_1 + d_2)]}{d_1^2 \lambda^2} \exp\left[-\frac{ikM}{2d_{10}}(x_3^2 + y_3^2)(1 + M)\right]$$

$$\iiint\int_{-\infty}^{\infty} P(x_2, y_2) o(x_1, y_1) \exp\left[-\frac{ik}{2}\left(\frac{1}{d_{10}}\right)^2 (z_1 - M^2 z_3)(x_2^2 + y_2^2)\right]$$

$$\exp\left\{\frac{ik}{d_{10}}\left[x_2(x_1 + Mx_3) + y_2(y_1 + My_3)\right]\right\} dx_1 dy_1 dx_2 dy_2, \tag{2.3.7}$$

Eq. (2.3.7) gives the image field in the plane at z_3 when a thin object is placed in the plane at z_1. If the object has a finite thickness, i. e., $o(x, y, z)$, each of the vertical sections in the thick object at a given position z_1 produces an image in the image plane placed at z_3. The total field in the image plane at z_3 is the superposition of the contributions from each section. The superposition principle holds because the first Born approximation[2.4] has been assumed. This assumption implies that secondary diffraction in a thick object is neglected and that the object is semi-transparent. Thus the image of an object with finite thickness, i. e., the image of a 3-D object, is the integration of Eq. (2.3.7) with respect to z_1:

$$U_3(x_3, y_3, z_3) = \frac{M \exp[-ik(d_{10} + d_{20})]}{d_1^2 \lambda^2} \exp\left[-\frac{ikM}{2d_{10}}(x_3^2 + y_3^2)(1 + M)\right]$$

$$\iiint\int_{-\infty}^{\infty} P(x_2, y_2) o(x_1, y_1, z_1) \exp[ik(z_1 - z_3)] \exp\left[-\frac{ik}{2}\left(\frac{1}{d_{10}}\right)^2 (z_1 - M^2 z_3)(x_2^2 + y_2^2)\right]$$

$$\exp\left\{\frac{ik}{d_{10}}\left[x_2(x_1 + Mx_3) + y_2(y_1 + My_3)\right]\right\} dx_1 dy_1 dz_1 dx_2 dy_2. \tag{2.3.8}$$

Here we have expressed the image field U_3 explicitly as a function of z_3. In Eq. (2.3.8), the linear phase factor $\exp[ik(z_1 - z_3)]$ originates from the factor $\exp[-ik(d_1 + d_2)]$ in Eq. (2.3.7) when Eq. (2.3.2) is employed. Eq. (2.3.8) can be rewritten as

$$U_3(x_3, y_3, z_3) = \exp[-ik(d_{10} + d_{20})]\exp\left[-\frac{ikM}{2d_{10}}(x_3^2 + y_3^2)(1+M)\right]$$

$$\iiint_{-\infty}^{\infty} o(x_1, y_1, z_1)\exp[ik(z_1 - z_3)]h(x_1 + Mx_3, y_1 + My_3, z_1 - M^2 z_3)dx_1 dy_1 dz_1,$$

(2.3.9)

where

$$h(x, y, z) = \frac{M}{d_1^2 \lambda^2}\iint_{-\infty}^{\infty} P(x_2, y_2)\exp\left[-\frac{ik}{2}\left(\frac{1}{d_{10}}\right)^2 z(x_2^2 + y_2^2)\right]$$

(2.3.10)

$$\exp\left[\frac{ik}{d_{10}}(x_2 x + y_2 y)\right]dx_2 dy_2.$$

The pre-phase $\exp(ikz_1)$ in Eq. (2.3.9) is caused by the thick object, representing a defocus phase because of the thickness. If the product $o(x_1, y_1, z_1)\exp(ikz_1)$ is called the effective object function, thus the image field is the 3-D convolution of the effective object function with $h(x, y, z)$. Imaging is therefore 3-D space-invariant with a transverse magnification factor $1/M$ and an axial magnification $-1/M^2$.

For a single point object, $o(x_1, y_1, z_1) = \delta(x_1)\delta(y_1)\delta(z_1)$, so that the image field is

$$U_3(x_3, y_3, z_3) = \exp[-ik(d_{10} + d_{20})]\exp\left[-\frac{ikM}{2d_{10}}(x_3^2 + y_3^2)(1+M)\right]$$

(2.3.11)

$$\exp(-ikz_3)h(Mx_3, My_3, -M^2 z_3).$$

Its intensity is thus

$$I_3(x_3, y_3, z_3) = \left|h(Mx_3, My_3, -M^2 z_3)\right|^2.$$

(2.3.12)

Therefore the significance of $h(x, y, z)$ is that it represents a 3-D field distribution of the image for a single point and can be called the 3-D amplitude point spread function as mentioned in the last section.

For a circular lens of radius a, the image of a single point can be expressed as

$$U_3(v,u) = \frac{M \exp\left[-ik\left(d_{10} + d_{20} + \frac{d_{20}^2 u}{ka^2}\right)\right]}{d_1^2 \lambda^2} \exp\left[-\frac{iv^2}{4\pi N}(1+M)\right]$$

(2.3.13)

$$\int_0^1 P(\rho)\exp\left(\frac{iu}{2}\rho^2\right)J_0(\rho v)2\pi\rho d\rho,$$

where $P(\rho)$ is the pupil function with a normalized radius and is given by Eq. (2.1.26) for a circular uniform pupil. Here

$$v = \frac{2\pi}{\lambda}r_3\frac{a}{d_{20}} \approx \frac{2\pi}{\lambda}r_3\sin\alpha_i,$$

$$u = \frac{2\pi}{\lambda}z_3\frac{a^2}{d_{20}^2} \approx \frac{8\pi}{\lambda}z_3\sin^2\frac{\alpha_i}{2},$$

(2.3.14)

$$N = \frac{a^2}{\lambda d_{20}},$$

where $\sin\alpha_i$ is the numerical aperture of the lens in the image space. For an imaging system, the relation $N \gg v^2/4$ is valid and we can neglect the corresponding phase term in Eq. (2.3.13). The constant phase term associated with $d_{10} + d_{20}$ can also be omitted. Finally, Eq. (2.3.13) becomes

$$U_3(v,u) = \frac{M \exp\left(-\frac{iu}{4\sin^2(\alpha_i/2)}\right)}{d_1^2 \lambda^2}\int_0^1 P(\rho)\exp\left(\frac{iu}{2}\rho^2\right)J_0(\rho v)2\pi\rho d\rho,$$

(2.3.15)

which has the same distribution as Eq. (2.1.34) except the pre-factors.

The 3-D amplitude point spread function in Eq. (2.3.10) accordingly becomes

$$h(v,u) = \int_0^1 P(\rho)\exp\left(-\frac{iu}{2}\rho^2\right)J_0(\rho v)2\pi\rho d\rho,$$

(2.3.16)

where a constant has been omitted and

$$v = \frac{2\pi}{\lambda} r \frac{a}{d_{10}} \approx \frac{2\pi}{\lambda} r \sin \alpha_o,$$

(2.3.17)

$$u = \frac{2\pi}{\lambda} z \frac{a^2}{d_{10}^2} \approx \frac{8\pi}{\lambda} z \sin^2 \frac{\alpha_o}{2}.$$

Here $\sin \alpha_o$ is the numerical aperture of the lens in the object space. Note that the definitions of v and u are different between Eqs. (2.3.14) and (2.3.17). The former represents the coordinates in the image space, while the latter in the object space. A defocused pupil function can be introduced in Eq. (2.2.16), which has the following circularly symmetric form:

$$P(\rho, u) = P(\rho) \exp\left(-\frac{iu\rho^2}{2}\right).$$

(2.3.18)

2. 4   Three-Dimensional Coherent Transfer Function

The function of an optical imaging system such as a microscope is to provide a magnified image of an object in which details are too fine to be seen by naked eye. It is desirable that the imaging system should have an ability to reproduce the details in the images. As we have mentioned in Chapter 1, any optical imaging system is a low pass filter which transmits only low spatial frequencies corresponding to slow variations in the object. The fine details of the object are represented by high spatial frequencies. These high spatial frequencies may not be imaged because the optical system has a cut-off spatial frequency. Further, the efficiency with which the periodic components are transmitted are dependent on the optical system. These properties can be analysed in terms of the Fourier transformation.[2.1, 2.5]

Suppose that the 3-D transmittance of a thick object is $o(x, y, z)$, which is called the 3-D object function hereafter. The distribution of the periodic components of the thick object, $O(m)$, which is called the 3-D amplitude spectrum of the object, can be found from the 3-D inverse Fourier transform of $o(r)$:

$$O(m) = \int_{-\infty}^{\infty} o(r) \exp(-2\pi i r \bullet m) dr.$$

(2.4.1)

In comparison with the normal definition of the Fourier transform,[2.1] a constant has been neglected for convenience. We use the word 'inverse' to refer to the negative sign in the exponent in Eq. (2.4.1) (see Appendix 1). Here r denotes the position vector with x, y, and z components, and m is the spatial frequency vector including m, n, and s components in x, y, and z directions, respectively. dr represents the 3-D integration with respect to dx, dy, and dz, and dm includes dm, dn, and ds. The object function is thus given by a 3-D Fourier transform:

$$o(r) = \int_{-\infty}^{\infty} O(m) \exp(2\pi i r \bullet m) dm. \tag{2.4.2}$$

Substituting Eq. (2.4.2) into Eq. (2.3.9) and neglecting the constant phase term and the quadratic phase term in the image plane for large values of the Fresnel number, one can express the image field as

$$U_3(x_3, y_3, z_3) = \int\int\int\int\int\int_{-\infty}^{\infty} O(m, n, s) \exp[2\pi i (mx_1 + ny_1 + sz_1)]$$

$$\tag{2.4.3}$$

$$\exp[ik(z_1 - z_3)] h(x_1 + Mx_3, y_1 + My_3, z_1 - M^2 z_3) dx_1 dy_1 dz_1 dm dn ds.$$

Performing the integration in the object space yields

$$U_3(x_3, y_3, z_3) = \exp(-ikz_3) \int\int\int_{-\infty}^{\infty} O(m, n, s) c(m, n, s + 1/\lambda)$$

$$\tag{2.4.5}$$

$$\exp\{-2\pi i [mMx_3 + nMy_3 - (s + 1/\lambda) M^2 z_3]\} dm dn ds.$$

Here the image field is given by the 3-D inverse Fourier transform of the 3-D object spectrum multiplied by the function $c(m, n, s+1/\lambda)$, implying that the image has been resolved into a superposition of a series of periodic components transmitted through the imaging system. The function $c(m, n, s+1/\lambda)$ gives the strength with which each periodic component in the object is imaged and is called the transfer function of the imaging system. Since it operates on the amplitude periodic components in coherent imaging, it is therefore termed the 3-D coherent transfer function (CTF) for an imaging system of a single lens and is given by

$$c(m) = \int_{-\infty}^{\infty} h(r) \exp(2\pi i r \bullet m) dr. \tag{2.4.6}$$

It is seen that the 3-D CTF is the 3-D Fourier transform of the 3-D APSF given in Eq. (2.3.10).

As an example, we consider a circular thin lens of radius a. The 3-D APSF is given by Eq. (2.3.16). Substituting Eq. (2.3.17) into Eq. (2.4.6) results in a circularly symmetric 3-D CTF:

$$c(l,s) = \int\int_{-\infty}^{\infty} h(v,u)J_0(lv)\exp(ius)vdvdu, \tag{2.4.7}$$

where some constant factors have been neglected. Hence we have used $c(l, s)$ to represent the 3-D CTF, where $l = (m^2 + n^2)^{1/2}$ is the radial spatial frequency normalized by

$$\sin\alpha_o / \lambda \tag{2.4.8}$$

and s is the axial spatial frequency normalized by

$$4\sin^2(\alpha_o / 2) / \lambda . \tag{2.4.9}$$

Using Eq. (2.3.16), one can finally derive the 3-D CTF as

$$c(l,s) = P(l)\delta(s - l^2 / 2), \tag{2.4.10}$$

which has been normalized by the value at $l = s = 0$. According to Eq. (2.1.26), it is cut off at $l = 1$ and $s = 1/2$, so that Eq. (2.4.8) is the transverse cut-off spatial frequency of the 3-D CTF. Thus the 3-D CTF for 3-D coherent imaging in Eq. (2.4.5) becomes

$$c(l,s + s_0) = P(l)\delta(s + s_0 - l^2 / 2), \tag{2.4.11}$$

where s_0 denotes the constant shift of the axial spatial frequency, defined as

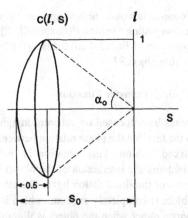

Fig. 2. 4.1   3-D coherent transfer function (CTF) for a single circular lens in coherent imaging. The 3-D CTF is axially shifted, has an axial bandwidth of 0.5, and cuts off at $l = 1$ in the transverse direction. s_0 denotes the axial shift and α_o is the semi-angular aperture of the lens in the object space.

$$s_0 = 1 / [4\sin^2(\alpha_o / 2)].\tag{2.4.12}$$

It is seen that the 3-D CTF for a single lens is axially shifted and is a cap of a paraboloid of revolution about the s axis with a weighting function $P(l)$. Recalling the discussion in Section 1.4, the 3-D CTF for a circular lens should be a cap of a sphere (i. e. Ewald sphere).[2.6] The difference is caused by the use of the paraxial approximation. A cap of a paraboloid is an approximate form of a cap of the sphere. Eq. (2.4.11) is shown in Fig. 2.4.1, which is cut off at $l = 1$ and has an axial bandwidth of $1/2$.

If the lens has an annular pupil function which has a central obstruction of radius a_ε:

$$P(\rho) = \begin{cases} 1 & , & \varepsilon < \rho < 1, \\ \\ 0 & , & otherwise, \end{cases}\tag{2.4.13}$$

where $\varepsilon = a_\varepsilon/a$, its 3-D CTF is given by a strip on the paraboloid of revolution about the s axis.

When the lens experiences aberration, the defocused pupil function can be expressed as[2.1]

$$P(\rho,\theta,u) = P(\rho)\exp\left[-\frac{iu\rho^2}{2} + i\Phi(\rho,\theta)\right],\tag{2.4.14}$$

where ρ and θ are the polar coordinates over the lens aperture. In this case, the 3-D CTF is a cap of a paraboloid with a weighting function $P(l)\exp[i\Phi(l, \theta)]$.

As is expected, the projection of Eq. (2.4.10) leads to the 2-D CTF, $P(l)\exp[i\Phi(l, \theta)]$, for in-focus imaging of a thin object.[2.1]

2. 5 Three-Dimensional Optical Transfer Function

So far, our discussion has been focused on coherent imaging of a thin lens. This imaging process is based on the fact that the phase relation between any two points in the object has a completely fixed fashion. This situation is true when the object is illuminated by a coherent field and the interaction of the object with the incident light does not destroy the coherence of the illumination light. The other limiting case is that the phase relation in the object is completely random, which is called the incoherent object. This process can happen either when the object is illuminated by an incoherent field or when the sample is labelled with fluorescent materials like those in biological studies. For each point in the incoherent object, it produces an image intensity distribution given by Eq. (2.3.12). Thus the total image intensity of a thick object is the superposition of the intensity contributed from each point in the object:

$$I(x_3, y_3, z_3) = \iiint_{-\infty}^{\infty} |o(x_1, y_1, z_1)|^2 |h(x_1 + Mx_3, y_1 + My_3, z_1 - M^2 z_3)|^2 \, dx_1 dy_1 dz_1.$$

(2.5.1)

Here the object function $|o(x, y, z)|^2$ should be recognized as either the intensity transmittance of the object or the fluorescence strength of the sample in fluorescence imaging. Eq. (2.5.1) describes 3-D image formation in the case of incoherent imaging. Note that Eq. (2.5.1) is the 3-D convolution of $|o(x, y, z)|^2$ with $|h(x, y, x)|^2$. $|h(x, y, x)|^2$ is therefore termed the 3-D intensity point spread function (IPSF) for a single thin lens.

If the object is resolved into a superposition of a series of periodic components in terms of the Fourier transform:

$$O_i(m) = \int_{-\infty}^{\infty} |o(r)|^2 \exp(-2\pi i r \bullet m) dm,$$

(2.5.2)

where $O_i(m)$ is the 3-D intensity spectrum of the object function and represents the strength of the periodic components, Eq. (2.5.1) can be rewritten as

$$I(x_3, y_3, z_3) = \iiint_{-\infty}^{\infty} O_i(m, n, s) C(m, n, s) \exp[-2\pi i(mMx_3 + nMy_3 - sM^2 z_3)] dm dn ds,$$

(2.5.3)

where

$$C(m) = \int_{-\infty}^{\infty} |h(r)|^2 \exp(2\pi i r \bullet m) dr.$$

(2.5.4)

Eq. (2.5.3) is the 3-D Fourier transform of the 3-D intensity spectrum of the object multiplied by the function $C(m)$. Therefore $C(m)$ gives the strength with which each periodic component in an incoherent object is imaged and thus is the 3-D transfer function for incoherent imaging with a single lens. Historically, it is termed the optical transfer function (OTF) rather than the incoherent transfer function.[2.1] It is noted that Eq. (2.5.4) is the Fourier transform of the 3-D intensity point spread function, as pointed out by Frieden,[2.7] and operates on the intensity spectrum of the object.

According to the convolution theorem of the Fourier transform[2.1, 2.5] and Eq. (2.4.6), the 3-D OTF can be expressed as

$$C(m) = c(m) \otimes_3 c^*(-m),$$

(2.5.5)

where $\otimes_3$ denotes the 3-D convolution in spatial frequency space and thus the 3-D OTF is mathematically given by the 3-D convolution of the 3-D CTF $c(m)$ with its inversed conjugate function $c^*(-m)$.

For a circular lens, we have

$$C(l,s) = K \int\int_{-\infty}^{\infty} |h(v,u)|^2 J_0(lv) \exp(ius) v \, dv \, du,$$ (2.5.6)

where l and s have been normalized by Eqs. (2.4.8) and (2.4.9). K is a constant of normalization. With the help of Eqs. (2.4.10) and (2.5.5), one has

$$C(l,s) = Kc(l,s) \otimes_3 c(l,-s)$$
$$= KP(l)\delta(s - l^2/2) \otimes_3 P(l)\delta(s + l^2/2).$$ (2.5.7)

The above 3-D convolution relation is shown in Fig. 2.5.1.

It is difficult to evaluate Eq. (2.5.7) directly. An alternative method for getting a solution of Eq. (2.5.7) is to derive first the corresponding defocused OTF $C(l, u)$:

$$C(l,u) = K \int_{-\infty}^{\infty} |h(v,u)|^2 J_0(lv) v \, dv.$$ (2.5.8)

Then the 3-D OTF can be obtained by performing the Fourier transform with respect to u.[2.8] Making use of Eq. (2.3.16) in Eq. (2.5.8) yields

$$C(l,u) = K[P(l,u) \otimes_2 P(l,-u)],$$ (2.5.9)

where $P(l, u)$ is the defocused pupil function defined in Eq. (2.3.18), and $\otimes_2$ denotes the 2-D convolution of $P(l, u)$ with $P(l, -u)$. The 2-D convolution in Eq. (2.5.9) can be evaluated, according to Fig. 2.5.2, as

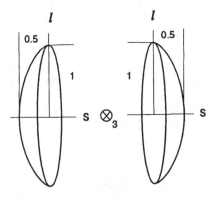

Fig. 2.5.1   3-D convolution of the 3-D CTF shown in Fig. 2.4.1 with its inverted conjugate function.

$$C(l,u) = K \iint_\sigma P_1(\rho_1)P_2(\rho_2)\exp\left(i\frac{\rho_1^2 u - \rho_2^2 u}{2}\right)\rho' \, d\rho' \, d\theta',$$

(2.5.10)

where σ represents the area overlapped by the two pupil functions as shown in Fig. 2.5.2. ρ' and θ' are the polar coordinates with an origin at O (see Fig. 2.5.2). ρ_1 and ρ_2 are given by

$$\rho_1^2 = \rho'^2 + \left(\frac{l}{2}\right)^2 + \rho' l \cos\theta',$$

(2.5.11)

$$\rho_2^2 = \rho'^2 + \left(\frac{l}{2}\right)^2 - \rho' l \cos\theta'.$$

Accordingly, we have

$$C(l,u) = K \iint_\sigma \exp(-i\rho' \, lu\cos\theta')\rho' \, d\rho' \, d\theta'.$$

(2.5.12)

It should be noticed that $C(l = 0, u)$ is, in fact, not a function of the defocus distance u at $l = 0$. This accordingly results in a singularity at the origin of the spatial frequency space so that the 3-D OTF cannot be normalized to unity at the origin. By performing the Fourier transform of $C(l, u)$ with respect to u, the 3-D OTF is evaluated as

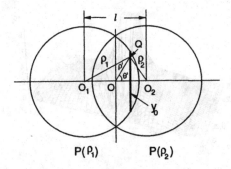

Fig. 2.5.2   2-D convolution of the two defocused pupil functions $P(\rho, u)$ and $P(\rho, -u)$. $P(\rho)$ is the pupil function normalized by the radius a. ρ' and θ are the polar coordinates with an origin at O. The lengths of $O_1 Q$ and $O_2 Q$ are denoted by ρ_1 and ρ_2. The thick vertical line is the path along which the integration in Eq. (2.5.13) is performed.

$$C(l,s) = \iint_{\sigma} \delta(s - \rho' \, l \cos\theta') \rho' \, d\rho' \, d\theta'. \tag{2.5.13}$$

Here the δ-function in Eq. (2.5.13) implies that the integration is taken along a vertical straight line on the $\rho' - \theta'$ plane as shown Fig. 2.5.2, the length of which is determined by the value of s and l. An analytical expression for Eq. (2.5.13) can be derived as follows:

$$C(l,s) = \frac{2}{l} \mathrm{Re}\left[\sqrt{1 - \left(\frac{|s|}{l} + \frac{l}{2}\right)^2} \, \right], \tag{2.5.14}$$

where Re[ ] denotes the real part of its argument. The result in Eq. (2.5.14) is an exact solution of the 3-D OTF for a single lens. It has a considerable difference from the 3-D OTF[2.8] based on the Stockseth's approximation.[2.9]

Fig. 2.5.3 shows the 3-D OTF of Eq. (2.5.14). It is seen that the 3-D OTF has a singularity at the origin and cuts off at $l = 2$ and $|s| = 1/2$, respectively. The non-zero region of the 3-D OTF is given by $|s| \leq l \, (1 - l/2)$. The 3-D OTF also exhibits a missing cone of spatial frequencies around the origin, implying that the information in this region cannot be imaged. This phenomenon is associated with the "wings" (streaking along the shadow edge) of the corresponding 3-D point spread function shown in Fig. 2.1.3b. The complete 3-D optical transfer function is radially symmetric about the s axis.

The projection of $C(l, s)$ in the plane at $s = 0$, i. e., in the focal plane, gives rise to

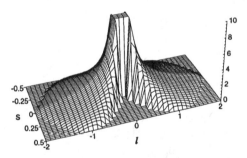

Fig. 2.5.3   3-D optical transfer function (OTF) for a single circular lens in incoherent imaging. The OTF is radially symmetrical about the s axis and has a singularity at the origin. According to Section 3.2, it is also the 3-D coherent transfer function in a confocal bright-field microscope consisting of a point source and a point detector and of two equal lenses.

$$C_2(l) = \frac{2}{\pi}\left[\cos^{-2}\left(\frac{l}{2}\right) - \frac{l}{2}\sqrt{1 - \left(\frac{l}{2}\right)^2}\right], \tag{2.5.15}$$

which is depicted in Fig. 2.5.4 and is the 2-D OTF $C_2(l)$, an description for in-focus imaging of a thin object (Chapter 10).[2.1, 2.10, 2.11]

For a lens with an annular pupil function (see Eq.(2.4.13)), the corresponding 3-D OTF can be derived, using the same way as described above, as[2.10]

$$C(l,s) = \frac{2}{l}\left\{\text{Re}\left[\sqrt{1 - \left(\frac{|s|}{l} + \frac{l}{2}\right)^2}\right] - \text{Re}\left[\sqrt{\varepsilon^2 - \left(\frac{|s|}{l} - \frac{l}{2}\right)^2}\right]\right\}. \tag{2.5.16}$$

It is seen that when $\varepsilon = 0$, Eq. (2.5.16) reduces to (2.5.14). The transverse cut-off spatial frequency remains 2 but the axial cut-off becomes $s = (1 - \varepsilon^2)/2$ and eventually is zero when $\varepsilon = 1$, so that the 3-D OTF gives non-zero values only on the plane of $s = 0$. This means that no axial information can be imaged for a very thin annular lens.[2.10] The depth of the focal length of the lens consequently becomes longer than that for a circular lens ($\varepsilon = 0$). The projection of Eq. (2.5.16) leads to the 2-D OTF for an annular lens (Fig. 2.5.4) as discussed by many people.[2.10, 2.11] The expression for the 2-D OTF is complicate[2.11, 2.12] but the corresponding 3-D OTF is simple and compact. Thus the specification of the 2-D OTF as a projection of the 3-D OTF avoids the necessity of defining various regions of applicability.

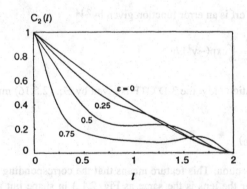

Fig. 2.5.4   2-D in-focus optical transfer function (OTF) for a single annular lens in incoherent imaging. ε represents the radius of the central obstruction normalized by the outer radius of the lens. When $\varepsilon = 0$, the 2-D OTF for a normal circular lens is obtained. According to Section 3.2, these curves also represent the 2-D in-focus coherent transfer function in a confocal bright-field microscope consisting of a point source and a point detector and of two equal lenses.

The 3-D OTF for a single lens can be degraded by the introduction of aberration. For example, if the lens has defocus and primary spherical aberration,[2.12, 2.13] the defocused pupil function can be expressed as

$$P(\rho,u) = P(\rho)\exp\left(-\frac{iu\rho^2}{2} + ikW_{020}\rho^2 + ikW_{040}\rho^4\right)$$

(2.5.17)

where W_{020} and W_{040} are the defocusing coefficient and the third-order (primary) spherical aberration coefficient.[2.2, 2.13] The aberrated 3-D OTF for an annular lens can be generalized, according to Eq. (2.5.9), to

$$C(l,s) = \frac{\sqrt{\pi}}{\sqrt{2kW_{040}}(1-i)l}\exp\left\{4ikW_{040}\left[\frac{l^2}{4} + \left(\frac{s}{l}\right)^2\right]\right\}$$

$$\left\{\text{erf}\left\{\sqrt{2kW_{040}}(1-i)\text{Re}\left[\sqrt{1-\left(\frac{|s|}{l}+\frac{l}{2}\right)^2}\right]\right\} - \right.$$

(2.5.18)

$$\left.\text{erf}\left\{\sqrt{2kW_{040}}(1-i)\text{Re}\left[\sqrt{\varepsilon^2-\left(\frac{|s|}{l}-\frac{l}{2}\right)^2}\right]\right\}\right\}$$

for $W_{020} = 0$. Here erf is an error function given by[2.14]

$$\text{erf}(x) = \frac{2}{\sqrt{\pi}}\int_0^x \exp(-y^2)dy.$$

(2.5.19)

For defocus aberration only, the 3-D OTF is given by Eq. (2.5.16) multiplied by a linear phase factor

$$\exp(2iksW_{020})$$

(2.5.20)

along the axial direction. This feature means that the corresponding 3-D intensity point spread function for the lens is the same as Fig. 2.1.3 in shape but axially shifted by a constant distance $2kW_{020}$.

2. 6 Three-Dimensional Imaging in a Conventional Microscope

Image formation in a conventional microscope is complicate.[2.4, 2.15, 2.16] Usually two lenses are involved in the imaging system. The first lens, called the condenser, focuses (or images) a finite-sized incoherent source on the sample. The function of this lens is to provide an illumination source and to control the degree of coherence for illumination. The second lens, called the objective, is responsible for imaging properties and the power of imaging resolution.

If the sample is labelled with fluorescent materials, the coherence of the light is completely destroyed by the sample and accordingly the imaging process becomes incoherent as the image intensity is simply given by the superposition of intensity of each point in the object. Therefore the discussions presented in Section 2.5 are applicable to 3-D incoherent imaging in a conventional microscope.

In general, the degree of coherence in conventional microscopy is dependent on the ratio, $\tilde{\gamma}$, of the condenser aperture to the objective aperture.[2.4] If $\tilde{\gamma}$ approaches zero, corresponding to the case that the objective is much larger than the condenser, imaging is thus purely coherent. As a result, the results presented in Sections 2.2, 2.3 and 2.4 can be used for the description of 3-D coherent imaging in a conventional microscope. For the other limiting case when $\tilde{\gamma} \rightarrow \infty$, i. e., when the condenser is much larger than the objective, imaging in the conventional microscope is incoherent and we can use those results in Section 2.5. For a finite value of $\tilde{\gamma}$, 3-D image formation in a conventional microscope is partially-coherent, so that the 3-D transmission cross-coefficient (TCC)[2.2, 2.4, 2.15, 2.16] must be used and in particular a 3-D weak-object transfer function (WOTF) is needed when scattering of light in the object is weak compared with the unscattered component.[2.15, 2.16] In Chapter 4, it will be seen that imaging in a partially-coherent conventional microscope is equivalent to that in a confocal microscope when a large detector is used.

References

2.1.   J. W. Goodman, *Introduction to Fourier Optics* (McGraw-Hill, New York, 1968).

2.2.   M. Born and E. Wolf, *Principles of Optics* (Pergamon, New York, 1980).

2.3.   M. Gu, *J. Opt. Soc. Am. A*, **12** (1995) 1602.

2.4.   C. J. R. Sheppard and X. Q. Mao, *J. Opt. Soc. Am. A*, **6** (1989) 1260 .

2.5.   B. R. Bracewell, *The Fourier Transform and Its Applications* (McGraw-Hill, New York, 1965).

2.6.   C. J. R. Sheppard and M. Gu, *J. Microscopy*, **165** (1992) 377.

2.7.   B. R. Frieden, *J. Opt. Soc. Am.*, **57** (1967) 56.

2.8.   C. J. R. Sheppard and M. Gu, *J. Opt. Soc. Am. A*, **8** (1991) 692.

2.9.   P. A. Stockseth, *J. Opt. Soc. Am.*, **59** (1969) 1314.

2.10.   C. J. R. Sheppard and M. Gu, *Opt. Commun.*, **81** (1991) 276.

2.11.   E. L. O'Nell, *J. Opt. Soc. Am.*, **46** (1956) 285.

2.12.   D. G. A. Jackson, M. Gu, and C. J. R. Sheppard, *J. Opt. Soc. Am. A*, **11** (1994) 1758.

2.13.   S. Wang and B. R. Frieden, *Applied Optics*, **29** (1990) 2424.

2.14.   I. S. Gradstein and I. M. Ryshik, *Tables of Series, Products and Integrals* (Harri Deutsch, Frankfurt, 1981).

2.15.   N. Streibl, *J. Opt. Soc. Am. A*, **2** (1985) 121.

2.16.   N. Streibl, *Optik*, **66** (1984) 341.

Chapter 3

CONFOCAL BRIGHT-FIELD MICROSCOPY
WITH A POINT DETECTOR

One of the key elements in a confocal scanning microscope is the pinhole mask placed in front of the detector. In order to achieve the true confocal imaging mode, the size of the pinhole should be chosen to be as small as possible. In the limiting case of a point-like detector, imaging is purely coherent in confocal bright-field microscopy providing that a specimen is non-fluorescent, and the system demonstrates a strong optical sectioning property which allows one to perform three-dimensional (3-D) imaging of thick objects.

Confocal bright-field microscopy includes two imaging modes in practice, confocal reflection-mode and confocal transmission-mode. For a point detector, image formation in these two cases is purely coherent as the detected intensity can be expressed as a form of superposition of amplitude. The 3-D imaging performance in confocal bright-field microscopy with a point detector is analysed in the present chapter. Section 3.1 presents the detailed derivation of 3-D image formation, showing the purely-coherent nature of the system. Section 3.2 introduces the 3-D coherent transfer function (CTF) for both reflection- and transmission-mode systems consisting of two equal lenses. Section 3.3 considers the influence of defocus and of primary spherical aberration, as well as the aberration balance in terms of the 3-D CTF. The aberrated 3-D CTF is then used to discuss axial and transverse resolution of the confocal system. The 3-D CTFs with detector offset are discussed in Section 3.4. The effect of the pupil functions including annular pupils on the 3-D CTF is analysed in Section 3.5.

3. 1   Coherent Image Formation

Schematic diagrams of the reflection-mode and the transmission-mode optical confocal scanning microscopes are shown in Fig. 3.1.1. A confocal reflection-mode microscope (Fig. 3.1.1a) can be considered to be a folded transmission system (Fig. 3.1.1b) in the backward direction. In general, there are two lenses included in a confocal system for imaging. The first lens is called the objective, the function of which is to provide a diffraction-limited spot on the sample, and the second lens, called the collector, is responsible for delivering the signal to a small detector. In a practical confocal

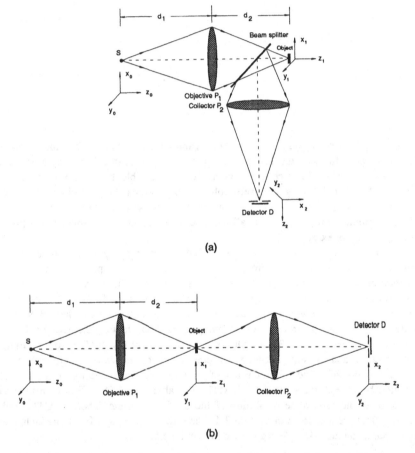

(a)

(b)

Fig. 3.1.1   Schematic diagram of confocal reflection-mode (a) and transmission-mode (b) microscopes.

reflection system,[3.1, 3.2] only one lens is used. In this case, a single lens plays the dual roles of illumination and collection. For an easy analysis, we explicitly draw two lenses in Fig. 3.1.1a. Let P_1 and P_2 denote the objective and collector lenses, respectively. Assume that the incident wavelength is λ. The incident light from a point source is focused by the objective lens onto a 3-D object. The signal from the illuminated spot is collected by the collection lens and recorded by a point detector. If the thick object is scanned along both axial and transverse directions, a map of the 3-D information, i. e., a 3-D image, may be obtained.

To analyse the image formation, three coordinate systems are created in source, object, and detector spaces (see Fig. 3.1.1) and represented by x_j, y_j, and z_j ($j = 0, 1, 2$). Position vectors in these spaces are denoted by r_j with three components x_j, y_j, and z_j.

The 3-D amplitude point spread functions (APSF) for the objective and collector lenses can be expressed, according to Eq. (2.3.10) in Chapter 2, as

$$h_q(x,y,z) = \iint_{-\infty}^{\infty} P_q(\xi,\eta,z) \exp\left[\frac{ik}{d_q}(\xi x + \eta y)\right] d\xi d\eta, \quad (q=1,2), \tag{3.1.1}$$

where $q = 1$ corresponds to the objective and $q = 2$ to the collector. In Eq. (3.1.1), pre-factors have been omitted as they do not affect the result in the case of continuous wave (CW) illumination. d_1 and d_2 are the distances from the objective to the source and to the object plane, respectively. They are also the distances from the collector lenses to the detector and to the object and satisfy the lens law:

$$\frac{1}{d_1} + \frac{1}{d_2} - \frac{1}{f} = 0, \tag{3.1.2}$$

where f is the focal length of the objective and collector, which implies that the objective and the collector have the same focal length. This assumption applies throughout the book because this is usually a practical situation. The functions $P_q(\xi_2\, \eta, z)$ in Eq. (3.1.1) represent the defocused pupil functions for the objective and the collector and are given, according to Eq. (2.3.10), by

$$P_q(\xi,\eta,z) = P_q(\xi,\eta)\exp\left[-\frac{ik}{2d_q^2}z(\xi^2 + \eta^2)\right]. \tag{3.1.3}$$

Here $P_1(\xi,\eta)$ and $P_2(\xi,\eta)$ are the pupil functions for the objective and collection lenses, respectively.

Consider the position vectors to be column vectors. If a point source is placed on the axis, the light field incident upon the object is given, in terms of Eq. (2.3.9), by

$$\int_{-\infty}^{\infty} \delta(r_0) \exp[ik(z_0 - z_1)] h_1(r_0 + M_1 r_1) dr_0,$$ (3.1.4)

where the constant pre-factors have been neglected as the Fresnel number has been assumed to be large. The delta function represents the point source and M_1 is a diagonal matrix of the magnification factors of the objective lens defined as

$$M_1 = \begin{bmatrix} M_1 & 0 & 0 \\ 0 & M_1 & 0 \\ 0 & 0 & -M_1^2 \end{bmatrix}.$$ (3.1.5)

Here the parameter $M_1 = d_1/d_2$ is the de-magnification of the objective lens. The 3-D object function is denoted by $o(r)$, which is the 3-D amplitude reflectivity for the reflection-mode microscope or the 3-D amplitude transmittance[3.3] for the transmission-mode microscope. Suppose that the object is scanned to a position $r_s(x_s, y_s, z_s)$. The light field from the object is given by

$$\left[\int_{-\infty}^{\infty} \delta(r_0) \exp[ik(z_0 - z_1)] h_1(r_0 + M_1 r_1) dr_0 \right] o(r_s - r_1).$$ (3.1.6)

The light field in detector space is given, according to Eq. (2.3.9), by the convolution of Eq. (3.1.6) with the 3-D APSF for the collection lens:

$$U(r_2, r_s) = \int_{-\infty}^{\infty} \left[\int_{-\infty}^{\infty} \delta(r_0) \exp[ik(z_0 - z_1)] h_1(r_0 + M_1 r_1) dr_0 \right] o(r_s - r_1)$$

(3.1.7)

$$\exp[ik(\pm z_1 - z_2)] h_2(r_1 + M_2 r_2) dr_1,$$

in which the positive sign in the exponent corresponds to the transmission case and the negative sign to the reflection case. The negative sign is caused by the fact that the light is reflected in the backward direction of the incident light. Here M_2 is a diagonal matrix of the magnification factors of the collection lens given by

$$M_2 = \begin{bmatrix} M_2 & 0 & 0 \\ 0 & M_2 & 0 \\ 0 & 0 & -M_2^2 \end{bmatrix},$$ (3.1.8)

where the parameter $M_2 = d_2/d_1$ is the magnification of the collection lens. The intensity in the detector space is the modulus squared of the field $U(r_2, r_s)$. If a point detector is placed at $r_{20}(x_{20}, y_{20}, z_{20})$ in the detector space, the detected intensity is

$$I(r_s) = \int_{-\infty}^{\infty} |U(r_2, r_s)|^2 \delta(r_2 - r_{20}) dr_2,$$ (3.1.9)

or

$$I(r_s) = \int \int \int \int \int_{-\infty}^{\infty} \delta(r_0) h_1(r_0 + M_1 r_1) o(r_s - r_1) h_2(r_1 + M_2 r_2)$$

$$\delta(r_0') h_1^*(r_0' + M_1 r_1') o^*(r_s - r_1') h_2^*(r_1' + M_2 r_2)$$ (3.1.10)

$$\exp[ik(z_0 - z_1 \pm z_1 - z_2)] \exp[-ik(z_0' - z_1' \pm z_0' - z_2)]$$

$$\delta(r_2 - r_{20}) dr_0 dr_0' dr_1 dr_1' dr_2,$$

where the bold letter r_j ($j = 0, 1, 2$) represents the position vector with components x_j', y_j', and z_j', and the symbol * denotes the conjugate operation.

It should be mentioned that we have used 3-D space-invariance in Eqs. (3.1.4) and (3.1.7) because it holds for the paraxial approximation as discussed in Section 2.3. When there is no detector offset, the 3-D space-invariance holds even for a high-aperture system as long as the confocal system is an object scanning system.

According to Fourier transform relations,[3.4, 3.5] Eq. (3.1.10) can be rewritten as

$$I(r_s) = |h_a(r_s) \otimes_3 o(r_s)|^2,$$ (3.1.11)

where $\otimes_3$ denotes the 3-D convolution operation. Eq. (3.1.11) represents a superposition principle of amplitude of the light field, meaning that the microscope behaves as a coherent microscope with a 3-D effective APSF given by

$$h_a(r) = \exp[ik(-z \pm z)] h_1(M_1 r) h_2(r + M_2 r_2),$$ (3.1.12)

which is the image field of a single point object. Without losing generality, we have used r and r_2 to replace r_s and r_{20}, respectively. The former denotes a position vector in object space, while the latter represents a position vector in detector space. Eq. (3.1.12) is a general form of the 3-D APSF in a confocal bright-field microscope with an offset point detector and an on-axis point source. It is applicable to transmission and reflection microscopes as well as to microscopes in the presence of aberration. In other words, if a point detector is utilized, imaging in confocal bright-field microscopy is purely coherent even for a system experiencing aberration. It is important to note that there is a linear phase term along the axial direction in Eq. (3.1.12). It disappears in transmission but causes the dark-field nature along the axial direction in confocal reflection microscopy as will be discussed in the next section.

When the detector offset is zero, i. e., when $r_2 = 0$, Eq. (3.1.12) reduces to

$$h_a(r) = \exp[ik(-z \pm z)]h_1(M_1r)h_2(r), \tag{3.1.13}$$

which applies to a confocal system that is perfectly aligned.[3.6]

For circular objective and collector lenses of radii a and b, respectively, introducing the optical coordinates to normalize Eq. (3.1.12) and using Eqs. (3.1.1) and (3.1.3), we have

$$h_a(v_x, v_y, u) = h_1(v, u)h_2(v_x, v_y, u), \tag{3.1.14}$$

where $h_1(v, u)$ and $h_2(v_x, v_y, u)$, expressed in the object space, are the amplitude point spread functions for the point source and the point detector, respectively, and given by

$$h_1(v, u) = \exp(-is_0u)\int_0^1 P_1(\rho, u)J_0(v\rho)\rho d\rho \tag{3.1.15}$$

and

$$h_2(v_x, v_y, u) = \exp(\pm is_0u)\int_0^{b/a} P_2(\rho, u)\exp(iu_2\rho^2/2)$$

$$\tag{3.1.16}$$

$$\exp\{i[(v_x + v_{2x})x_\rho + (v_y + v_{2y})y_\rho]\}dx_\rho dy_\rho.$$

Note that Eq. (3.1.16) does not have the circular symmetry due to the transverse detector offset and that s_0 has the same definition as in Eq. (2.4.12). The radial coordinate $\rho = (x_\rho^2 + y_\rho^2)^{1/2}$ has been normalized by a, so that x_ρ and y_ρ are two orthogonal coordinates over the lens aperture. v and u are the transverse and axial optical coordinates in the object space and $v = (v_x^2 + v_y^2)^{1/2}$. They are defined by

$$(v, v_x, v_y) = \frac{2\pi}{\lambda}(r, x, y)\frac{a}{d_2} \approx \frac{2\pi}{\lambda}(r, x, y)\sin\alpha_o,$$

$$\tag{3.1.17}$$

$$u = \frac{2\pi}{\lambda}z\frac{a^2}{d_2^2} \approx \frac{8\pi}{\lambda}z\sin^2(\alpha_o/2),$$

where $\sin\alpha_o$ is the numerical aperture of the objective in the object space. The coordinates for detector offset are denoted by v_{2x}, v_{2y} and u_2:

$$(v_2, v_{2x}, v_{2y}) = \frac{2\pi}{\lambda}(r_2, x_2, y_2)\frac{a}{d_1} \approx \frac{2\pi}{\lambda}(r_2, x_2, y_2)\sin\alpha_d,$$

(3.1.18)

$$u_2 = \frac{2\pi}{\lambda}z_2\frac{a^2}{d_1^2} \approx \frac{8\pi}{\lambda}z_2\sin^2(\alpha_d / 2).$$

Here $\sin\alpha_d$ is the numerical aperture of the collector in the detector space and $v_2 = (v_{2x}^2 + v_{2y}^2)^{1/2}$ and $r_2 = (x_2^2 + y_2^2)^{1/2}$.

Eq. (3.1.14) is not of circular symmetry when there exists detector offset. But as long as the transverse detector offset disappears, Eq. (3.1.14) reduces to a form of circular symmetry:

$$h_a(v, u) = h_1(v, u)h_2(v, u).$$

(3.1.19)

3. 2 Coherent Transfer Function

Eq. (3.1.11) means that the total amplitude detected by a point detector is the superposition of the amplitude contributed from different parts of the sample. On the other hand, one can consider the object to be a superposition of a series of periodic components. If the efficiency with which each component is imaged is derived, the performance of the imaging system can be understood. In order to obtain the efficiency, let us introduce a 3-D inverse Fourier transform of the object function $o(r)$ (the definition of the inverse Fourier transform is given in Appendix 1):

$$O(m) = \int_{-\infty}^{\infty} o(r)\exp(-2\pi i r \bullet m)dr$$

(3.2.1)

with the Fourier transform giving the 3-D object function:

$$o(r) = \int_{-\infty}^{\infty} O(m)\exp(2\pi i r \bullet m)dm,$$

(3.2.2)

where m represents the spatial frequency vector with two transverse components m and n, and one axial component s. Substituting Eq. (3.2.2) into Eq. (3.1.11) gives

$$I(r_s) = \left|\int_{-\infty}^{\infty} c(m)O(m)\exp(2\pi i r_s \bullet m)dm\right|^2.$$

(3.2.3)

Here $c(m)$ is given by

$$c(m) = \int_{-\infty}^{\infty} h_a(r)\exp(-2\pi ir \bullet m)dr, \tag{3.2.4}$$

which is the 3-D inverse Fourier transform of the 3-D effective APSF $h_a(r)$ and can be explicitly expressed as

$$c(m) = \int_{-\infty}^{\infty} \exp[ik(-z \pm z)]h_1(M_1r)h_2(r + M_1r_2)\exp(-2\pi ir \bullet m)dr. \tag{3.2.5}$$

It is seen that the product $c(m)O(m)$ in Eq. (3.2.3) represents the strength of the periodic components in the image. Thus $c(m)$ is the ratio of the image spectrum to the object spectrum and called the 3-D coherent transfer function (CTF) for a confocal microscope with a point detector. It is noted from Eq. (3.2.3) that the imaging property is fully determined by the 3-D CTF. Once the 3-D CTF is given, images of any object can be derived. The property of the 3-D CTF is only dependent on the optical system parameters such as aberrations, pupil functions, detector positions and so on. It is therefore understood that the 3-D CTF provides the information regarding the behaviour of the imaging system.

Suppose that the objective and the collector have circular pupils, so that $P_1(x, y, z) = P_1(r, z)$ and $P_2(x, y, z) = P_2(r, z)$, where $r = (x^2 + y^2)^{1/2}$. Substituting Eq. (3.1.1) into Eq. (3.2.5) and taking the cylindrical symmetry into account, we have

$$c(m,n,s) = \int_{-\infty}^{\infty} c(m,n,z)\exp(-2\pi izs)dz, \tag{3.2.6}$$

where $c(m, n, z)$ is the two-dimensional (2-D) defocused coherent transfer function given by

$$c(m,n,z) = \exp[ik(-z \pm z)]\Big\{P_1(\lambda d_1 l, -M_1^2 z) \otimes_2$$

$$\tag{3.2.7}$$

$$\Big\{P_2(\lambda d_2 l, z - M_2^2 z_2)\exp[2\pi iM_2(x_2 m + y_2 n)]\Big\}\Big\}.$$

Here $l = (m^2 + n^2)^{1/2}$ denotes the radial spatial frequency and $\otimes_2$ represents the 2-D convolution operation. While Eq. (3.2.5) is a general form of the 3-D CTF, Eq. (3.2.7) is only applicable to both reflection and transmission confocal microscopes using circular lenses.

Using Eqs. (3.1.15) and (3.1.16), we can rewrite Eqs. (3.2.5), (3.2.6) and (3.2.7) as

$$c(m,n,s) = \iiint_{-\infty}^{\infty} h(v,u)h(v_x,v_y,u)\exp[-i(v_x m + v_y n + us)]dv_x dv_y du, \tag{3.2.8}$$

$$c(m,n,s) = \int_{-\infty}^{\infty} c(m,n,u)\exp(-ius)du, \tag{3.2.9}$$

and

$$c(m,n,u) = \exp[is_0(-u \pm u)]\left\{ P_1(l,u) \otimes_2 \left\{ P_2(l,u)\exp(iu_z l^2 / 2)\exp[i(v_{2x}m + v_2 n)]\right\}\right\}, \tag{3.2.10}$$

where l, m and n have been normalized by Eq. (2.4.8) and s by Eq. (2.4.9). s_0 is given by Eq. (2.4.12). The factor in the inner curl brackets in Eq. (3.2.10) can be termed the effective defocused pupil function for the collector lens when the system has detector offset.

3. 2. 1 Reflection

Let us consider that the objective and collection lenses are aberration-free and have the same radius a, and that there is no detector offset. Eq. (3.2.10) reduces to a circularly symmetric form:

$$c_r(l,u) = K_r \exp(-2is_0 u)[P_1(l,u) \otimes_2 P_2(l,u)]. \tag{3.2.11}$$

K_r is a constant of normalization in reflection. The defocused pupil function for the objective lens in Eq. (3.2.11) can be expressed, in terms of Eqs. (3.1.3) and (2.3.18), as

$$P_1(\rho,u) = P(\rho)\exp\left(\frac{iu\rho^2}{2}\right), \tag{3.2.12}$$

where $P(\rho)$ is the pupil function for the objective. The positive sign in the quadratic phase term results from the fact that u is defined in the object space as shown in Fig. 3.1.1.

In the reflection confocal system, the signal is collected in the backward direction of the incident light, so that the defocus distance is the same for both objective and collector lenses. Accordingly,

$$P_2(\rho, u) = P_1(\rho, u). \tag{3.2.13}$$

Substituting $P_1(\rho, u)$ and $P_2(\rho, u)$ into Eq. (3.2.11) and performing the inverse Fourier transform of $c_r(l, u)$ with respect to u yields

$$c_r(l,s) = K_r[P(l)\delta(s+s_0 - l^2/2)] \otimes_3 [P(l)\delta(s+s_0 - l^2/2)]. \qquad (3.2.14)$$

It is noted that Eq. (3.2.14) is a convolution of two identical functions, each of which is the 3-D CTF for a single lens discussed in Eq. (2.4.10). Thus the 3-D CTF for a reflection confocal microscope is the auto-convolution of the 3-D CTF for conventional coherent microscopy, as shown in Fig. 3.2.1. Eq. (3.2.14) can be rewritten as

$$c_r(l,s) = K_r\delta(s+2s_0) \otimes_1 [P(l)\delta(s - l^2/2)] \otimes_3 [P(l)\delta(s - l^2/2)], \qquad (3.2.15)$$

which clearly indicates that the 3-D CTF is axially shifted by $2s_0$.

In order to evaluate Eq. (3.2.15), we return to the defocused CTF given by Eq. (3.2.11). With the help of Fig. 3.2.2., it can be expressed as

$$c_r(l,u) = K_r\iint_\sigma P(\rho_1)P(\rho_2)\exp\left(iu\frac{\rho_1^2+\rho_2^2}{2}\right)\rho'\,d\rho'\,d\theta', \qquad (3.2.16)$$

where ρ_1 and ρ_2 are given by Eq. (2.5.11) and σ is the area overlapped by the two pupil functions as shown Fig. 3.2.2. ρ' and θ' are the polar coordinates with an origin at O. Eq. (3.2.16) can be rewritten as

$$c_r(l,u) = K_r\exp(-2is_0u)\exp(iul^2/4)\int_0^{\pi/2}\int_0^{\rho_0}\exp(iu\rho'^2)\rho'\,d\rho'\,d\theta', \qquad (3.2.17)$$

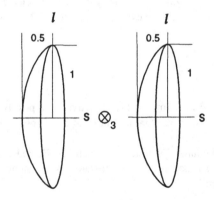

Fig. 3.2.1   3-D auto-convolution of the 3-D CTF for a conventional coherent microscope shown in Fig. 2.4.1.

where

$$\rho_0 = -\frac{l|\cos\theta'|}{2} + \sqrt{1 - \frac{l^2 \sin^2\theta'}{4}}, \tag{3.2.18}$$

as shown in Fig. 3.2.2

Performing the inverse Fourier transform of $c_r(l, u)$ with respect to u yields the 3-D coherent transfer function for the reflection-mode confocal scanning microscope:

$$c_r(l,s) = K_r \int_0^{\pi/2} \int_0^{\rho_0} \delta(s + 2s_0 - \rho'^2 - l^2/4)\rho' \, d\rho' \, d\theta', \tag{3.2.19}$$

where the δ-function implies that the above integral can be evaluated along a curve in the $\rho'-\theta'$ plane, as shown in Fig. 3.2.2. In the present case, the curve is a circle or part of a circle, depending on the value of ρ_0. After mathematical manipulations,[3.6] the 3-D coherent transfer function $c_r(l, s)$ normalized by the value of $c_r(l = 0, s = -2s_0)$ can be finally derived as

$$c_r(l,s) = \begin{cases} 1, & \frac{l^2}{4} \le \bar{s} \le 1 - l\left(1 - \frac{l}{2}\right), \\ \dfrac{2}{\pi}\sin^{-1}\dfrac{1 - \bar{s}}{l\sqrt{\bar{s} - \dfrac{l^2}{4}}}, & 1 - l\left(1 - \frac{l}{2}\right) \le \bar{s} \le 1, \\ 0, & otherwise, \end{cases} \tag{3.2.20}$$

where

$$\bar{s} = s + s0. \tag{3.2.21}$$

Obviously, the non-zero area of $c_r(l, s)$ is $l^2/4 < \bar{s} < 1$. It is seen that the 3-D CTF in Eq. (3.2.20) has a constant spatial frequency shift $s0$ along the s direction, given by

$$s0 = 2s_0 = \frac{1}{2\sin^2(\alpha_o/2)}. \tag{3.2.22}$$

The fact that the 3-D CTF has a single side band with a constant spatial frequency offset $s0$ along the z axis means that low axial spatial frequencies cannot be imaged and that the system exhibits an axial dark-field nature, so that axial phase variations can be

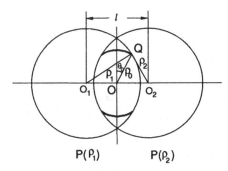

Fig. 3.2.2   2-D auto-convolution of the defocused pupil function $P(\rho, u)$. $P(\rho)$ is the pupil function normalized by the radius a. The lengths of O_1Q and O_2Q are ρ_1 and ρ_2. The thick curves are the paths along which the integration in Eq. (3.2.19) is performed.

detected. The spatial frequency offset can be also associated with the wings of the reflection 3-D amplitude point spread function.

In Fig. 3.2.3, the numerical 3-D plot of Eq. (3.2.20) gives a complete view of the 3-D CTF $c_r(l, s)$ in the reflection-mode confocal scanning microscope. Here, the discrete peaks result from the fact that $c_r(l, s)$ near $l = 2$ is too narrow to be continuously plotted when the plotting points are limited by the computer storage. Because of the constant spatial frequency shift $s0$, the value of the 3-D CTF is zero in the region where $1 - s0 \leq s \leq 0$. $c_r(l, s)$ remains non-zero only in the region of $l^2/4 \leq \bar{s} \leq 1$ with a transverse cut-off

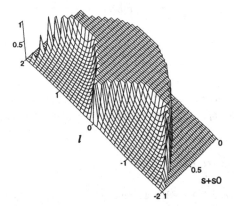

Fig. 3.2.3   3-D coherent transfer function for a confocal bright-field microscope with a point source and a point detector. The 3-D CTF is radially symmetrical about the s axis.

spatial frequency of 2 and an axial bandwidth of 1, both of which are twice as large as those in conventional coherent imaging presented in Fig. 2.4.1 The non-zero region is divided into three sub-regions.

Fig. 3.2.4 is the contour distribution of Fig. 3.2.4. It is obvious to find the three non-zero regions of $c_r(l, s)$. The 3-D CTF is a constant in one of the non-zero regions, while that in the other ones decays to zero as $s + s0 = 1$. The left border of the non-zero region (Fig. 3.2.4) is a parabolic curve, $s = l^2/4$. This feature results from the assumed paraxial approximation. But when the numerical aperture of the objective becomes large, it should be represented by a circular curve (Chapter 9).

It is important to notice that the distribution of $c_r(l = 0, s)$ is a square function which therefore results in a good optical sectioning effect in the microscope. To understand this effect, we consider a perfect reflector scanned in the axial direction. As explained in Chapter 1, the response to the reflector gives the strength of the optical sectioning effect. Since the object does not exhibit any transverse structure, the amplitude of the axial response is determined by $c_r(l = 0, u)$, which is the Fourier transform of $c_r(l = 0, s)$ according to Eq. (3.2.3) and can be analytically expressed as

$$U(u) = K_r \exp(-is0u)\exp(iu/2)\frac{\sin(u/2)}{u/2}. \tag{3.2.23}$$

The modulus squared of Eq. (3.2.23) gives rise to the intensity:

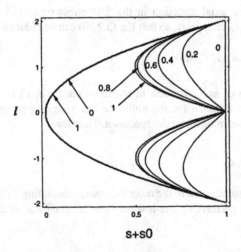

Fig. 3.2.4   Contours of the 3-D CTF in reflection, shown in Fig. 3.2.3.

$$I(u) = \left[\frac{\sin(u/2)}{(u/2)} \right]^2, \tag{3.2.24}$$

which has been normalized to unity at $u = 0$.

It is interesting to consider a case in which the object is a thin layer located in the focal plane ($u = 0$). Its object spectrum is $O(m) = O(m, n)$. In terms of Eq. (3.2.3), imaging in this case is described by the in-focus coherent transfer function, $c_r(l, u = 0)$, which is determined by the integration of the 3-D CTF, $c_r(l, s)$, with respect to the variable s and can be analytically expressed as

$$c_r(l, u = 0) = c_2(l) = \frac{2}{\pi} \left[\cos^{-1}\left(\frac{l}{2}\right) - \frac{l}{2}\sqrt{1 - \frac{l^2}{4}} \right]. \tag{3.2.25}$$

Eq. (3.2.25) has been normalized to unity at $l = 0$. It is not surprising that Eq. (3.2.25) is the well-known result for 2-D imaging of the confocal scanning microscope[3.1, 3.2, 3.7] and is identical to the in-focus 2-D OTF, depicted in Fig. 2.5.4, for a single lens in incoherent imaging. The general relationship of the 3-D CTF to the 2-D in-focus CTF will be discussed further in Chapter 10.

3. 2. 2   Transmission

In the case of the transmission-mode confocal scanning microscope (Fig. 3.1.1b), the linear phase in the axial direction in the 3-D effective amplitude point spread function (see Eq. 3.1.13) disappears, so that Eq. (3.2.10) can be reduced to

$$c_t(l, u) = K_t \left[P_1(l, u) \otimes_2 P_2(l, u) \right]. \tag{3.2.26}$$

Here K_t is a constant of normalization in transmission. $P_1(\rho, u)$ is the same as Eq. (3.2.12). The defocus distance for the collector has a sign opposite to that for the objective because the beam traverses the specimen. Therefore

$$P_2(\rho, -u) = P_1(\rho, u). \tag{3.2.27}$$

For the same lenses as those used in the reflection case, substituting $P_1(\rho, u)$ and $P_2(\rho, u)$ into Eq. (3.2.26) and performing the inverse Fourier transform of Eq. (3.2.26) with respect to u yields

$$c_t(l, s) = K_t [P(l)\delta(s - l^2/2)] \otimes_3 [P(l)\delta(s + l^2/2)]. \tag{3.2.28}$$

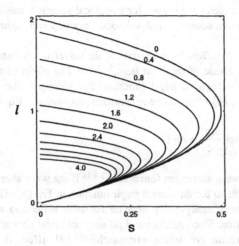

Fig. 3.2.5   Contours of the 3-D CTF in transmission, shown in Fig. 2.5.3.

Notice that Eq. (3.2.28) is the convolution of the 3-D CTF for a single lens, discussed in Eq. (2.4.10), with its axially inverted function, as shown in Fig. 2.5.1. Therefore, Eq. (3.2.28) is identical to Eq. (2.5.7). Thus we conclude that the 3-D CTF for an *aberration-free* transmission confocal microscope with two equal lenses is the same as the 3-D OTF for a single lens in incoherent imaging.[3.8, 3.9] Therefore the analytical solution of Eq. (3.2.28) is given by (2.5.14), which is shown in Fig. 2.5.3. Fig. 3.2.5 shows the contours of $c_t(l, s)$ in the region of $0 < l < 2$ and $0 < s < 1/2$.

The behaviour of the 3-D CTF in transmission differs from that in reflection. The former exhibits a missing cone of spatial frequencies around the origin, implying that the spatial frequencies in this region cannot be imaged. This conclusion leads to the fact that there is no optical sectioning effect in the transmission confocal microscope as $c_t(l = 0, s)$ is given by a delta function and its Fourier transform is thus constant. Another difference of the 3-D CTF between the two confocal systems is that there is no axial shift in the 3-D CTF for transmission, whereas the 3-D CTF in reflection is axially shifted, exhibiting an axial dark-field mode.

3. 3 Effects of Spherical Aberration

So far, we have derived the 3-D CTFs for confocal reflection and transmission microscopes without the presence of aberration. In practice, as a laser beam is focused into a specimen of finite thickness, the mismatching of the refractive indices between immersion materials and the specimen or the use of a cover slip of incorrect thickness results in spherical aberration[3.10, 3.11] even for a confocal system consisting of aberration-free lenses. This aberration source includes defocus, primary and higher-order spherical aberration.[3.10, 3.11]

If an optical system suffers from aberration, the wavefront is distorted because the aberration changes the phase of the light field.[3.12-3.14] The effect of aberration can be incorporated into the defocused pupil function. For example, the defocused pupil function $P_1(\rho, u)$ for the objective can be generalized, according to Eq. (2.4.14), to

$$P_1(\rho, u) = P(\rho) \exp\left(\frac{iu\rho^2}{2} + i\Phi\right),$$ (3.3.1)

where Φ is called the wave aberration function.[3.12-3.14] If the wave aberration function Φ is zero, $P_1(\rho, u)$ reduces to the defocused pupil function in Eq. (3.2.12), in which case, the 3-D CTFs have been analytically derived for both reflection and transmission systems in the last section. Since defocus and primary spherical aberration such as those resulting from the refractive index mismatch[3.10, 3.11] affect the axial imaging performance strongly, we may express the wave aberration function as

$$\Phi = kW_{020}\rho^2 + kW_{040}\rho^4.$$ (3.3.2)

Here W_{020} and W_{040} are called the defocusing coefficient and the third-order spherical aberration coefficient,[3.13, 3.14] respectively.

It is noted from Eqs. (3.3.1) and (3.3.2) that the phase aberration variation related to ρ^2 is given by $(u\rho^2/2 + kW_{020}\rho^2)$: the first term denotes the effect of the defocus distance u of the 3-D object, while the second represents the defocus aberration which may result from the refractive index mismatch of the specimen.[3.10, 3.11] Consider that the objective and collection lenses are identical. We further assume that both the objective and the collector have the same defocus aberration and third-order spherical aberration coefficient, so that Eq. (3.3.2) also applies to the collector lens. The effect of the defocus aberration W_{020} and the primary spherical aberration W_{040} on confocal bright-field imaging can be understood from the aberrated 3-D CTF discussed in the following.[3.15]

3. 3. 1   *Aberrated Coherent Transfer Function in Reflection*

As has been pointed out in the last section, we have $P_2(\rho, u) = P_1(\rho, u)$ for a reflection-mode system. It is, accordingly, seen that the defocus aberration W_{020} represents a shift in coordinates of the object for both the objective and the collector, in terms of Eqs. (3.3.1) and (3.3.2). By inserting the defocused pupil functions $P_1(\rho, u)$ and $P_2(\rho, u)$ including the defocus and primary aberration into Eq. (3.2.11) and taking the inverse Fourier transform with respect to u, the 3-D CTF is thus given by[3.15]

$$c_r(l,s) = K_r \exp(2ikW_{020}\bar{s})\int_0^{\pi/2}\int_0^{\rho_0} \delta(\bar{s}-\rho'^2-l^2/4)\exp[ikW_{040}(2\rho'^4+l^4/8+$$

$$\text{(3.3.3)}$$

$$\rho'^2 l^2(1+2\cos^2\theta'))]\rho'\,d\rho'\,d\theta',$$

where ρ_0 and $\bar{s}$ are given by Eqs. (3.2.18) and (3.2.21), respectively. In a similar way to solving Eq. (3.2.19), the 3-D CTF $c_r(l, s)$, normalized by the value of $c_r(l = 0, s = -s0)$, can be finally expressed as

$$c_r(l,s) = \exp\left\{2ikW_{020}\bar{s} + ikW_{040}\left[2\left(\bar{s}-\frac{l^2}{4}\right)^2 + \frac{l^4}{8} + \left(\bar{s}-\frac{l^2}{4}\right)l^2\right]\right\}f_1(l,s) \qquad \text{(3.3.4)}$$

with

$$f_1(l,s) = \begin{cases} \dfrac{2}{\pi}\displaystyle\int_0^{\pi/2} \exp\left[2ikW_{040}\left(\bar{s}-\dfrac{l^2}{4}\right)l^2\cos^2\theta'\right]d\theta', & \dfrac{l^2}{4}\le\bar{s}\le 1-l\left(1-\dfrac{l}{2}\right), \\[4mm] \dfrac{2}{\pi}\displaystyle\int_{\pi/2-\theta_0}^{\pi/2} \exp\left[2ikW_{040}\left(\bar{s}-\dfrac{l^2}{4}\right)l^2\cos^2\theta'\right]d\theta', & 1-l\left(1-\dfrac{l}{2}\right)\le\bar{s}\le 1, \\[4mm] 0, & \textit{otherwise,} \end{cases}$$

$$\text{(3.3.5)}$$

where

$$\theta_0 = \sin^{-1}\frac{1-\bar{s}}{l\sqrt{\bar{s}-\dfrac{l^2}{4}}}, \qquad \text{(3.3.6)}$$

which is depicted in Fig. 3.2.2.

Note that for an aberration-free reflection confocal system, i. e., for $W_{040} = W_{020} = 0$, the 3-D CTF $c_r(l, s)$ shown in Eq. (3.3.5) reduces to the analytical, real solution of the 3-D CTF (see Eq. (3.2.20)), while it is a complex function when either $W_{040} \neq 0$ or $W_{020} \neq 0$.

If there is no defocus ($W_{020} = 0$), the effects of primary spherical aberration on the modulus of the 3-D CTF in a reflection confocal system are shown in Fig. 3.3.1 for various values of the spherical aberration coefficient W_{040}. It is confirmed that when $W_{040} = 0.25\lambda$, corresponding to Rayleigh's quarter wavelength rule,[3.13] the 3-D CTF (Fig. 3.3.1a) shows little change compared with that for the aberration-free case (Fig.

3.2.3). The 3-D CTF reveals a significant change at $W_{040} = \lambda$ (Fig. 3.3.1b): a decrease of the 3-D CTF appears in a region around $\bar{s} = 0.5$ and $l = 1$. This decrease develops into a dip with increasing amounts of aberration. Between $W_{040} = \lambda$ and $W_{040} = 2\lambda$ (Fig. 3.3.1c), the dip becomes zero. After that point, the 3-D CTF in the region around $\bar{s} = 0.5$ and $l = 1$ increases again and becomes oscillatory (see the round peak in Fig. 3.3.1d or Fig. 3.3.2a), while the values of the 3-D CTF in the rest region keep decaying.

The variation of the CTF around $\bar{s} = 0.5$ and $l = 1$ with primary spherical aberration is further displayed in Fig. 3.3.2, in which Fig. 3.3.2a corresponds to the

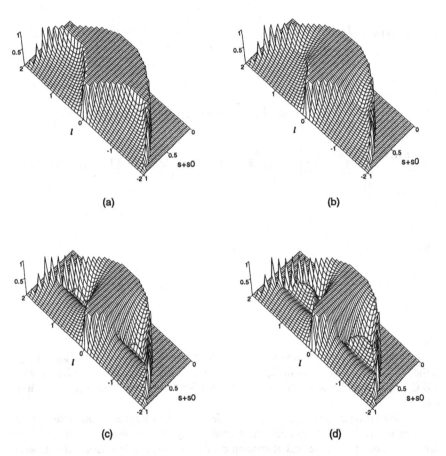

Fig. 3.3.1   Modulus of the 3-D CTF in reflection for various values of W_{040} (in units of wavelength) when $W_{020} = 0$: (a) $W_{040} = 0.25$; (b) $W_{040} = 1$; (c) $W_{040} = 2$; (d) $W_{040} = 3$.

cross-section of the modulus of the 3-D CTF, while Fig. 3.3.2b to the variation of the phase of the 3-D CTF on this section. It is seen that the phase of the 3-D CTF undergoes a π-jump, for example, at points A and A' when $W_{040} = 2\lambda$, and at points B and B' when $W_{040} = 3\lambda$ (Fig. 3.3.2b). From the numerical calculations, the phase jump appears only when $W_{040} \geq 1.55$. At $W_{040} = 1.55$, the dip of the modulus of the CTF decays to zero (see Fig. 3.3.2a), so that the phase jump represents a change in sign of the complex amplitude of the CTF. Note that the variation in phase in the s direction is monotonically increasing, except for the π-jump, (Fig. 3.3.3b), while that in the l direction is not (Fig.

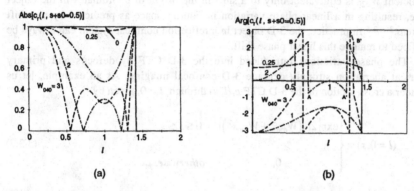

Fig. 3.3.2   Modulus (a) and phase (b) of the transverse cross-section of the 3-D CTF in reflection for different values of W_{040} (in units of wavelength) when $W_{020} = 0$. The phase variation is plotted within a range from $-\pi$ to π.

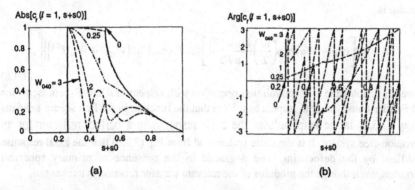

Fig. 3.3.3   Modulus (a) and phase (b) of the axial cross-section of the 3-D CTF in reflection for different values of W_{040} (in units of wavelength) when $W_{020} = 0$. The phase variation is plotted within a range from $-\pi$ to π.

3.3.2b).

If, on the other hand, there is no primary spherical aberration ($W_{040} = 0$) in the system, the 3-D CTF is reduced to a form which has a modulus, $f_1(l, s)$, identical to the 3-D CTF for an aberration-free confocal reflection system (Eq. (3.2.20)), and multiplied by a phase of $\exp[2ik(s+s0)W_{020}]$. In other words, the defocusing coefficient W_{020} generally does not affect the modulus of the 3-D CTF and only causes an extra phase shift. On the cross-section of the CTF at $\bar{s} = 0$, this phase shift disappears, implying that imaging for a thick object with only transverse change in strength is not affected by its defocusing. This can be understood by the fact that the presence of the defocusing coefficient W_{020} is equivalent only to a shift in the origin of coordinates in the object space, resulting in a linear phase variation in Fourier space as predicted by the shift theorem.[3.4, 3.5] In practice, the 3-D object in a reflection confocal system can always be redefined to remove this linear phase shift.

The phase variations introduced into the 3-D CTF by defocus and primary spherical aberration strongly degrade 3-D confocal imaging. As an example, let us consider a cross-section of the 3-D CTF $c_r(l, s)$ through $l = 0$, given by

$$c_r(l = 0, s) = \begin{cases} \exp\left[2ik\left(W_{020}\bar{s} + W_{040}\bar{s}^2\right)\right], & 0 \le \bar{s} \le 1, \\ \\ 0, & \textit{otherwise.} \end{cases}$$

(3.3.7)

It is seen that the modulus of $c_r(l = 0, s)$ is equal to unity, identical to the 3-D CTF for an aberration-free reflection confocal microscope. However, the amplitude of the axial response to a perfect reflector scanned along the axial direction is the Fourier transform of $c_r(l = 0, s)$ with respect to s (see Eq. (3.2.3)). Accordingly, the intensity of the axial response is

$$I(u) = K_r\left[\exp(-is0u)\exp\left(\frac{iu}{2}\right)\frac{\sin(u/2)}{u/2}\right]\otimes_1 F_1\left\{\exp\left[2ik\left(W_{020}\bar{s} + W_{040}\bar{s}^2\right)\right]\right\}\Big|^2,$$

(3.3.8)

where $\otimes_1$ denotes the 1-D convolution operation with respect to u and F_1 represents the 1-D Fourier transform with respect to s. Note that the function in the first square brackets of Eq. (3.3.8) is the amplitude of the axial response to a perfect reflector for an aberration-free system. It is therefore understood from Eq. (3.3.8) that the axial response is shifted by the defocusing, and degraded in the presence of primary spherical aberration, even though the modulus of the relevant transfer function is unchanged.

3. 3. 2 *Aberrated Coherent Transfer Function in Transmission*

For a transmission system, we therefore have $P_2(\rho, u) = P_1(\rho, -u)$. According to Eqs. (3.3.1) and (3.3.2), the defocusing coefficient W_{020} represents equivalently in this case an error in the separation between the two lenses. There are thus two degrees of freedom, u and W_{020}, which together define the defoci of the objective and collector lenses. Substituting $P_1(\rho, u)$ and $P_2(\rho, u)$ into Eq. (3.2.26) and evaluating the 2-D convolution and taking the inverse Fourier transform yields[3.15]

$$c_t(l,s) = K_t \int_0^\pi \int_0^{\rho_0} \delta(s - \rho' l \cos\theta') \exp\left\{ 2ikW_{020}\left(\frac{l^2}{4} + \rho'^2\right) + \right.$$

$$\left. ikW_{040}\left[2\rho'^4 + \frac{l^4}{8} + \rho'^2 l^2(1 + 2\cos^2\theta') \right] \right\} \rho' \, d\rho' \, d\theta'. \tag{3.3.9}$$

When either $W_{020} \neq 0$ or $W_{040} \neq 0$, Eq. (3.3.9) can be derived as

$$c_t(l,s) = \frac{2}{l}\exp\left\{ 2ikW_{020}\left[\frac{l^2}{4} + \left(\frac{s}{l}\right)^2\right] + ikW_{040}\left[3s^2 + 2\left(\frac{s}{l}\right)^2 + \frac{l^4}{8} \right] \right\}$$

$$\int_0^{y_0} \exp\left\{ 2ikW_{020}y^2 + ikW_{040}\left[\left(l^2 + 4\left(\frac{s}{l}\right)^2\right)y^2 + 2y^4 \right] \right\} dy, \tag{3.3.10}$$

where y_0, denoting the half length of the straight line along which the integration is evaluated (see Fig. 2.5.2), is given by

$$y_0 = \sqrt{1 - \left(\frac{|s|}{l} + \frac{l}{2}\right)^2}, \qquad \left(\frac{|s|}{l} + \frac{l}{2}\right)^2 \leq 1. \tag{3.3.11}$$

As is expected, a singularity exists at $l = 0$. The 3-D CTF $c_t(l, s)$ exhibits the following symmetry: $c_t(l, -s) = c_t(l, s)$.

Let us first consider the case when $W_{020} = 0$. Comparing Eq. (3.3.10) with Eq. (2.5.18), we find that the 3-D CTF in Eq. (3.3.10) for $W_{020} = 0$ is different from the 3-D optical transfer function (OTF) for a conventional incoherent system with primary aberration.[3.12] The reason is that the generalized defocused pupil function $P_2(\rho,-u)$ is not identical to $P_1^*(\rho,u)$ for a transmission confocal system *in the presence of primary* *spherical aberration even for two identical lenses.* In fact, only for an aberration-free system, i. e., for $\Phi = 0$, is the 3-D CTF for a transmission confocal system identical to the form of the 3-D OTF for a conventional incoherent imaging system.

It is also noted that the 3-D CTF $c_t(l, s)$ in Eq. (3.3.10) is a complex function. In the presence of primary spherical aberration, the modulus of the 3-D CTF $c_t(l, s)$ is shown in Fig. 3.3.4. A comparison of Fig. 3.3.4a with Fig. 2.5.3 shows that a quarter wavelength of aberration again causes only a weak effect on the 3-D CTF, but it is now stronger than the effect in the reflection case. As W_{040} increases, the modulus of the CTF displays an oscillatory behaviour nearly along the s direction. The period of the oscillation gradually becomes smaller as the aberration is enhanced (see Fig. 3.3.6). The 3-D CTF suffers from a strong attenuation in the region of high spatial frequencies in the presence of the aberration. This attenuation is clear in Fig. 3.3.6a. From Figs. 3.3.5 and 3.3.6, it is seen that as s and l increase, there is no appearance of the phase jumps

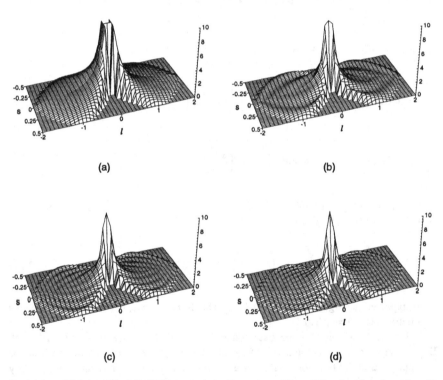

(a)   (b)

(c)   (d)

Fig. 3.3.4   Modulus of the 3-D CTF in transmission for various values of W_{040} (in units of wavelength) when $W_{020} = 0$: (a) $W_{040} = 0.25$; (b) $W_{040} = 1$; (c) $W_{040} = 2$; (d) $W_{040} = 3$.

because the complex amplitude of the 3-D CTF does not undergo a sign change. The phase variation is monotonically increasing in both l and s directions (see Figs. 3.3.5b and 3.3.6b).

If the transmission confocal system does not exhibit primary spherical aberration, i. e., $W_{040} = 0$ in Eq. (3.3.10), the 3-D CTF $c_t(l, s)$ in Eq. (3.3.10) can be analytically derived, if $W_{020} \neq 0$, as

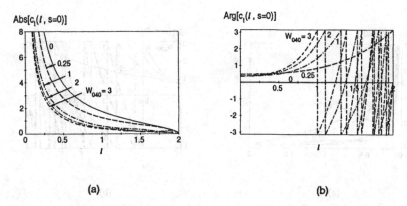

(a)   (b)

Fig. 3.3.5   Modulus (a) and phase (b) of the transverse cross-section of the 3-D CTF in transmission for different values of W_{040} (in units of wavelength) when $W_{020} = 0$. The phase variation is plotted within a range from $-\pi$ to π.

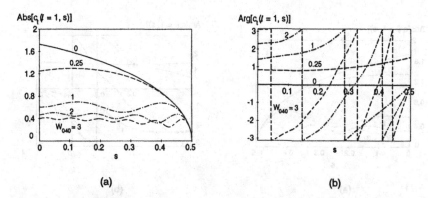

(a)   (b)

Fig. 3.3.6   Modulus (a) and phase (b) of the axial cross-section of the 3-D CTF in transmission for different values of W_{040} (in units of wavelength) when $W_{020} = 0$. The phase variation is plotted within a range from $-\pi$ to π.

$$c_t(l,s) = \frac{\sqrt{\pi}}{\sqrt{kW_{020}}\,(1-i)l}\exp\left\{2ikW_{020}\left[\frac{l^2}{4}+\left(\frac{s}{l}\right)^2\right]\right\}\mathrm{erf}\left[\sqrt{kW_{020}}\,(1-i)\,\mathrm{Re}(y_0)\right],$$

(3.3.12)

where Re( ) denotes the real part of its argument, which becomes zero when values of the argument are negative. Eq. (3.3.12) is identical to Eq. (2.5.18) with $\varepsilon = 0$ if W_{020} in Eq. (3.3.12) is replaced by $2W_{040}$. erf represents the error function[3.16] defined as

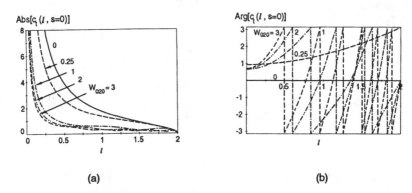

(a)   (b)

Fig. 3.3.7   Modulus (a) and phase (b) of the transverse cross-section of the 3-D CTF in transmission for different values of W_{020} (in units of wavelength) when $W_{040} = 0$. The phase variation is plotted within a range from $-\pi$ to π.

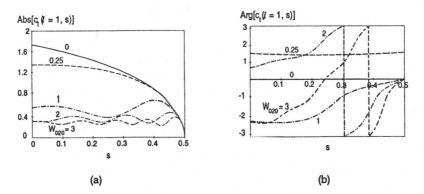

(a)   (b)

Fig. 3.3.8   Modulus (a) and phase (b) of the axial cross-section of the 3-D CTF in transmission for different values of W_{020} (in units of wavelength) when $W_{040} = 0$. The phase variation is plotted within a range from $-\pi$ to π.

$$\text{erf}(x) = \frac{2}{\sqrt{\pi}} \int_0^x \exp(-y^2) dy. \tag{3.3.13}$$

Now the defocusing coefficient W_{020}, because it is present in both lenses and is not equivalent to a shift in coordinates of the object, affects both the modulus and the phase of the 3-D CTF in transmission. In the case of conventional incoherent imaging with a single lens, W_{020} causes only a linear phase along the axial direction (see Eq. (2.5.20)).[3.17]

Numerical calculations of the variation of the 3-D CTF with W_{020} are shown in Figs. 3.3.7 and 3.3.8. It is seen that the variations in both modulus and phase are similar to those in Figs. 3.3.5 and 3.3.6. Therefore, the corresponding 3-D plots of the CTF, which are not shown here, are also similar to those in Fig. 3.3.4 if we set $W_{020} = W_{040}$.

3. 3. 3 Aberration Balance

As we pointed out in Section 2.5, the defocusing coefficient W_{020} does not affect the 3-D intensity distribution for a single lens system but causes an axial shift, which is the so-called displacement theorem (page 462 of Ref. 3.13). The combination of defocus and primary spherical aberration can be appropriately chosen, so that the total contribution of the aberration can be reduced.[3.17] In a similar way, the 3-D CTF in confocal microscopy can be also optimized.

If the confocal system includes both defocus and primary spherical aberration and $W_{040} = -W_{020}$, which is called the mid-focus setting condition, the primary spherical aberration is compensated by the defocus aberration W_{020} at the edge of the pupils. As we have shown, the modulus of the 3-D CTF in a reflection confocal system is not affected by the defocusing coefficient W_{020} but the phase of the 3-D CTF is shifted by $2kW_{020}\bar{s}$. Accordingly, under the condition for the balance ($W_{040} = -W_{020}$), Fig. 3.3.1, Fig. 3.3.2a and 3.3.3a still hold, while the phases shown in the Fig. 3.3.2b are shifted by a constant phase $-kW_{020}$ along the vertical axis as $\bar{s}$ is constant. However, on the section at $l = 1$, the phase variation is shown in Fig. 3.3.9. Comparing it with Fig. 3.3.3b, we find that the phase in the former case changes more slowly than that in the latter due to the balance of the two sources of the aberration. Note that the phase in Fig. 3.3.9 gradually decreases to zero at the cut-off spatial frequency because of the complete compensation of the aberration at the edge of the pupils.

On the other hand, it is seen from Eq. (3.3.10) that the variation of the 3-D CTF in transmission is complicate when $W_{040} = -W_{020}$. The 3-D CTFs on two sections at $s = 0$ and $l = 1$ are shown in Figs. 3.3.10 and 3.3.11. It is seen that the phase variations in Figs. 3.3.10b and 3.3.11b are slower compared with those in Figs. 3.3.5b and 3.3.6b. In particular, for the quarter wavelength aberration, the phase variations are nearly constant except for the region at the cut-off spatial frequency, which, accordingly, results in very little decrease in the modulus of the 3-D CTF in comparison with the 3-D CTF for an aberration-free case (see Figs. 3.3.10a and 3.3.11a). As a result of the balanced CTF, the image quality may be improved, which will be discussed in Section 3.3.4.

An alternative way for compensating for the primary aberration is to introduce another source of spherical aberration. For example, in practice, the primary spherical aberration caused by the mismatching of the refractive indices can be compensated for by alteration of the effective tube length.[3.10, 3.11] In this case the next order of the spherical aberration, i. e., the fifth-order spherical aberration term, i. e., $W = W_{060}\rho^6$ plays an important role in 3-D confocal imaging. Its effect on the 3-D CTF can be

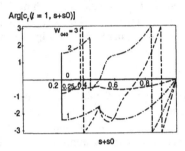

Fig. 3.3.9   Phase of the axial cross-section of the 3-D CTF in reflection for different values of W_{040} (in units of wavelength) under the condition of balance ($W_{040} = -W_{020}$). The phase variation is plotted within a range from $-\pi$ to π.

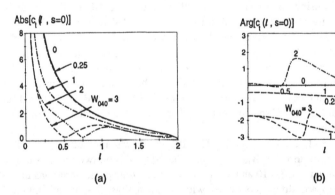

(a)   (b)

Fig. 3.3.10   Modulus (a) and phase (b) of the transverse cross-section of the 3-D CTF in transmission for different values of W_{040} (in units of wavelength) under the condition of balance ($W_{040} = -W_{020}$). The phase variation is plotted within a range from $-\pi$ to π.

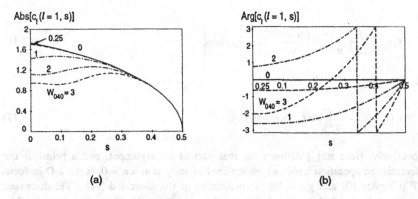

Fig. 3.3.11 Modulus (a) and phase (b) of the axial cross-sections of the 3-D CTF in transmission for different values of W_{020} (in units of wavelength) under the condition of balance ($W_{040} = -W_{020}$). The phase variation is plotted within a range from $-\pi$ to π.

investigated using the method presented in this section. In fact, the method described in this section is also applicable to analysing the effects of any aberration if the factor $k(W_{020}\rho^2 + W_{040}\rho^4)$ in Eq. (3.3.4) is given by the wave aberration function Φ.

3. 3. 4 Image of a Sharp Edge

The improvement in the 3-D CTF under the condition of $W_{040} = -W_{020}$ accordingly results in better images. For example, the sharpness of the images of a straight edge can be increased by using the balance condition.[3.18] This is importantly related to improving transverse resolution.

Let us consider a straight edge placed to be perpendicular to the x axis and initially located at $x = 0$. Its object function and Fourier transform can be expressed as

$$o(x,y,z) = \delta(z)\begin{cases} 1, & x \geq 0, \\ \\ 0, & x < 0, \end{cases} \qquad (3.3.14)$$

and

$$O(m,n,s) = \left[\frac{1}{2}\delta(m) + \frac{1}{2\pi i m}\right]\delta(n), \qquad (3.3.15)$$

respectively. Substituting Eq. (3.3.15) into Eq. (3.2.3) yields the image intensity of the edge and its gradient at the edge:

$$I(v_x) = \left| \frac{1}{2} + \frac{1}{\pi} \int_0^{\infty} \frac{c_2(m, n = 0)}{m} \sin(v_x m) dm \right|^2 \tag{3.3.16}$$

and

$$\Gamma'(v_x = 0) = \frac{1}{\pi} \int_0^{\infty} \mathrm{Re}[c_2(m, n = 0)] dm, \tag{3.3.17}$$

respectively. Here Re[ ] denotes the real part of its argument, and a prime is the differentiation operation. $c_2(m, n)$, normalized to unity at $m = n = 0$, is the 2-D in-focus CTF (Chapter 10) and given by the projection of the aberrated 3-D CTF, discussed above, into the focal plane. It is noted that the intensity at the edge is one quarter of its value far from the edge, as is expected from the normal coherent imaging,[3.7] and that the gradient of the intensity at the edge is given by the area under the real part of the aberrated 2-D in-focus CTF,[3.18] $c_2(m, n = 0)$. For a circularly symmetric system, $c_2(m, n = 0) = c_2(l)$.

For an aberration-free system, substituting Eq. (3.2.25) into Eq. (3.3.17) leads to

$$\Gamma'(v_x = 0) = \frac{8}{3\pi^2}. \tag{3.3.18}$$

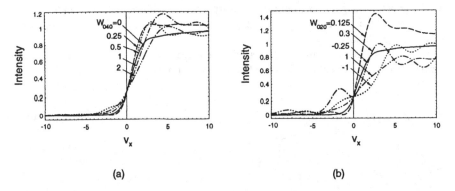

(a)   (b)

Fig. 3.3.12   Images of a sharp edge in the presence of aberration (in units of wavelength): (a) $W_{020} = 0$; (b) $W_{040} = 0.25$.

We first consider a confocal system in which there is no defocus aberration. Fig. 3.3.12a gives the images of the edge as a function of W_{040} when $W_{020} = 0$. Note that as the primary spherical aberration is gradually enhanced, the image of the edge first becomes sharper and then broader, leading to a maximum gradient at the edge for a particular amount of spherical aberration. This maximum gradient (see the dashed curve of Fig. 3.3.12a) is approximately 54% larger than that for the aberration-free case.[3.18] Once the gradient begins decreasing, the edge response becomes non-smooth (see Fig. 3.3.12a). The improvement in gradient at the edge was also found in the case of small defocus aberration.[3.18, 3.19]

In general, a confocal system may suffer from both defocus and primary spherical aberration. If the system has a certain amount of primary spherical aberration, the edge response can be made sharper by changing the defocus coefficient W_{020}. For example, Fig. 3.3.12b shows that for a given value of $W_{040} = 0.25$, the edge response for $W_{020} = 0.125$ is sharper than that for $W_{020} = 0$ but there is a strong overshoot at the edge. If the defocusing coefficient is too large, the edge response is then degraded. Notice that with primary spherical aberration the system does not show a dark-field behaviour and that the gradient of the intensity at the edge is asymmetric[3.18] with respect to the positive and negative values of W_{020}. In particular, when $W_{040} = 0.25$ and $W_{020} = -0.25$, i. e., under the mid-focus setting condition, the image of an edge is less affected by the aberrations and has an intensity gradient of 0.2730, almost the same as the gradient of $8/3\pi^2$ for an aberration-free system (see Eq. (3.3.18)).[3.7]

3. 4 Imaging with Detector Offset

An ideal confocal microscope requires that a point detector is placed on the axis and that there is no defocus of the detector. In practice, the detector may be offset as a result of improper alignment and accordingly the image quality can be altered. Under some conditions, an offset detector can lead to an improvement in image properties. For example, the sharpness of the image intensity of a straight edge can be improved when a detector is offset either in the transverse direction [3.20, 3.21] or in the axial direction.[3.18, 3.19] The 3-D CTF for a confocal bright-field microscope with detector offset is discussed in the present section.

3. 4. 1 Offset in the Transverse Direction

If the detector offset is only presented in the transverse direction, we can set $z_2 = 0$ in Eq. (3.2.7). Let us consider that x_2 and y_2 are the transverse coordinates of the offset detector. The defocused pupil function for the objective lens is given by Eq. (3.2.12) and the effective defocused pupil function for the collection lens is, according to Eqs. (3.1.16) and (3.2.10),

$$P_2(x_\rho, y_\rho, u) = P(\rho)\exp\left(\frac{iu\rho^2}{2} + iv_{2x}x_\rho + iv_{2y}y_\rho\right) \tag{3.4.1}$$

for reflection and

$$P_2(x_\rho, y_\rho, u) = P(\rho)\exp\left(-\frac{iu\rho^2}{2} + iv_{2x}x_\rho + iv_{2y}y_\rho\right) \tag{3.4.2}$$

for transmission. Here two equal circular lenses are assumed and $v_2 = (v_{2x} + v_{2y})^{1/2}$, defined in Eq. (3.1.18). The 3-D CTFs in two imaging modes can be expressed, in terms of the method presented in Section 3.2, as

$$c_r(m,n,s) = \exp\left[\frac{iv_2 l\cos(\phi + \phi')}{2}\right]f_2(m,n,s) \tag{3.4.3}$$

for reflection and

$$c_t(m,n,s) = \frac{2}{l}\exp\left[iv_2\left(\frac{l}{2} - \frac{s}{l}\right)\cos(\phi + \phi')\right]\mathrm{sinc}\left\{v_2\left[1 - \left(\frac{|s|}{l} + \frac{l}{2}\right)^2\right]\sin(\phi + \phi')\right\}$$

$$\mathrm{Re}\left[1 - \left(\frac{|s|}{l} + \frac{l}{2}\right)^2\right] \tag{3.4.4}$$

for transmission. Here

$$f_2(m,n,s) = \begin{cases} J_0\left(v_2\sqrt{\bar{s} - \frac{l^2}{4}}\right), & \frac{l^2}{4} \le \bar{s} \le 1 - l\left(1 - \frac{l}{2}\right), \\ \frac{1}{\pi}\int_{\pi/2-\theta_0}^{\pi/2+\theta_0}\exp\left[iv_2\sqrt{\bar{s} - \frac{l^2}{4}}\cos(\theta + \phi + \phi')\right]d\theta, & 1 - l\left(1 - \frac{l}{2}\right) \le \bar{s} \le 1, \\ 0, & otherwise, \end{cases} \tag{3.4.5}$$

$$\mathrm{sinc}(x) = \frac{\sin x}{x}, \tag{3.4.6}$$

$$\phi = \tan^{-1}\frac{n}{m},$$ (3.4.7)

$$\phi' = \tan^{-1}\frac{y_2}{x_2},$$ (3.4.8)

where θ_0 is the same as Eq. (3.3.6) and J_0 is a zero-order Bessel function of the first kind. The angle ϕ' is defined in a polar coordinate in the detector plane and ϕ is the angle in the m-n spatial frequency plane. The integration in Eq. (3.4.5) can be expressed by an Anger function.[3.22, 3.23] As is expected, the 3-D CTFs are complex and are not of circular symmetry due to the detector offset. If the detector offset is zero, Eqs. (3.4.3) and (3.4.4) reduce to Eqs. (3.2.20) and (2.5.14). When the detector offset equals the first minimum of the Airy function in the detector plane, the confocal system behaves as a dark-field imaging system.[3.20] The 3-D CTF for transmission is now asymmetric with respect to s.

Axial resolution in reflection confocal microscopy can be degraded because the modulus squared of the Fourier transform of $c_r(0, 0, s)$ becomes

$$I(u) = K\left[\exp(-is0u)\exp\left(\frac{iu}{2}\right)\frac{\sin(u/2)}{u/2}\right] \otimes_1 F_1\left[J_0\left(v_2\sqrt{s}\right)\right]^2.$$ (3.4.9)

3. 4. 2   Offset in the Axial Direction

When the detector has an offset only along the axial direction,[3.18, 3.19, 3.24] we have $x_2 = y_2 = 0$ in Eq. (3.3.7). With the help of Eq. (3.2.10), the effective defocused pupil function for the collection lens is

$$P_2(\rho,u) = P(\rho)\exp\left(\frac{iu\rho^2}{2} + \frac{iu_2\rho^2}{2}\right)$$ (3.4.10)

for reflection and

$$P_2(\rho,u) = P(\rho)\exp\left(-\frac{iu\rho^2}{2} + \frac{iu_2\rho^2}{2}\right)$$ (3.4.11)

for transmission. Here u_2 is the axial offset defined in Eq. (3.1.18).

Comparing Eqs. (3.4.10) and (3.4.11) with Eqs. (3.3.1) and (3.3.2), we can set

$$kW_{020} = u_2/2$$ (3.4.12)

for the collection lens. It is clear that the axial detector offset introduces a defocus aberration in the collection lens. Therefore, the 3-D CTF for reflection confocal microscopy with an axial detector offset is

$$c_r(l, s) = \exp(iu_2 \bar{s}/2) f_3(l, s) \tag{3.4.13}$$

with

$$f_3(l,s) = \begin{cases} J_0\left(\dfrac{u_2 l}{2}\sqrt{\bar{s} - \dfrac{l^2}{4}}\right), & \dfrac{l^2}{4} \le \bar{s} \le 1 - l\left(1 - \dfrac{l}{2}\right), \\[4ex] \dfrac{1}{\pi}\displaystyle\int_{\pi/2-\theta_0}^{\pi/2+\theta_0} \exp\left[-\dfrac{iu_2 l}{2}\sqrt{\bar{s} - \dfrac{l^2}{4}}\cos(\theta')\right] d\theta', & 1 - l\left(1 - \dfrac{l}{2}\right) \le \bar{s} \le 1, \\[4ex] 0, & otherwise, \end{cases} \tag{3.4.14}$$

and the 3-D CTF for transmission confocal microscopy with axial offset is

$$c_t(l,s) = \frac{\sqrt{\pi}}{\sqrt{u_2/4}(1-i)l}\exp\left(-\frac{iu_2 s}{2}\right)\exp\left\{\frac{iu_2}{2}\left[\frac{l^2}{4}+\left(\frac{s}{l}\right)^2\right]\right\} \tag{3.4.15}$$

$$\mathrm{erf}\left[\sqrt{u_2/4}(1-i)\mathrm{Re}(y_0)\right].$$

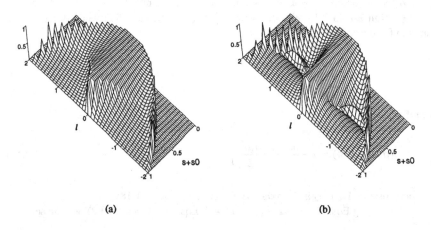

(a)   (b)

Fig. 3.4.1   Modulus of the 3-D CTF in reflection with axial detector offset u_2: (a) $u_2 = 6$; (b) $u_2 = 4\pi$.

The 3-D CTFs in both cases experience a linear phase change along the axial direction. In transmission, the 3-D CTF is of the same form as Eq. (3.3.12) except the linear phase term along s. This is because the defocus aberration exists only in the collection lens due to the axial detector offset. As a result , Figs. 3.3.7 and 3.3.8 are applicable in the present case, if we set $2kW_{020} = u_2/2$. Axial resolution in reflection is not degraded as $c_r(0, s) =$ $\exp(iu_2\bar{s}/2)$. But the complete modulus of the 3-D CTF changes (see Fig. 3.4.1), compared with that for the case without detector offset (see Fig. 3.2.3). Fig. 3.4.1b corresponds to the case in which the detector is axially placed in the first minimum of the 3-D intensity point spread function given by the modulus squared of Eq. (3.1.19).

3. 5 Influence of Pupil Functions

In the preceding sections, the 3-D CTFs have been derived for a confocal system with two equal pupil functions. In practice, the pupil functions for the objective and collector lenses may be different from each other because there may be different spatial filters placed in entrance and exit paths. For example, a central obstruction is often placed on either entrance or exit pupils of a microscope[3.25-3.27] to form effectively annular objective and collector lenses, so that the transverse resolution can be improved and the effects of aberrations reduced. Further, using an annular objective only may result in improved signal level compared with using two annular pupils.[3.25]

For an annular lens, the pupil function can be considered to be a circular pupil function obstructed by a circular central mask (see Eq. (2.4.13)),[3.26] so that the defocused pupil functions for annular objective and collector lenses P_1 and P_2 can be expressed as

$$P_1(\rho, u) = P_{1a}(\rho, u) - P_{1b}(\rho, u),$$

(3.5.1)

$$P_2(\rho, u) = P_{2a}(\rho, u) - P_{2b}(\rho, u).$$

(3.5.2)

Here we have assumed that the focal length of the objective is the same as that of the collector. By substituting Eqs. (3.5.1) and (3.5.2) into Eq. (3.2.10), the defocused CTF can be thus derived, if there is no detector offset, as

$$c(l,u) = \exp[is_0(-u \pm u)]\left[P_{1a}(l,u) \otimes_2 P_{2a}(l,u) + P_{1b}(l,u) \otimes_2 P_{2b}(l,u)\right.$$

(3.5.3)

$$\left. - P_{1a}(l,u) \otimes_2 P_{2b}(l,u) - P_{1b}(l,u) \otimes_2 P_{2a}(l,u)\right].$$

Note that the defocused CTF is finally given by four terms each of which represents the defocused CTF for two circular lenses of different radii.

Assume that the defocused pupil function for the annular objective with inner and outer radii of $\varepsilon_1 a$ and a $(0 < \varepsilon_1 < 1)$ can be expressed as

$$P_{1a}(\rho, u) = \exp\left(\frac{iu\rho^2}{2}\right), \qquad 0 \leq \rho \leq 1, \tag{3.5.4}$$

$$P_{1b}(\rho, u) = \exp\left(\frac{iu\rho^2}{2}\right), \qquad 0 \leq \rho \leq \varepsilon_1. \tag{3.5.5}$$

For a reflection-mode confocal microscope, if the annular collector has inner and outer radii of $\varepsilon_2 b$ and b $(0 < \varepsilon_2 < 1)$, its defocused pupil function is then given by

$$P_{2a}(\rho, u) = \exp\left(\frac{iu\rho^2}{2}\right), \qquad 0 \leq \rho \leq b/a, \tag{3.5.6}$$

$$P_{2b}(\rho, u) = \exp\left(\frac{iu\rho^2}{2}\right), \qquad 0 \leq \rho \leq \varepsilon_2 b/a. \tag{3.5.7}$$

By substituting Eqs. (3.5.4)-(3.5.7) into Eq. (3.5.3) and performing the inverse Fourier transform with respect to u, one has the 3-D CTF for the reflection-mode confocal microscope:

$$c_r(l,s) = K_r\left(\iint_{\sigma_1} + \iint_{\sigma_2} - \iint_{\sigma_3} - \iint_{\sigma_4}\right)\delta\left(\bar{s} - \rho'^2 - \frac{l^2}{4}\right)\rho'\,d\rho'\,d\theta'. \tag{3.5.8}$$

Here σ_1, σ_2, σ_3 and σ_4 are the non-zero areas overlapped by two unequal circular pupil functions with normalized radii $\rho_{1a} = 1$ and $\rho_{2a} = b/a$, $\rho_{1b} = \varepsilon_1$ and $\rho_{2b} = \varepsilon_2 b/a$, $\rho_{1a} = 1$ and $\rho_{2b} = \varepsilon_2 b/a$, and $\rho_{1b} = \varepsilon_1$ and $\rho_{2a} = b/a$, respectively. The delta function in Eq. (3.5.8) implies that the double integral can be reduced to a single integral with respect to θ'. Each of four terms represents the 3-D CTF for two unequal circular pupil functions.

In the case of the transmission-mode confocal scanning microscope, the defocused pupil function for the collector now becomes

$$P_{2a}(\rho, u) = \exp\left(-\frac{iu\rho^2}{2}\right), \qquad 0 \leq \rho \leq b/a, \tag{3.5.9}$$

$$P_{2b}(\rho, u) = \exp\left(-\frac{iu\rho^2}{2}\right), \qquad 0 \leq \rho \leq \varepsilon_2 b/a. \tag{3.5.10}$$

Using Eqs. (3.5.4), (3.5.5), (3.5.9) and (3.5.10) in Eq. (3.5.3), we can derive the 3-D CTF:

$$c_t(l,s) = K_t \left(\iint_{\sigma_1} + \iint_{\sigma_2} - \iint_{\sigma_3} - \iint_{\sigma_4} \right) \delta(s - \rho' l \cos\theta') \rho' \, d\rho' \, d\theta'. \qquad (3.5.11)$$

where the meanings of σ_1, σ_2, σ_3, and σ_4 remain the same as those in the reflection case. It should be noticed that $c_t(l = 0, s)$ has a singularity at the origin of the spatial frequency space, so that the 3-D CTF cannot be normalized to unity at the origin. Like the reflection case, the 3-D CTF in transmission is composed of four terms. Each of them denotes the 3-D CTF for the transmission-mode confocal scanning microscope with two unequal circular pupil functions.

3. 5. 1 *Two Unequal Circular Pupils*

Let us consider one of the four terms in Eqs. (3.5.8) and (3.5.11). It corresponds to the 3-D CTF for two unequal circular lenses in reflection or transmission systems. Without losing generality, we assume that the two pupils have normalized radii ρ_1 and ρ_2, respectively.

The 3-D CTF in the reflection case can be expressed as

$$c_r(l,s) = K_r \iint_{\sigma} \delta\left(\bar{s} - \rho'^2 - \frac{l^2}{4} \right) \rho' \, d\rho' \, d\theta'. \qquad (3.5.12)$$

Here σ denotes the non-zero area overlapped by the two circular pupil functions with normalized radii ρ_1 and ρ_2. The 3-D CTF $c_r(l, s)$, normalized by the value of $c_r(l = 0, s = -s0)$, can be derived analytically as follows:[3.26]

1) $\rho_1 > \rho_2$

1a) $0 < l < \rho_1 - \rho_2$

i) $2\rho_2 < l < \rho_1 - \rho_2$

$$c_r(l,s) = \begin{cases} \dfrac{1}{\pi}\cos^{-1}\dfrac{\bar{s}-\rho_2^2}{l\sqrt{\bar{s}-\dfrac{l^2}{4}}}, & \rho_2^2 - l\left(\rho_2-\dfrac{l}{2}\right) \le \bar{s} \le \rho_2^2 + l\left(\rho_2+\dfrac{l}{2}\right), \\[6mm] 0, & \text{otherwise.} \end{cases}$$

(3.5.13)

ii) $0 < l < 2\rho_2$

$$c_r(l,s) = \begin{cases} 1, & \dfrac{l^2}{4} \le \bar{s} \le \rho_2^2 - l\left(\rho_2-\dfrac{l}{2}\right), \\[6mm] \dfrac{1}{\pi}\cos^{-1}\dfrac{\bar{s}-\rho_2^2}{l\sqrt{\bar{s}-\dfrac{l^2}{4}}}, & \rho_2^2 - l\left(\rho_2-\dfrac{l}{2}\right) \le \bar{s} \le \rho_2^2 + l\left(\rho_2+\dfrac{l}{2}\right), \\[6mm] 0, & \text{otherwise.} \end{cases}$$

(3.5.14)

1b) $\rho_1 - \rho_2 < l < \rho_1 + \rho_2$

i) $2\rho_2 < l < \rho_1 + \rho_2$

$$c_r(l,s) = \begin{cases} \dfrac{1}{\pi}\cos^{-1}\dfrac{\bar{s}-\rho_2^2}{l\sqrt{\bar{s}-\dfrac{l^2}{4}}}, & \rho_2^2 - l\left(\rho_2-\dfrac{l}{2}\right) \le \bar{s} \le \rho_1^2 - l\left(\rho_1-\dfrac{l}{2}\right), \\[8mm] \dfrac{1}{\pi}\left(\cos^{-1}\dfrac{\bar{s}-\rho_2^2}{l\sqrt{\bar{s}-\dfrac{l^2}{4}}} - \cos^{-1}\dfrac{\rho_1^2-\bar{s}}{l\sqrt{\bar{s}-\dfrac{l^2}{4}}}\right), & \rho_1^2 - l\left(\rho_1-\dfrac{l}{2}\right) \le \bar{s} \le \dfrac{\rho_1^2+\rho_2^2}{2}, \\[8mm] 0, & \text{otherwise.} \end{cases}$$

(3.5.15)

ii) $\rho_1 - \rho_2 < l < 2\rho_2$

$$
c_r(l,s) = \begin{cases}
1, & \dfrac{l^2}{4} \le \bar{s} \le \rho_2^2 - l\left(\rho_2 - \dfrac{l}{2}\right), \\[3mm]
\dfrac{1}{\pi}\cos^{-1}\dfrac{\bar{s} - \rho_2^2}{l\sqrt{\bar{s} - \dfrac{l^2}{4}}}, & \rho_2^2 - l\left(\rho_2 - \dfrac{l}{2}\right) \le \bar{s} \le \rho_1^2 - l\left(\rho_1 - \dfrac{l}{2}\right), \\[3mm]
\dfrac{1}{\pi}\left(\cos^{-1}\dfrac{\bar{s} - \rho_2^2}{l\sqrt{\bar{s} - \dfrac{l^2}{4}}} - \cos^{-1}\dfrac{\rho_1^2 - \bar{s}}{l\sqrt{\bar{s} - \dfrac{l^2}{4}}}\right), & \rho_1^2 - l\left(\rho_1 - \dfrac{l}{2}\right) \le \bar{s} \le \dfrac{\rho_1^2 + \rho_2^2}{2}, \\[3mm]
0, & \text{otherwise.}
\end{cases}
$$

$$(3.5.16)$$

2) $\rho_1 < \rho_2$

The expressions for the 3-D CTF in this case are similar to Eqs. (3.5.13)-(3.5.16) with the following replacements:

$$\rho_1 \Rightarrow \rho_2, \qquad\qquad (3.5.17)$$

$$\rho_2 \Rightarrow \rho_1. \qquad\qquad (3.5.18)$$

Eqs. (3.5.17) and (3.5.18) imply that the 3-D CTF in reflection remains unchanged when the objective and collector lenses are exchanged.

For transmission, we can express the 3-D CTF as

$$c_t(l,s) = K_t \iint_\sigma \delta(s - \rho' l\cos\theta')\rho'\, d\rho'\, d\theta'. \qquad\qquad (3.5.19)$$

An analytical expression for Eq. (3.5.19) can be derived as follows:

1) $\rho_1 > \rho_2$

1a) $0 < l < \rho_1 - \rho_2$

$$c_t(l,s) = \begin{cases} \dfrac{2}{l}\mathrm{Re}\left[\sqrt{\rho_2^2 - \left(\dfrac{s}{l} - \dfrac{l}{2}\right)^2}\right], & -l\left(\rho_2 - \dfrac{l}{2}\right) \le s \le l\left(\rho_2 + \dfrac{l}{2}\right), \\ \\ 0, & otherwise. \end{cases}$$

(3.5.20)

1b) $\rho_1 - \rho_2 < l < \rho_1 + \rho_2$

$$c_t(l,s) = \begin{cases} \dfrac{2}{l}\mathrm{Re}\left[\sqrt{\rho_1^2 - \left(\dfrac{s}{l} + \dfrac{l}{2}\right)^2}\right], & \dfrac{\rho_1^2 - \rho_2^2}{2} \le s \le l\left(\rho_1 - \dfrac{l}{2}\right), \\ \\ \dfrac{2}{l}\mathrm{Re}\left[\sqrt{\rho_2^2 - \left(\dfrac{s}{l} - \dfrac{l}{2}\right)^2}\right], & -l\left(\rho_2 - \dfrac{l}{2}\right) \le s \le \dfrac{\rho_1^2 - \rho_2^2}{2}, \\ \\ 0, & otherwise. \end{cases}$$

(3.5.21)

Here Re [ ] remains the same meaning as before.

2) $\rho_1 < \rho_2$

In this case, the 3-D CTF has a similar form to Eqs. (3.5.20) and (3.5.21) using the replacements:

$$\rho_2 \Rightarrow \rho_1,$$

(3.5.22)

$$\rho_1 \Rightarrow \rho_2,$$

(3.5.23)

$$s \Rightarrow -s.$$

(3.5.24)

It is obvious that imaging in the axial direction is asymmetric when exchanging the objective and collection lenses.

Fig. 3.5.1 gives the 3-D CTFs for a confocal reflection system with two unequal circular pupil functions. Comparing them with the 3-D CTF for two equal circular pupil functions ($\rho_{1a} = \rho_{2a} = 1$) depicted in Fig. 3.2.3, we find that the cut-off at high true axial spatial frequencies (i. e., $s + s0$) is zero, the same as that of the 3-D CTF for two equal circular pupil functions, while that at low true axial spatial frequencies in the former case is $(1 + \rho_{2a}^2)/2$ rather than unity as occurs in the latter case. This decrease in the cut-off results in a reduced region where the 3-D CTF is not zero, degrading resolution in both transverse and axial directions. In particular, as ρ_{2a} decreases, the 3-D CTFs in Fig. 3.5.1 demonstrate a missing cone of spatial frequencies (i. e., $s + s0 = 1$) at $l = 0$, meaning that

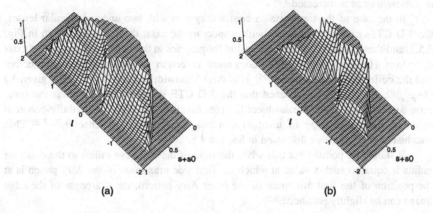

(a)   (b)

Fig. 3.5.1   3-D CTF in reflection with two unequal circular pupils ($\rho_{1a} = 1$, $\rho_{1b} = \rho_{2b} = 0$): (a) $\rho_{2a} = 0.75$; (b) $\rho_{2a} = 0.5$.

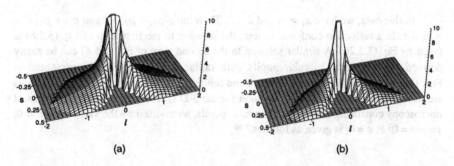

(a)   (b)

Fig. 3.5.2   3-D CTF in transmission with two unequal circular pupils ($\rho_{1a} = 1$, $\rho_{1b} = \rho_{2b} = 0$): (a) $\rho_{2a} = 0.75$; (b) $\rho_{2a} = 0.5$.

the cut-off at low true axial spatial frequencies at $l = 0$ is less than that for $l \neq 0$. This property directly affects optical sectioning because the strength of optical sectioning is given by the Fourier transform of the 3-D CTF at $l = 0$. Therefore, the use of two unequal circular pupils causes a reduction of the optical sectioning strength. Note that when $\rho_{2a} = 1$, the 3-D CTF is bounded by $l^2/4$ at high true axial spatial frequencies. This is, however, no longer true if $\rho_{2a} < 1$. Eventually, the border at $s + s0 = 0$ is given by $l^2/2$ for $\rho_{2a} = 0$. The corresponding 3-D CTF is expressed by Eq. (2.4.10), which is the 3-D CTF for a single lens. This property implies that when the role of the collector is completely degraded, the imaging property reduces to that in a conventional coherent system. Although the 3-D imaging ability becomes poor in this case, the sharpness of the image of a straight edge is increased.[3.28]

In the case of the transmission confocal system with two unequal circular lenses, the 3-D CTFs exhibit asymmetry with respect to the axial direction, as shown in Fig. 3.5.2, and thus give different cut-off spatial frequencies in the positive and negative axial directions. However, the 3-D CTF does show inversion symmetry when the objective and the collector are exchanged[3.26]. The axial bandwidth of the 3-D CTF is given by $(1+\rho_{2a}^2)/2$. It should be mentioned that the 3-D CTF in the present case has the same form as that for the 3-D weak-object transfer function (WOTF) for partially-coherent conventional microscopy of unequal condenser and objective lenses.[3.26, 3.29] This conclusion will be further discussed in Section 4.6.

It should be pointed out that when the ratio of the objective radius to the collector radius is equal to such a value at which the first side maximum of the Airy patten is at the position of the first minimum of the other Airy pattern, the sharpness of the edge image can be slightly enhanced.[3.26]

3. 5. 2   Two Equal Annular Pupils

In this case, we have $\varepsilon_1 = \varepsilon_2$ and $a = b$. Therefore, $\rho_{1a} = \rho_{2a} = 1$ and $\rho_{1b} = \rho_{2b} = \varepsilon_1 = \varepsilon_2 = \varepsilon$. In a reflection confocal system, the solution to the first term of Eq. (3.5.8) is given by Eq. (3.2.20). A similar solution to the second term of Eq. (3.5.8) can be easily derived for two equal circular pupils with radius ε. The expressions presented in Eqs. (3.5.13) - (3.5.15) apply to the third and fourth terms when appropriate substitutions are used. Finally an analytical expression for the 3-D CTF in reflection-mode confocal microscopy consisting of two equal annular pupils, normalized by the value of $c_r(l = 0, s = -s0 = 0)$ at $\varepsilon = 0$, is given as follows.[3.30]

1) $\varepsilon \le 1/3$

i) $\varepsilon + 1 \le l \le 2$

$$c_r(l,s) = \begin{cases} 1, & \dfrac{l^2}{4} \le \bar{s} \le 1 - l\left(1 - \dfrac{l}{2}\right), \\[4mm] \dfrac{2}{\pi}\sin^{-1}\dfrac{1 - \bar{s}}{l\sqrt{\bar{s} - \dfrac{l^2}{4}}}, & 1 - l\left(1 - \dfrac{l}{2}\right) \le \bar{s} \le 1, \\[4mm] 0, & otherwise. \end{cases} \qquad (3.5.25)$$

ii) $1 - \varepsilon \le l \le \varepsilon + 1$

$$c_r(l,s) = \begin{cases} 1, & \dfrac{l^2}{4} \le \bar{s} \le \varepsilon^2 - l\left(\varepsilon - \dfrac{l}{2}\right), \\[4mm] \dfrac{2}{\pi}\sin^{-1}\dfrac{\bar{s} - \varepsilon^2}{l\sqrt{\bar{s} - \dfrac{l^2}{4}}}, & \varepsilon^2 - l\left(\varepsilon - \dfrac{l}{2}\right) \le \bar{s} \le \dfrac{1 + \varepsilon^2}{2}, \\[4mm] \dfrac{2}{\pi}\sin^{-1}\dfrac{1 - \bar{s}}{l\sqrt{\bar{s} - \dfrac{l^2}{4}}}, & \dfrac{1 + \varepsilon^2}{2} \le \bar{s} \le 1, \\[4mm] 0, & otherwise. \end{cases} \qquad (3.5.26)$$

iii) $2\varepsilon \le l \le 1 - \varepsilon$

$$c_r(l,s) = \begin{cases} 1, & \dfrac{l^2}{4} \le \bar{s} \le \varepsilon^2 - l\left(\varepsilon - \dfrac{l}{2}\right), \\[2ex] \dfrac{2}{\pi}\sin^{-1}\dfrac{\bar{s}-\varepsilon^2}{l\sqrt{\bar{s}-\dfrac{l^2}{4}}}, & \varepsilon^2 - l\left(\varepsilon - \dfrac{l}{2}\right) \le \bar{s} \le \varepsilon^2 + l\left(\varepsilon + \dfrac{l}{2}\right) \\[2ex] 1, & \varepsilon^2 + l\left(\varepsilon + \dfrac{l}{2}\right) \le \bar{s} \le 1 - l\left(1 - \dfrac{l}{2}\right), \\[2ex] \dfrac{2}{\pi}\sin^{-1}\dfrac{1-\bar{s}}{l\sqrt{\bar{s}-\dfrac{l^2}{4}}}, & 1 - l\left(1 - \dfrac{l}{2}\right) \le \bar{s} \le 1, \\[2ex] 0, & \text{otherwise.} \end{cases}$$

(3.5.27)

iv) $0 \le l \le 2\varepsilon$

$$c_r(l,s) = \begin{cases} \dfrac{2}{\pi}\sin^{-1}\dfrac{\bar{s}-\varepsilon^2}{l\sqrt{\bar{s}-\dfrac{l^2}{4}}}, & \varepsilon^2 \le \bar{s} \le \varepsilon^2 + l\left(\varepsilon + \dfrac{l}{2}\right), \\[2ex] 1, & \varepsilon^2 + l\left(\varepsilon + \dfrac{l}{2}\right) \le \bar{s} \le 1 - l\left(1 - \dfrac{l}{2}\right), \\[2ex] \dfrac{2}{\pi}\sin^{-1}\dfrac{1-\bar{s}}{l\sqrt{\bar{s}-\dfrac{l^2}{4}}}, & 1 - l\left(1 - \dfrac{l}{2}\right) \le \bar{s} \le 1, \\[2ex] 0, & \text{otherwise.} \end{cases}$$

(3.5.28)

2) $\varepsilon \ge 1/3$

i) $\varepsilon + 1 \le l \le 2$

The 3-D CTF is the same as Eq. (3.5.25).

ii) $2\varepsilon \le l \le \varepsilon + 1$

The 3-D CTF is the same as Eq. (3.5.26)

iii) $1 - \varepsilon \le l \le 2\varepsilon$

$$c_r(l,s) = \begin{cases} \dfrac{2}{\pi} \sin^{-1} \dfrac{\bar{s} - \varepsilon^2}{l\sqrt{\bar{s} - \dfrac{l^2}{4}}}, & \varepsilon^2 \le \bar{s} \le \dfrac{1+\varepsilon^2}{2}, \\[4mm] \dfrac{2}{\pi} \sin^{-1} \dfrac{1 - \bar{s}}{l\sqrt{\bar{s} - \dfrac{l^2}{4}}}, & \dfrac{1+\varepsilon^2}{2} \le \bar{s} \le 1, \\[4mm] 0, & otherwise. \end{cases} \tag{3.5.29}$$

iv) $0 \le l \le 1 - \varepsilon$

The 3-D CTF is the same as Eq. (3.5.28).

It is seen that when $\varepsilon = 0$, Eqs. (3.5.25) - (3.15.29) reduce to Eq. (3.2.20) for the confocal system with two equal circular lenses.

Fig. 3.5.3 gives the complete view of the 3-D CTF in a reflection-mode confocal scanning microscope for different radii of the central obstruction of the annular lenses. Because of the constant spatial frequency shift $s0$, the value of the 3-D CTF is zero in the region where $-s0 +1 \le s \le 0$ so that the image information in this region is missing. It is noted from Fig. 3.5.3 that the transverse cut-off spatial frequency of the 3-D CTF is two, independent of the radius of the central obstruction.

It is important to notice that the distribution of $c_r(l = 0, s)$ is a square function which therefore results in a good optical sectioning effect in such a reflection-mode confocal scanning microscope. Though the shape of the variation $c_r(l = 0, s)$ is independent of ε, its width is given by $1 - \varepsilon^2$. Thus, as ε increases, $c_r(l = 0, s)$ becomes narrower, which implies that the optical sectioning property is degraded. This is because the intensity of the axial response, $I(u)$, to a perfect reflector is determined, according to Eq. (3.2.3), by

$$I(u) = \left\{ \frac{\sin[(1 - \varepsilon^2)u / 2]}{(1 - \varepsilon^2)u / 2} \right\}^2. \tag{3.5.30}$$

When $\varepsilon = 0$, the border of the non-zero region of the 3-D CTF at true high true axial spatial frequencies (i. e., $s+s0 = 0$) is a parabolic curve, $s = l^2/4$ (see Figs. 3.2.3 and 3.2.4), for circular lenses, while it is a straight line for annular lenses.[3.30]

It is noted that the 3-D CTF exhibits a dip along the transverse direction for non-zero values of ε. Accordingly, the values of the 3-D CTF at higher transverse spatial frequencies become considerably larger as ε increases and the transverse resolution for a confocal system using annular lenses is improved.

If the object is a thin plate located on the focal plane, the in-focus CTF of the thin object, $c_r(l, u = 0)$, is determined by the integration of the 3-D CTF $c_r(l, s)$ with respect to the variable s (Chapter 10) and is identical to the 2-D OTF for a single lens for incoherent imaging (see Fig. 2. 5.4).[3.30]

The 3-D CTF $c_t(l, s)$ for transmission confocal microscopy with two equal annular lenses can be derived according to Eq. (3.5.11) and Eqs. (3.5.20)-(3.5.24) and is found to be identical to the 3-D OTF for conventional incoherent imaging, as shown in Eq. (2.5.16). Note that $c_t(l, s)$ approaches infinity at $l = 0$. When $\varepsilon = 0$, i. e., in the case of two circular lenses, the second term in Eq. (2.5.16) is equal to zero, so that Eq. (2.5.16) reduces to Eq. (2.5.14), including the first term only. It can also be seen that the cut-off transverse spatial frequency is determined by the first term, so that $c_t(l, s)$ is cut off at $l = 2$. The cut-off axial spatial frequency is given by the condition that $c_t(l, s)$ must be larger than or equal to zero, because $c_t(l, s)$ is identical to the length of a line within the overlapped area (see Fig. 2.5.2). We thus find that the axial spatial frequency of $c_t(l, s)$ is cut off at $s = (1 - \varepsilon^2)/2$, which is smaller than that for two circular lenses.

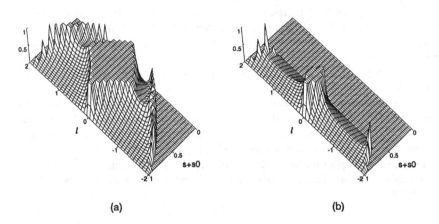

(a)   (b)

Fig. 3.5.3   3-D CTF in reflection with two equal annular pupils ($\varepsilon_1 = \varepsilon_2 = \varepsilon$, $\rho_{1a} = \rho_{2a} = 1$): (a) $\varepsilon = 0.25$; (b) $\varepsilon = 0.75$.

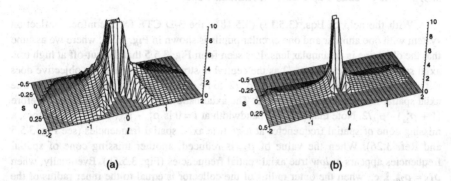

Fig. 3.5.4   3-D CTF in transmission with two equal annular pupils ($\varepsilon_1 = \varepsilon_2 = \varepsilon$, $\rho_{1a} = \rho_{2a} = 1$): (a) $\varepsilon = 0.25$; (b) $\varepsilon = 0.75$.

For general cases, the dependence of the 3-D CTF $c_t(l, s)$ on the parameter ε of the annular lens is plotted in Fig. 3.5.4. Within the non-zero region of $c_t(l, s)$, $c_t(l, s)$ reveals two parts:[3.8] one in which the values of $c_t(l, s)$ are unchanged with varying ε, while in the other $c_t(l, s)$ is dipped compared with that when $\varepsilon = 0$. The unchanged region of $c_t(l, s)$ and the depth of the dip become smaller and deeper, respectively, as ε increases. This can be understood from Eq. (2.5.16). For given values of ε, the first and the second terms in Eq. (2.5.16) give non-zero values in areas where $(|s|/l + l/2)^2 \leq 1$ and $(|s|/l - l/2)^2 \leq \varepsilon^2$, respectively. The latter area approaches zero as ε becomes zero. In the region overlapped by both areas, $c_t(l, s)$ is contributed by the difference between the two terms of Eq. (2.5.16), so that the 3-D CTFs are dipped. Outside the overlapped region, $c_t(l, s)$ is only given by the first term.

It is noted that the size variation of the central obstruction of the annular lenses does not change the transverse cut-off spatial frequency. Interestingly, because of the existence of the dip, the transfer function becomes narrow for the low transverse spatial frequency part, and smooth for the high transverse spatial frequency part. This property implies that more energy fills up the outer rings of the image of the point object, which may lead to a decreased transverse width of the central peak of the point image, and accordingly an improved transverse resolution may be obtained.[3.8]

As expected, the projection of $c_t(l, s)$ gives rise to the 2-D in-focus CTF which is the same as the result given by the integration of the 3-D CTF for reflection and is depicted in Fig. 2.5.4.

3. 5. 3   One Circular Pupil and One Annular Pupil

With the help of Eqs. (3.5.13)-(3.5.18), the 3-D CTF for a confocal reflection system with one annular and one circular pupils is shown in Fig. 3.5.5, where we assume that the objective is an annular lens. It is seen from Fig. 3.5.5 that the cut-off at high true axial spatial frequencies is $\rho_{1b}^2/2$ as the central obstruction of the annular objective does not allow spatial frequencies less than $\rho_{1b}^2/2$ to be passed, and that the cut-off at low true axial spatial frequencies is $(1 + \rho_{2a}^2)/2$. The axial bandwidth of the 3-D CTF is therefore $(1 + \rho_{2a}^2) - \rho_{1b}^2/2$. Note that the axial bandwidth at $l = 0$ is $\rho_{2a}^2 - \rho_{1b}^2$. There is, therefore, a missing cone of spatial frequencies at *high* true axial spatial frequencies (see Fig. 3.5.5 and Ref. 3.26). When the value of ρ_{2a} is reduced, another missing cone of spatial frequencies appears at *low* true axial spatial frequencies (Fig. 3.5.5b). Eventually, when $\rho_{1b} = \rho_{2a}$, i. e., when the outer radius of the collector is equal to the inner radius of the objective, the axial bandwidth at $l = 0$ becomes zero (Fig. 3.5.5b), implying that the microscope behaves as a dark–field microscope because low transverse spatial frequencies cannot be imaged.

In the limiting case in which either a thin annular objective or a thin annular collector is used, we can derive an analytical expression[3.23] for $c_r(l, s)$ as

$$c_r(l,s) = \frac{2}{l} \frac{1}{\mathrm{Re}\left[\sqrt{1 - \left(\frac{1-\bar{s}}{l} + \frac{l}{2}\right)^2}\right]}, \qquad (3.5.31)$$

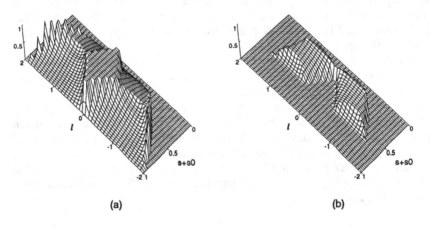

(a)   (b)

Fig. 3.5.5   3-D CTF in reflection with one circular pupil and one annular pupil ($\rho_{1a} = 1$, $\rho_{1b} = 0.5$, $\rho_{2b} = 0$): (a) $\rho_{2a} = 1$; (b) $\rho_{2a} = 0.5$.

as shown in Fig. 3.5.6. It is straightforward to show that the 2-D in-focus CTF of Eq. (3.5.31) is

$$c_2(l) = \cos^{-1}(l/2).\tag{3.5.32}$$

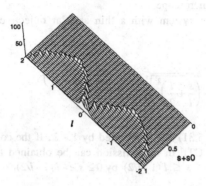

Fig. 3.5.6   3-D CTF in reflection with one circular pupil and one thin annular pupil ($\rho_{1a} = 1$, $\rho_{1b} = 0$, $\rho_{2a} \rightarrow 1$, $\rho_{2b} \rightarrow 1$).

The 3-D CTF for one circular pupil and one annular pupil in transmission can be obtained from Eqs. (3.5.20)-(3.5.24). Use of one circular pupil and one annular pupil in the transmission confocal microscope results in the reduction of cut-off spatial frequencies in either positive or negative axial directions. For example, for a system with

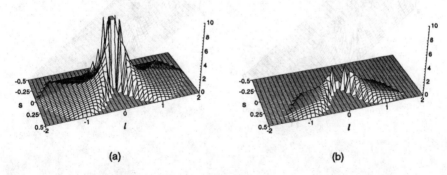

(a)   (b)

Fig. 3.5.7   3-D CTF in transmission with one circular pupil and one annular pupil ($\rho_{1a} = 1$, $\rho_{1b} = 0.5$, $\rho_{2b} = 0$): (a) $\rho_{2a} = 1$; (b) $\rho_{2a} = 0.5$.

an annular objective, the cut-off in the negative axial direction (Fig. 3.5.7a) is smaller than that in the positive direction. In addition, the reduction in the outer radius of the collector causes a further decrease in the cut-off in the negative axial direction (Fig. 3.25a). Finally, when the outer radius of the collector is equal to the inner radius of the objective, the 3-D CTF splits into two parts (see Fig. 3.5.7b), implying that the system behaves as a dark-field microscope.

For a transmission system with a thin annular objective lens and a circular collection lens, we have[3.23]

$$c_t(l,s) = \frac{2}{l} \frac{1}{\text{Re}\left[\sqrt{1 - \left(\frac{s}{l} + \frac{l}{2}\right)^2}\right]}, \qquad 0 \le s \le l\left(1 - \frac{l}{2}\right). \tag{3.5.32}$$

It is the same as Eq. (3.5.31) if s is replaced by $1 - \bar{s}$. If the collection lens is a thin annular pupil, the 3-D CTF in transmission can be obtained from Eq. (3.5.32) by replacing s by $|s|$, and $0 \le s \le l(1 - l/2)$ by $0 \ge s \ge -l(1 - l/2)$.

3.5.4 Two Unequal Annular Pupils

By applying Eqs. Eqs. (3.5.13)-(3.5.18) to each term in Eq. (3.5.8), and Eqs. (3.5.20)-(3.5.24) to Eq. (3.5.11), the final expressions for the 3-D CTFs for two unequal annular pupils in reflection and transmission can be expressed in analytical forms. Notice

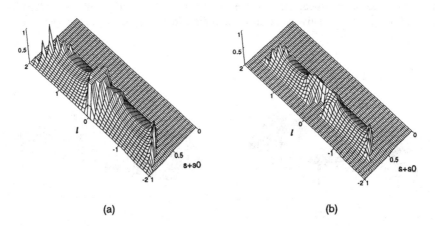

(a)   (b)

Fig. 3.5.8   3-D CTF in reflection with two unequal pupils ($\rho_{1a} = 1$, $\rho_{1b} = 0.75$, $\rho_{2b} = 0.5$): (a) $\rho_{2a} = 1$; (b) $\rho_{2a} = 0.75$.

that the transverse cut-off spatial frequency is given by $1+\rho_{2a}$ rather than 2.

The 3-D CTFs for a reflection confocal system consisting of two unequal annular lenses are shown in Fig. 3.5.8. Since two annular lenses are adopted, the cut-off at high true axial spatial frequencies is now determined by two central obstructions, i. e., by $(\rho_{1b}^2+\rho_{2b}^2)/2$. Note that the axial bandwidth at $l=0$ is given by $(\rho_{2a}^2-\rho_{1b}^2)$ (if $\rho_{1b}>\rho_{2b}$), which is smaller than that, i. e., $(1+\rho_{2a}^2-\rho_{1b}^2-\rho_{2b}^2)/2$, when $l \neq 0$. Thus a missing cone of spatial frequencies is formed at high true axial spatial frequencies (see Figs 3.5.8 and Ref. 3.26). If the outer radius of the collector decreases, the cut-off at low true axial spatial frequencies is decreased by $1 - (1+\rho_{2a}^2)/2$, and another missing cone of spatial frequencies appears again at low true axial spatial frequencies (Fig. 3.5.8). The condition for the dark-field microscope is $\rho_{1b}=\rho_{2a}$ (Fig. 3.5.8b).

Fig. 3.5.9 gives the 3-D CTFs for the transmission confocal system containing two unequal annular lenses. The two annular lenses produce different axial cut-offs in two directions and thus the bandwidth in the axial direction is $(1+\rho_{2a}^2-\rho_{1b}^2-\rho_{2b}^2)/2$. The 3-D CTF when $\rho_{2a}=\rho_{1b}$, corresponding to a dark-field case, is shown in Fig. 3.5.9b.

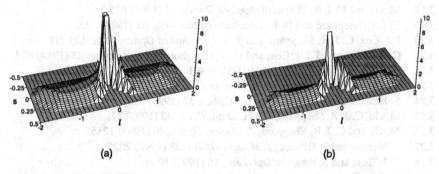

(a)   (b)

Fig. 3.5.9   3-D CTF in transmission with two unequal pupils ($\rho_{1a}= 1$, $\rho_{1b}= 0.75$, $\rho_{2b}= 0.5$): (a) $\rho_{2a}= 1$; (b) $\rho_{2a}= 0.75$.

References

3.1.   C. J. R. Sheppard, Scanning optical microscopy, in *Advances in Optical and Electron Microscopy*, Vol. 10, eds. R. Barer & V. E. Cosslett (Academic, London, 1987).

3.2.   T. Wilson, *Confocal Microscopy* (Academic, London, 1990).

3.3   E. Wolf, *Opt. Commun.*, 1 (1969) 153.

3.4.   J. W. Goodman, *Introduction to Fourier Optics* (McGraw-Hill, New York, 1968).

3.5.   B. R. Bracewell, *The Fourier Transform and Its Applications* (McGraw-Hill, New York, 1965).

3.6.   C. J. R. Sheppard, M. Gu, and X. Q. Mao, *Opt. Commun.*, **81** (1991) 281.

3.7.   T. Wilson and C. J. R. Sheppard, *Theory and Practice of Scanning Optical Microscopy* (Academic Press, London, 1984).

3.8.   C. J. R. Sheppard and M. Gu, *Opt. Commun.*, **81** (1991) 276.

3.9.   C. J. R. Sheppard and M. Gu, *J. Opt. Soc. Am. A*, **8** (1991) 692.

3.10. C. J. R. Sheppard and M. Gu, *Applied Optics*, **30** (1991) 3560.

3.11. C. J. R. Sheppard and M. Gu, *Opt. Commun.*, **88** (1992) 180.

3.12. S. Wang and B. R. Frieden, *Applied Optics*, **29** (1990) 2424.

3.13. M. Born and E. Wolf, *Principles of Optics* (Pergamon, New York, 1980).

3.14. V. N. Mahajan, *Applied Optics*, **19** (1983) 3035.

3.15. M. Gu and C. J. R. Sheppard, *Applied Optics*, **31** (1992) 2541.

3.16. I. S. Gradstein and I. M. Ryshik, *Tables of Series, Products and Integrals* (Harri Deutsch, Frankfurt, 1981).

3.17. D. G. A. Jackson, M. Gu, and C. J. R. Sheppard, *J. Opt. Soc. Am. A*, **11** (1994) 1758.

3.18. M. Gu and C. J. R. Sheppard, *Applied Optics*, **33** (1994) 625.

3.19. C. J. R. Sheppard and D. K. Hamilton, *Optica Acta*, **31** (1984) 723.

3.20. I. J. Cox, C. J. R. Sheppard, and T. Wilson, *Aplied Optics*, **21** (1982) 778.

3.21. C. J. R. Sheppard, I. J. Cox, and D. K. Hamilton, *Applied Optics*, **23** (1984) 657.

3.22. T. Wilson, *J. Modern Optics*, **40** (1993) 401.

3.23. M. Gu and C. J. R Sheppard, *J. Modern Optics*, **41** (1994) 1701.

3.24. S. Kimura and T. Wilson, *Applied Optics*, **32** (1993) 2257.

3.25. M. Gu, C. J. R. Sheppard, and H. Zhou, *Optik*, **93** (1993) 87.

3.26. M. Gu and C. J. R. Sheppard, *J. Modern Optics*, **40** (1993) 1255.

3.27   T. Wilson and S. Hewlett, *J. Modern Optics*, **37** (1990) 2025.

3.28. T. Wilson and S. Hewlett, *Opt. Lett.*, **16** (1991) 1062.

3.29. C. J. R. Sheppard and X. Q. Mao, *J. Opt. Soc. Am. A*, **6** (1989) 1260.

3.30. M. Gu and C. J. R Sheppard, *J. Modern Optics*, **39** (1992) 783.

Chapter 4

CONFOCAL BRIGHT-FIELD MICROSCOPY
WITH A FINITE-SIZED DETECTOR

The aim of this chapter is to develop a proper theory of image formation of three-dimensional (3-D) objects for confocal bright-field microscopes with a finite-sized detector, and thus to understand the effects of the detector size on 3-D partially-coherent confocal imaging.[4.1, 4.2] Although it is assumed that a point source is used, the effect of a finite-sized source can be dealt with in terms of the methods presented in the current chapter.[4.3, 4.4] In Section 4.1, 3-D partially-coherent image formation in confocal reflection-mode and transmission-mode microscopes, consisting of a finite-sized pinhole mask in front of the detector, is analysed. The dependence of the image of a point object on the detector size is discussed in Section 4.2. The 3-D transmission cross-coefficient (TCC) is derived for both systems with either a circular detector or a slit detector in Section 4.3. Section 4.4 discusses the effect of detector size on axial resolution, where a method for the improvement in axial resolution based on the combination of a finite-sized detector and annular lenses is demonstrated. In Section 4.5, transverse resolution in the presence of a finite-sized detector is analysed, which gives rise to the edge-setting criteria. Finally, for a weakly-scattering object, the 3-D weak-object transfer function (WOTF), a special form of the 3-D TCC, is in detail investigated in Sections 4.6.

4. 1 Partially-Coherent Image Formation

In the previous chapter, we have considered 3-D image formation in confocal bright-field microscopy containing a point source and a point detector. In practice, two pinhole masks of finite size are used in a confocal microscope. The first one acts as a spatial filter to provide a diffraction-limited illumination spot on the specimen. The size of the first pinhole is therefore not crucial to confocal imaging qualities. However, the second pinhole placed in front of an incoherent detector does have an important influence on confocal bright-field imaging. In addition, the size of the second pinhole is often opened up to a certain size in order to record enough signal. It has been shown that two-dimensional (2-D) confocal bright-field imaging with a finite-sized detector is partially coherent.[4.5-4.7] Furthermore, the strength of optical sectioning is degraded as the size of the pinhole increases.[4.8, 4.9]

The geometry of reflection and transmission confocal systems can be referred to Fig. 3.1.1, where P_1 and P_2 denote the aberration-free objective and collector lenses, respectively. We consider the source to be a point. Assume that the detector has the uniform intensity sensitivity $D(r_2)$ and that an object scanning method is used.

According to the discussion in Section 3.1, the light field in the detector space is given by Eq. (3.1.7). Therefore, the final image intensity detected by a finite-sized detector is the integration of the modulus squared of Eq. (3.1.7) over the detector aperture:

$$I(r_s) = \int_{-\infty}^{\infty} |U(r_2, r_s)|^2 D(r_2) dr_2, \tag{4.1.1}$$

which becomes

$$I(r_s) = \int_{-\infty}^{\infty} \left| \int_{-\infty}^{\infty} \left\{ \int_{-\infty}^{\infty} \delta(r_0) \exp[ik(z_0 - z_1)] h_1(r_0 + M_1 r_1) dr_0 \right\} \right. \tag{4.1.2}$$

$$\left. o(r_s - r_1) \exp[ik(\pm z_1 - z_2)] h_2(r_1 + M_2 r_2) dr_1 \right|^2 D(r_2) dr_2,$$

where the bold letters r_j ($j = 0, 1, 2$) represent the position vectors with components x_j, y_j and z_j (see Fig. 3.1.1), and the scan point is represented by $r_s(x_s, y_s, z_s)$. Parameters M_1 and M_2 are defined by Eqs. (3.1.5) and (3.1.8). The delta function in Eq. (4.1.2) represents a point source.

For a detector with a circular pinhole mask placed in front of the detector, its intensity sensitivity $D(r_2)$ can be written as

$$D(r_2) = \delta(z_2) \begin{cases} 1 & , \quad r_2 \leq r_d, \\ \\ 0 & , \quad otherwise, \end{cases} \tag{4.1.3}$$

where $r_2 = (x_2^2 + y_2^2)^{1/2}$ is the radial coordinate in the detector plane, r_d is the radius of the pinhole, and the delta function $\delta(z_2)$ means that the detector is placed in the plane at

$z_2 = 0$. For an infinitely-long slit detector, with width $2x_d$, placed along the x_2 axis, Eq. (4.1.3) should be replaced by

$$D(r_2) = \delta(z_2) \begin{cases} 1 & , & |x_2| \le x_d, \\ \\ 0 & , & otherwise, \end{cases} \qquad (4.1.4)$$

if the length of the slit is very long compared with the width $2x_d$.

For a point object, the measured intensity, i. e., the image of the point object, can be easily derived as

$$I(r_s) = |h_1(M_1 r_s)|^2 \int_{-\infty}^{\infty} |h_2(r_s + M_2 r_2)|^2 D(r_2) dr_2. \qquad (4.1.5)$$

This is the 3-D intensity point spread function (IPSF) in a confocal bright-field microscope with a finite-sized detector. As will be seen in Chapter 5, Eq. (4.1.5) may be also applicable to confocal single-photon fluorescence microscopy because only a point object is considered. Although Eq. (4.1.5) is the image intensity of a single point object, Eq. (4.1.2) is not the superposition of the intensity from different points in the 3-D object, so that it does not represent an incoherent imaging process. On the other hand, Eq. (4.1.2) also implies that due to the use of a finite-sized detector, the image intensity of any object cannot be expressed as a superposition of the amplitude from different points in the 3-D object. Therefore, Eq. (4.1.2) does not correspond to a coherent imaging process either. We conclude that the 3-D image formation represented by Eq. (4.1.2) is partially coherent.

4. 2 Intensity Point Spread Function

Before we introduce the transfer function in the partially-coherent confocal imaging process, it is worthwhile to discuss the image of a point object further, so that we can understand how the detector size affects the imaging performance. Rewriting Eq. (4.1.5) as an explicit convolution relation and denoting it by $h_i(r)$ yields

$$h_i(r) = |h_1(M_1 r)|^2 \left[|h_2(r)|^2 \otimes_3 D(r / M_2) \right]. \qquad (4.2.1)$$

Here $\otimes_3$ represents the 3-D convolution operation. Eq. (4.2.1) applies for both pinhole and slit detectors. As an example, we consider a system consisting of circular lenses and a circular pinhole detector. In this case, Eq. (4.2.1) has the following circularly symmetric form if Eq. (4.1.3) is used:

$$h_i(v,u) = |h_1(v,u)|^2 |h_2(v,u)|^2 \otimes_2 D(v), \tag{4.2.2}$$

where

$$D(v) = \begin{cases} 1 & , \quad v < v_d, \\ \\ 0 & , \quad otherwise, \end{cases} \tag{4.2.3}$$

and $\otimes_2$ is the two-dimensional (2-D) convolution operation in a vertical plane. Here v_d is the normalized radius of the detector, defined as

$$v_d = \frac{2\pi}{\lambda} r_d \sin \alpha_d, \tag{4.2.4}$$

where $\sin \alpha_d$ is the numerical aperture of the collector lens in the detection space (see Eq. (3.1.18)). $h_1(v, u)$ and $h_2(v, u)$ are, respectively, the 3-D amplitude point spread functions for objective and collector lenses in the object space and are given, according to Eqs. (3.1.15) and (3.1.16), by

$$h_1(v,u) = \exp(-is_0 u)\int_0^1 P_1(\rho,u)J_0(v\rho)\rho d\rho \tag{4.2.5}$$

and

$$h_2(v,u) = \exp(\pm is_0 u)\int_0^{b/a} P_2(\rho,u)J_0(v\rho)\rho d\rho. \tag{4.2.6}$$

In Eqs. (4.2.2), (4.2.3), (4.2.5) and (4.2.6), v and u are the transverse and axial optical coordinates in the object space, defined by Eq. (3.1.17). $P_1(\rho, u)$ and $P_2(\rho, u)$ are the defocused pupil functions for the objective and collector lenses, respectively. ρ is the radial coordinate of the pupil function of the objective lens, normalized by the radius a. The radius of the pupil function for the collector is b.

For a point detector, the image of a point object is therefore

$$h_i(v,u) = |h_1(v,u)|^2 |h_2(v,u)|^2, \tag{4.2.7}$$

which is, as expected, the modulus squared of the 3-D amplitude point spread function (APSF) given by Eq. (3.1.19). On the other hand, when an infinitely-large detector is used, the 3-D IPSF in Eq. (4.2.2) reduces to

$$h_i(v,u) = |h_1(v,u)|^2, \tag{4.2.8}$$

which is thus the 3-D IPSF for the objective lens. This result can be understandable because the role of the collector lens is completely degraded owing to the use of a large detector.

For a confocal system of two identical lenses, the effect of the detector radius on the 3-D IPSF is shown in Fig. 4.2.1, where contours of the 3-D IPSF normalized to unity at $v = u = 0$ are depicted. As the detector size increases, the size of the central bright spot

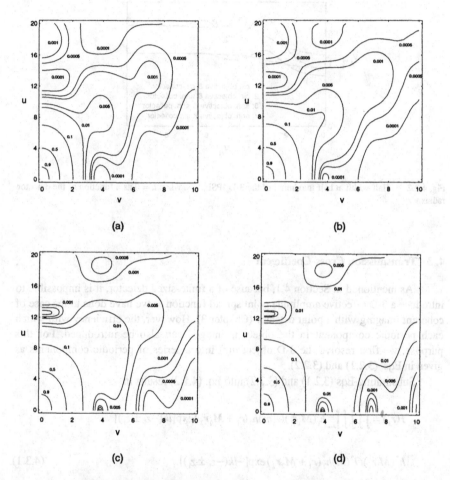

(a)   (b)

(c)   (d)

Fig. 4.2.1   Contours of the 3-D intensity point spread function (IPSF) for a confocal bright-field microscope comprising two identical circular lenses and a circular detector. The effect of the detector size is given by the normalized radius v_d: (a) $v_d = 0$; (b) $v_d = 2$; (c) $v_d = 5$; (d) $v_d \to \infty$.

of the 3-D IPSF becomes large, meaning that resolution of the imaging system is gradually degraded. The half width at half maximum (HWHM) of the 3-D IPSF, $v_{1/2}$, in the transverse direction, i. e., when $u = 0$, is depicted as a function of the detector radius v_d in Fig. 4.2.2, which includes the curves for the cases of annular lenses. Note that for a large detector, the function of the collector is completely degraded.

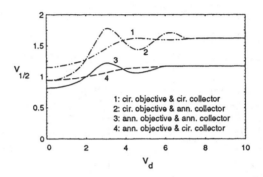

Fig. 4.2.2   Half width at half maximum of the 3-D IPSF, $v_{1/2}$, when $u = 0$ as a function of the detector radius v_d.

4. 3   Transmission Cross-Coefficient

As mentioned in Section 4.1, because of a finite-sized detector, it is impossible to introduce a 3-D effective amplitude point spread function as we have done in the case of coherent imaging with a point detector (Chapter 3). However, the efficiency with which each periodic component in the object is imaged can also be introduced. For this purpose, we first resolve the 3-D object $o(r)$ into a series of periodic components as given in Eqs. (3.2.1) and (3.2.2).

Substituting Eqs (3.2.1) and (3.2.2) into Eq. (4.1.2) results in

$$I(r_s) = \int\int\int\int\int_{-\infty}^{\infty} h_1(M_1 r_1)O(m)h_2(r_1 + M_2 r_2)\exp[ik(-z_1 \pm z_1)]$$

$$h_1^*(M_1 r_1')O^*(m)h_2^*(r_1' + M_2 r_2')\exp[-ik(-z_1' \pm z_1')] \qquad (4.3.1)$$

$$\exp[2\pi i r_s \bullet (m - m')]\exp[-2\pi i(r_1 \bullet m - r_1' \bullet m')]D(r_2)dmdm' \, dr_1 dr_1' dr_2,$$

where m (or m') represents the spatial frequency vector and $O(m)$ (or $O(m')$), the Fourier spectrum of the 3-D object, is given by the 3-D inverse Fourier transform of $o(r)$. Eq. (4.3.1) can be rewritten as

$$I(r_s) = \iint_{-\infty}^{\infty} C(m;m')O(m)O^*(m')\exp[2\pi i r_s \bullet (m-m')]dmdm'. \qquad (4.3.2)$$

Here $C(m; m')$ is defined as

$$C(m;m') = \iiint_{-\infty}^{\infty} h_1(M_1 r_1)h_2(r_1 + M_2 r_2)\exp[ik(-z_1 \pm z_1)]h_1^*(M_1 r_1')h_2^*(r_1' + M_2 r_2')$$

$$\exp[-ik(-z_1' \pm z_1')]\exp[-2\pi i(r_1 \bullet m - r_1' \bullet m')]D(r_2)dr_1 dr_1' dr_2,$$

$$(4.3.3)$$

which is a six-variable function. Generally, m and m' are not separable in $C(m; m')$. From Eq. (4.3.2), it can be seen that the significance of $C(m; m')$ is that it describes the imaging properties of the confocal system and gives the efficiency with which pairs of spatial frequencies m and m' in an object of interest are imaged. $C(m; m')$ is therefore the 3-D transfer function for 3-D confocal bright-field imaging with a finite-sized detector. Since the imaging performance is partially coherent, the transfer function $C(m; m')$ in this case is termed the 3-D transmission cross-coefficient (TCC), being consistent with the terminology of the early definition for partially-coherent imaging.[4.10, 4.11]

An alternative expression for $C(m; m')$ can be derived from Eq. (4.3.3):

$$C(m;m') = \int_{-\infty}^{\infty} c(m,r_2)c^*(m',r_2)D(r_2)dr_2, \qquad (4.3.4)$$

where

$$c(m,r_2) = \int_{-\infty}^{\infty} h_1(M_1 r_1)h_2(r_1 + M_2 r_2)\exp[ik(-z_1 \pm z_1)]\exp(-2\pi i r_1 \bullet m)dr_1. \qquad (4.3.5)$$

Here $c(m, r_2)$ is the 3-D CTF for confocal bright-field microscopy with an offset point detector at r_2 in the detector space and has been discussed in Section 3.4.

In general, when a finite-sized detector is employed, the 3-D TCC is a complex function and the variables m and m' in the 3-D TCC are not separable but the following symmetry property holds:

$$C(m; m') = C^*(m'; m). \qquad (4.3.6)$$

If there is no detector offset, the 3-D transmission cross-coefficient separates and can be simply given by

$$C(m; m') = c(m, 0)c^*(m', 0),$$ (4.3.7)

where $c(m, 0)$, the 3-D CTF without detector offset, is a real function for an aberration-free system. For a circularly symmetric system, an analytical expression for $c(m, 0)$ has been derived (see Eq. (3.2.20)). As a result, $C(m; m')$ is real and the variables m and m' are separable.

For a confocal system in the limiting case of an infinitely-large detector, one can obtain, if either Eq. (4.1.3) or Eq. (4.1.4) is used,

$$C(m;m') = \int_{-\infty}^{\infty} c(m,r_2)c^*(m',r_2)\delta(z_2)dr_2.$$ (4.3.8)

Using Eqs. (3.1.3) and (4.3.5), we can express the 3-D TCC as

$$C(m;m') = \iint_{-\infty}^{\infty} |P_2(\xi,\eta)|^2 P_1(m-\xi,n-\eta)P_1^*(m'-\xi,n'-\eta)$$

$$\delta\left(s - \frac{m^2+n^2}{2} + \xi m + \eta n\right)\delta\left(s' - \frac{m'^2+n'^2}{2} + \xi m' + \eta n'\right)d\xi d\eta$$ (4.3.9)

for transmission, and

$$C(m;m') = \iint_{-\infty}^{\infty} |P_2(\xi,\eta)|^2 P_1(m-\xi,n-\eta)P_1^*(m'-\xi,n'-\eta)$$

$$\delta\left(s - \frac{m^2+n^2}{2} + \xi m + \eta n - \xi^2 - \eta^2 + s0\right)$$

$$\delta\left(s' - \frac{m'^2+n'^2}{2} + \xi m' + \eta n' - \xi^2 - \eta^2 + s0\right)d\xi d\eta$$ (4.3.10)

for reflection. Here m, n and s (or m', n', s') represent the transverse and axial spatial frequencies of m (or m'), and normalized by Eqs. (2.4.8) and (2.4.9), respectively. $s0$ denotes the constant axial spatial frequency shift expressed in Eq. (3.2.22).

It is noted from Eqs. (4.3.9) and (4.3.10) that if the objectives are the same in confocal and conventional microscopes and if the collector in the confocal system is identical to the condenser in the conventional system, Eqs. (4.3.9) and (4.3.10) are the same as the 3-D TCCs for conventional imaging.[4.10] Thus it can be concluded that image formation in a confocal microscope with a large detector is identical to that in a conventional microscope with a condenser identical to the collector.

4. 3. 1 Circular Detector

We now first consider a confocal system consisting of equal circular objective and collector lenses and a circular pinhole, which is usually a case in practice. The analytical expression for $c(m, r_2)$ can be derived (see Eqs. (3.4.3) and (3.4.4)). Since the whole system is of circular symmetry, the 3-D TCC must be dependent only on l (or l') and s (or s'). Letting $n = 0$ in Eqs. (3.4.3) and (3.4.4) and $m = l$, one has $\phi = 0$. From Eq. (4.3.4) we can express the 3-D TCC as

$$C(l,s;l',s') = \frac{K_t}{ll'} \int_0^{v_d} \int_0^{2\pi} \exp\left[iv_2\left(\frac{l}{2} - \frac{s}{l}\right)\cos\phi'\right] \mathrm{sinc}\left[v_2\sqrt{1 - \left(\frac{|s|}{l} + \frac{l}{2}\right)^2}\sin\phi'\right]$$

$$\mathrm{Re}\left[\sqrt{1 - \left(\frac{|s|}{l} + \frac{l}{2}\right)^2}\right] \exp\left[-iv_2\left(\frac{l'}{2} - \frac{s'}{l'}\right)\cos\phi'\right] \mathrm{sinc}\left[v_2\sqrt{1 - \left(\frac{|s'|}{l'} + \frac{l'}{2}\right)^2}\sin\phi'\right]$$

$$\mathrm{Re}\left[\sqrt{1 - \left(\frac{|s'|}{l'} + \frac{l'}{2}\right)^2}\right] v_2 dv_2 d\phi'$$

$$(4.3.11)$$

for transmission, and

$$C(l,s;l',s') = K_r \int_0^{v_d} \int_0^{2\pi} \exp\left[iv_2 l(\cos\phi')/2\right] f_2(l,0,s)$$

$$(4.3.12)$$

$$\exp\left[-iv_2 l'(\cos\phi')/2\right] f_2(l',0,s') v_2 dv_2 d\phi'$$

for reflection. Here $l = (m^2 + n^2)^{1/2}$ and $l' = (m'^2 + n^2)^{1/2}$ are the radial spatial frequencies. K_t and K_r are constants of normalization in reflection and transmission, respectively. v_2 and ϕ' are the polar coordinates in the detector plane, where v_2 is given by Eq. (3.1.18). Re[ ] takes the real value of its argument. The function sinc is defined by

$$\frac{\sin x}{x} \qquad\qquad (4.3.13)$$

and the functions $f_2(l, 0, s)$ and $f_2(l', 0, s')$ are given by Eq. (3.4.5) for $\phi = 0$.

Although the 3-D TCC in Eqs. (4.3.11) and (4.3.12) becomes a four variable function $C(l, s; l', s')$ as a result of circular symmetry, l, s and l', s' are still not separable. In the reflection case, the cross-section of the 3-D TCC at $l = l' = 0$ is of particular significance: its Fourier transform gives rise to the strength of optical sectioning. Eq. (4.3.12) can be simplified, when $l = l' = 0$, as

$$C(0,s;0,s') = K_r \int_0^{v_a} J_0(v_2 \sqrt{s + s0}) J_0(v_2 \sqrt{s' + s0}) v_2 dv_2. \tag{4.3.14}$$

Here J_0 represents a Bessel function of the first kind of order zero. Thus the intensity of the axial response to a perfect reflector is, from Eq. (4.3.2),

$$I(u) = K_r \iint_{-\infty}^{\infty} C(0,s;0,s') \exp[iu(s - s')] ds ds'. \tag{4.3.15}$$

Substituting Eq. (4.3.14) into Eq. (4.3.15), performing the integration with respect to s', and setting $\sqrt{s} = \rho$ gives

$$I(u) = K_r \int_0^{v_a} \left| \int_0^1 \exp(iu\rho^2) J_0(v_2\rho) \rho d\rho \right|^2 v_2 dv_2, \tag{4.3.16}$$

which applies only to a system consisting two equal lenses. In general, for a system with two unequal pupil functions, Eq. (4.3.16) can be generalized to

$$I(u) = K_r \int_0^{v_a} \left| \int_0^1 P_1(\rho) P_2(\rho) \exp(iu\rho^2) J_0(v_2\rho) \rho d\rho \right|^2 v_2 dv_2, \tag{4.3.17}$$

where $P_1(\rho)$ and $P_2(\rho)$ are the pupil functions for the objective and collector lenses, respectively, and ρ can be understood to be the radial coordinate of the pupil function normalized by the radius, a, of the objective lens. We will discuss Eq. (4.3.17) in detail in the next section.

4. 3. 2   Slit Detector

Although most of commercial confocal microscopes adopt a circular pinhole detector, the limitation associated with it is the slow scanning speed. One way to overcome this difficulty is to use a finite-sized slit detector.[4.7, 4.8, 4.12, 4.13] This configuration has the great advantage that a whole line of information from the object under inspection can be imaged at one time, so that the scanning speed can be increased significantly and the signal level can also be improved.[4.7, 4.8]

The imaging performance in a confocal bright-field microscope with a finite-sized slit detector can be analysed by using a similar way presented in the early part of this section. Because the system except the detector has circular symmetry, we can assume the slit detector is parallel to the x axis without losing generality. The intensity sensitivity of the detector is given by Eq. (4.1.4).

According to Eqs. (3.4.3)-(3.4.7), the corresponding 3-D TCCs for transmission and reflection systems can be expressed as

$$C(m,n,s;m',n',s') = \frac{K_t}{ll'} \int_{-v_{dx}}^{v_{dx}} \int_{-\infty}^{\infty} \exp\left[iv_2\left(\frac{l}{2}-\frac{s}{l}\right)\cos(\phi+\phi')\right]$$

$$\mathrm{sinc}\left[v_2\sqrt{1-\left(\frac{|s|}{l}+\frac{l}{2}\right)^2}\sin(\phi+\phi')\right] \mathrm{Re}\left[\sqrt{1-\left(\frac{|s|}{l}+\frac{l}{2}\right)^2}\right]$$

$$\exp\left[-iv_2\left(\frac{l'}{2}-\frac{s'}{l'}\right)\cos(\phi+\phi')\right] \mathrm{sinc}\left[v_2\sqrt{1-\left(\frac{|s'|}{l'}+\frac{l'}{2}\right)^2}\sin(\phi+\phi')\right]$$

$$\mathrm{Re}\left[\sqrt{1-\left(\frac{|s'|}{l'}+\frac{l'}{2}\right)^2}\right] dv_{2x} dv_{2y}$$

$$(4.3.18)$$

and

$$C(m,n,s;m',n',s') = K_r \int_{-v_{dx}}^{v_{dx}} \int_{-\infty}^{\infty} \exp\{iv_2 l[\cos(\phi+\phi')]/2\} f_2(m,n,s)$$

$$(4.3.19)$$

$$\exp\{-iv_2 l'[\cos(\phi+\phi')]/2\} f_2(m',n',s') dv_{2x} dv_{2y},$$

respectively, where

$$v_{dx} = \frac{2\pi}{\lambda} x_d \sin\alpha_d \tag{4.3.20}$$

is the normalized width of the slit detector. As expected, Eqs. (4.3.18) and (4.3.19) are not of circular symmetry. Accordingly, the intensity of the axial response to a perfect reflector is, by generalizing Eq. (4.3.17),

$$I(u) = K_r \int_{-v_{dx}}^{v_{dx}} \int_{-\infty}^{\infty} \left| \int_0^1 P_1(\rho) P_2(\rho) \exp(iu\rho^2) J_0(v_2\rho) \rho d\rho \right|^2 dv_{2x} dv_{2y}. \tag{4.3.21}$$

It should be mentioned that imaging in a confocal bright-field microscope with a slit detector is partially coherent even when the width of the slit is zero.[4.7, 4.8] Axial resolution in this case is poorer than that for a point detector because the corresponding 3-D TCC is not separable in l, s and l', s', and thus Eq. (4.3.21) for a thin line detector includes the contribution from a series of off-axis point detectors (see Eq. (3.4.9)).

4. 4 Effect of Detector Size on Axial Resolution

For a confocal microscope having two circular pupils, the axial response to a perfect reflector is given by Eq. (4.3.17) for a circular detector or by Eq. (4.3.21) for a slit detector. It is the Fourier transform of the axial cross-section of the 3-D TCC with respect to s and s', giving the strength of optical sectioning: the narrower the axial response the stronger the optical sectioning effect and thus the higher the axial resolution. We will focus our discussion on the circular detector in this section, while results and conclusions can be easily generalized to the case of a slit detector.

From Eq. (4.3.17), the axial response is a function of the radius of the detector. For a point detector, in which case imaging is coherent, the axial response can be expressed by Eq. (3.2.24), which is the Fourier transform of the axial cross-section of the corresponding 3-D CTF. In the limiting case of an infinitely-large detector, the axial response becomes independent of the defocus distance u, i. e., it is a constant. This is

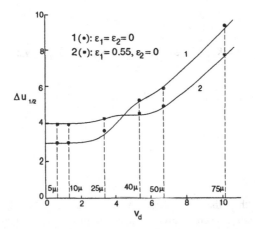

Fig. 4.4.1   Half width of the axial response to a perfect reflector, $\Delta u_{1/2}$, as a function of the detector radius v_d. The dark spots and squares correspond to experimental values. Curve 1 ($\varepsilon_1 = \varepsilon_2 = 0$) denotes the two circular lenses and curve 2 ($\varepsilon_1 = 0.55$, $\varepsilon_2 = 0$) represents an annular objective lens and a circular collector lens. The vertical lines denotes the position of the detector radii in experiments.

understood from the corresponding 3-D IPSF in Eq. (4.2.8): all the reflected energy from the reflector can be measured regardless of the position of the reflector if the detector is infinitely large. For a given value of the detector size, the dependence of the half width at half maximum (HWHM), $\Delta u_{1/2}$, of the axial response on the detector radius can be numerically calculated from Eq. (4.3.17), shown as curve 1 in Fig. 4.4.1. As expected, the axial response becomes broad as the detector radius increases, implying that the axial resolution is degraded compared with that for a point detector.

The degradation of axial resolution for a finite-sized detector is a practical problem as any pinhole has finite size. For weakly-scattering objects, the signal from the object is not strong so that the detector is opened up to a certain dimension in order to receive enough signal. One of the ways to solve this problem in practice is to use lenses with annular pupils rather than the circular pupils. In this case, the axial response is now a function of both the detector size and the size of the central obstruction of the annular lenses.[4.14-4.16] It is thus hopeful that the axial response can be optimized.[4.14]

According to Eq. (4.3.17) the axial response is determined by the smaller value of the outer radii of the annular pupils and the larger of the inner radii of the annular pupils. This property also implies that the axial response remains unchanged when exchanging the objective and the collector lenses. It is known that reducing the outer radius of a lens loses image information because the numerical aperture becomes smaller. Therefore, let us assume that the annular objective and collector lenses have the same outer radius of the annular pupils. The axial response becomes, from Eq. (4.3.17),

$$I(u) = K_r \int_0^{v_d} \left| \int_\varepsilon^1 \exp(iu\rho^2) J_0(v_2\rho)\rho d\rho \right|^2 v_2 dv_2, \qquad (4.4.1)$$

where ε ($0 \le \varepsilon \le 1$) is the larger radius of the two central obstructions, normalized by the outer radius of the objective. For $v_d = 0$, corresponding to a point detector, $I(u)$ can be analytically expressed as

$$I(u) = \frac{\sin^2\left[u(1-\varepsilon^2)/2\right]}{\left[u(1-\varepsilon^2)/2\right]^2}, \qquad (4.4.2)$$

which has been normalized to unity for $u = 0$, so that the HWHM $\Delta u_{1/2}$ is solved as

$$\Delta u_{1/2} = \frac{\Delta u_{1/2}^0}{1-\varepsilon^2}, \qquad (4.4.3)$$

where $\Delta u_{1/2}^0 = 2.78$ is the HWHM for $\varepsilon = 0$. It is obvious that the HWHM is increased with ε. Thus the axial response for $\varepsilon = 0$ is narrower than that for non-zero values of ε, as can be seen from Fig. 4.4.1. Fig. 4.4.2 shows the axial responses for $\varepsilon = 0$ and $\varepsilon = 0.55$

(solid curves). This result implies that using an annular pupil lens results in a decreased axial resolution when a point detector is used.

However, the situation changes when a finite-sized detector is used. Fig. 4.4.2 also gives the axial responses for $v_d = 6.7$, corresponding a radius of 50 μm for a pinhole in the experiment.[4.14] It can be seen that the axial response is narrower for $\varepsilon = 0.55$ than that for $\varepsilon = 0$ (dashed curves). This property becomes further clear in the dependence of $\Delta u_{1/2}$ on v_d for $\varepsilon = 0.55$, shown in Fig. 4.4.1. This demonstrates that the axial resolution

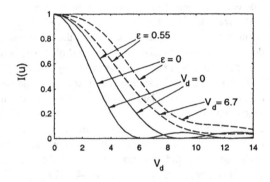

Fig. 4.4.2   Intensity of the axial response to a perfect reflector for $\varepsilon = 0$ and $\varepsilon = 0.55$ when $v_d = 0$ (solid curves) and $v_d = 6.7$ (dashed curves).

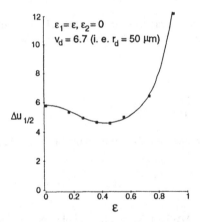

Fig. 4.4.3   Half width of the axial response to a perfect reflector, $\Delta u_{1/2}$, as a function of the radius of the central obstruction for a given value of $v_d (v_d = 6.7)$. The dark squares correspond to experimental values.

is indeed improved when the radius of the detector is larger than a certain value which depends on the given size of pupil obstruction. For a given value of v_d, the value of $\Delta u_{1/2}$ varies with ε. This means that for a given value of v_d, there is an optimum point ε_o at which the width of the axial response becomes the smallest. The relationship of $\Delta u_{1/2}$ to ε is shown in Fig. 4.4.3 for $v_d = 6.7$, in which the optimized value of ε is approximately 0.45. The optimized value of ε increases monotonically with the detector radius.[4.16]

The results shown in Figs. 4.4.1-4.4.3 were confirmed in an experimental confocal microscope with a wavelength of 0.633 μm and an objective of numerical aperture 0.8.[4.14] In Figs. 4.4.1 and 4.4.3, the dark squares and dots represent the experimental results. Although the improvement in axial resolution is the same when either an annular objective lens or an annular collector lens is used, making use of an annular objective only results in a significant improvement in signal level (Chapter 7).[4.14]

4. 5 Effect of Detector Size on Transverse Resolution - Edge-Setting Criteria

One of the ways to characterize transverse resolution is based on the image of a sharp straight edge: the sharper the image the higher the transverse resolution. It is also important to study the image of the edge for the application of optical microscopy in metrology of structures of various different forms. Confocal microscopy has been applied to investigate linewidths on integrated circuit masks and wafers,[4.17, 4.18] the diameters[4.19] of optical fibres and other fibres such as wool and mohair, dimensions of manufactured parts, and particle sizing in metallography, geology or biology.

In order to determine the edge location and the transverse resolution an edge-setting criterion is needed. However in conventional microscopy this criterion depends on the degree of coherence in the image formation process[4.20-4.22] A possible criterion is that the intensity displays an inflection point. However this property strictly only occurs at the edge for purely-incoherent illumination. Similarly, image shearing methods, which effectively determine the position where the intensity has dropped to one half of its value far from the edge, are strictly accurate only for purely-incoherent illumination. Incoherent illumination, however, requires the aperture of the condenser lens to be much larger than that of the objective, and this is obviously not possible when using a large aperture objective as is required for high resolution. In the case of coherent imaging, the intensity at the edge is one quarter of its value far from the edge in the bright side.[4.5, 4.24] For partially-coherent imaging, Watrasiewitz[4.22] calculated the intensity at the edge for different condenser apertures. In particular, the intensity at the edge is precisely one third for equal condenser and objective aperture.[4.23, 4.24] However, apart from the coherent and incoherent limits, the intensity at the edge depends on the presence of aberrations and apodization.

As discussed in Chapter 3, confocal imaging with a point detector is purely coherent, regardless of the presence of aberrations and apodization, and the intensity at the edge is one quarter. However, as a result of using a pinhole mask of finite size, the overall imaging operation is in practice partially coherent. We investigate, in this section, how the intensity at the edge varies with confocal aperture size and shape in terms of Eq. (4.3.2). For an edge given in Section 3.3.4, substituting Eq. (3.3.15) in to Eq. (4.3.2) results in the image of the edge:

$$I(v_x) = \frac{1}{4} + \frac{1}{2\pi} \text{Re}\left[\int_{-\infty}^{\infty} \frac{C_2(m, n = 0; m' = 0, n' = 0)}{m} \exp(iv_x m) dm\right] +$$

$$\text{(4.5.1)}$$

$$\frac{1}{4\pi^2} \int\int_{-\infty}^{\infty} \frac{C_2(m, n = 0; m', n' = 0)}{mm'} \exp[(iv_x(m - m'))] dm dm'.$$

Here $C_2(m, n; m', n')$, normalized to unity at $m = m' = n = n' = 0$, is the 2-D in-focus TCC for a thin object, given by the projection of the 3-D TCC in the focal plane (Chapter 10). The intensity and the gradient of the image at the edge can be derived from Eq. (4.5.1). Making use of Eqs. (4.3.11), (4.3.12), (4.3.18), and (4.3.19), the effect of the detector size on these two parameters can be evaluated. Alternatively, the intensity distribution in the detector space can be calculated in terms of Eqs. (3.1.7) and (3.3.14). By integrating the intensity within the aperture of the detector as a function of the scan position, the image of the edge can also be obtained.[4.25]

The image of the edge for different values of the detector radius is shown in Fig. 4.5.1. As can be seen, the gradient and the intensity at the edge change with the detector size. The resulting intensity at the edge as a function of v_d is shown in Fig. 4.5.2, where the intensity has been normalized to one quarter when $v_d = 0$, because imaging in this case is fully coherent. Fig. 4.5.1 indicates that the intensity at the edge increases with the detector radius. When the detector radius approaches infinity it becomes one third of its

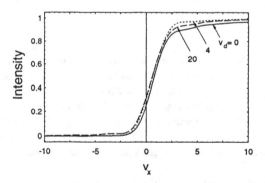

Fig. 4.5.1   Image of a sharp edge for different values of the detector radius, v_d.

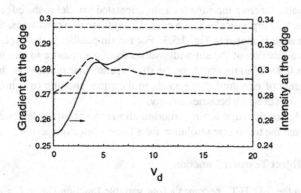

Fig. 4.5.2   Intensity and gradient at the edge as a function of the detector radius, v_d.

value far from the edge on the bright side, because imaging is equivalent to the equal aperture case for a conventional microscope (see Eqs. (4.3.9) and (4.3.10)). The variation of the gradient as a function of the detector radius v_d is not monotonic. For $v_d = 0$, it is equal to $8/3\pi^2$, as expected. Then, it reaches a maximum approximately at $v_d = 3.75$, which is close to the position of the first minimum of the Airy pattern in the detector plane. After the maximum, the gradient decreases, and eventually approaches $8/3\pi^2$ again, as predicated for a conventional microscope of full illumination.[4.23, 4.24]

For slit detectors, two important cases should be considered: the slit being oriented parallel and perpendicular to the edge.

According to Wilson and Hewlett,[4.26] when $v_{dx} = 0$, the intensity at the edge is one

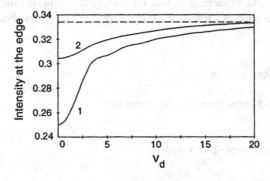

Fig. 4.5.3   Intensity at the edge as a function of the detector width, v_{dx}: (1) slit parallel to the edge; (2) slit perpendicular to the edge.

quarter of its value far from the edge for a slit orientated parallel to the edge and 0.30428 for a slit orientated perpendicular to the edge. Using these two conditions, the calculated gradient at the edge is shown in Fig. 4.5.3. For the slit parallel to the edge, the intensity at the edge as a function of the slit width varies from one quarter to one third, while it increases from 0.30428 to one third for the slit perpendicular to the edge. In both cases, the limiting value of one third corresponds to the equal aperture case in conventional imaging when the slit width becomes infinity.

Knowledge of the edge-setting criterion allows the accurate determination of the edge location and the transverse resolution for a known detector aperture.

4. 6 Weak-Object Transfer Function

Although the 3-D TCC becomes a four variable function $C(l, s; l', s')$ for a system of circular symmetry, it is still difficult to calculate $C(l, s; l', s')$ and then plot it in an illustrative form. In this section, we will discuss a useful and special form of $C(l, s; l', s')$, the weak-object transfer function (WOTF), if light scattering is weak.

4. 6. 1   Weak-Object Transfer Function in Transmission

For a transmission system, if the scattered light is much weaker than the unscattered light through the object, the transmittance of the object can be expressed as[4.1, 4.10]

$$o(r) = \delta(z) + o_1(r), \tag{4.6.1}$$

where the delta function represents the unscattered component in the forward direction and $o_1(r)$ denotes the scattered component which is weaker than the unscattered component. It should be mentioned that Eq. (4.6.1) is a general form of a weakly-scattering object function in transmission microscopy.[4.10] The inverse Fourier transform of Eq. (4.6.1) is

$$O(m) = \delta(m)\delta(n) + O_1(m), \tag{4.6.2}$$

where $O_1(m)$ is the 3-D spectrum of $o_1(r)$, given by

$$O_1(m) = \int_{-\infty}^{\infty} o_1(r)\exp(-2\pi i r \bullet m)dr. \tag{4.6.3}$$

Substituting Eq. (4.6.2) into Eq.(4.3.2) and neglecting the second-order terms, we obtain

$$I(r_s) = \int\int_{-\infty}^{\infty} C(0,0,s;0,0,s')\exp[2\pi i z_s(s-s')]dsds' +$$

$$\int\int_{-\infty}^{\infty} C(0,0,s;m')O_1^*(m')\exp(-2\pi i r_s \bullet m')\exp(2\pi i z_s s)dsdm' + \qquad (4.6.4)$$

$$\int\int_{-\infty}^{\infty} C(m;0,0,s')O_1(m)\exp(2\pi i r_s \bullet m'_s)\exp(-2\pi i z_s s')ds' dm,$$

which can be rewritten as

$$I(r_s) = \tilde{A} + \int_{-\infty}^{\infty} B_w(m')O_1^*(m')\exp(-2\pi i r_s \bullet m')dm' +$$

$$\qquad (4.6.5)$$

$$\int_{-\infty}^{\infty} C_w(m)O_1(m)\exp(2\pi i r_s \bullet m)dm,$$

where

$$\tilde{A} = \int\int_{-\infty}^{\infty} C(0,0,s;0,0,s')\exp[2\pi i z_s(s-s')]dsds', \qquad (4.6.6)$$

$$B_w(m') = \int_{-\infty}^{\infty} C(0,0,s;m')\exp(2\pi i z_s s)ds, \qquad (4.6.7)$$

$$C_w(m) = \int_{-\infty}^{\infty} C(m;0,0,s')\exp(-2\pi i z_s s')ds'. \qquad (4.6.8)$$

According to the symmetry relation in Eq. (4.3.6), we have

$$C_w^*(m) = B_w(m), \qquad (4.6.9)$$

and therefore

$$I(r_s) = \tilde{A} + 2\operatorname{Re}\left[\int_{-\infty}^{\infty} C_w(m)O_1(m)\exp(2\pi i r_s \bullet m)dm\right]. \qquad (4.6.10)$$

Here $\tilde{A}$, a constant, represents the contribution from the unscattered beam and $C_w(m)$ is called the 3-D weak-object transfer function (WOTF) in transmission.

For a system with a point-like detector on the axis,

$$C_w(m) = c(m, 0), \qquad (4.6.11)$$

which is identical to the 3-D CTF in confocal transmission without detector offset. In the case of a finite-sized pinhole detector, the 3-D WOTF may be calculated, for equal circular objective and collector lenses, from

$$C_w(l,s) = \frac{K_l}{l} \int_0^{v_d} \int_0^{2\pi} \exp\left[iv_2\left(\frac{l}{2} - \frac{s}{l}\right)\cos\phi' \right] \mathrm{sinc}\left[v_2\sqrt{1 - \left(\frac{|s|}{l} + \frac{l}{2}\right)^2}\sin\phi' \right]$$

(4.6.12)

$$\mathrm{Re}\left[\sqrt{1 - \left(\frac{|s|}{l} + \frac{l}{2}\right)^2} \right] J_1(v_2)dv_2d\phi' ,$$

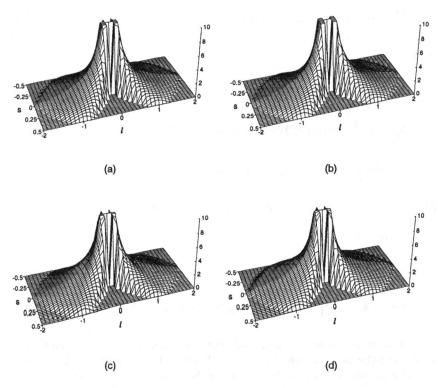

(a)   (b)

(c)   (d)

Fig. 4.6.1   3-D weak-object transfer function in transmission for different values of the detector radius v_d: (a) $v_d = 2$; (b) $v_d = 5$; (c) $v_d = 10$; (d) $v_d = 40$.

where J_1 is the Bessel function of the first kind of order unity. It is evident that, unlike the TCC, $C_w(l, s)$ is a real function regardless of the detector size.

It can be easily proved that the 3-D WOTF's, $C_w(l, s)$, for a point detector and for an infinitely-large detector can be expressed as

$$C_w(l,s) = \frac{2}{l} \text{Re}\left[\sqrt{1 - \left(\frac{|s|}{l} + \frac{l}{2}\right)^2}\right],$$ (4.6.13)

which is the 3-D CTF for confocal transmission without detector offset and is therefore the 3-D optical transfer function (OTF) for a single lens (see Eq. (2.5.14) and Fig. 2.5.3). This result, together with Eq. (4.3.9), means that, for a weakly-scattering object, a confocal transmission microscope behaves as a conventional imaging system.

For a given non-zero value of the pinhole size, the 3-D WOTFs are depicted in Fig. 4.6.1, where the WOTF has been normalized to the value of $C_w(l = 1, s = 0)$ for $v_d = 0$. As expected, the 3-D WOTF exhibits a missing cone of spatial frequencies, so that no optical sectioning can be achieved in transmission for weakly-scattering objects exhibiting only axial variations. When the pinhole radius is increased, the 3-D WOTF becomes asymmetric with respect to the plane at $s = 0$. The asymmetry can be clearly seen in Fig. 4.6.2b in which axial cross-sections of the 3-D WOTF through $l = 1$ are plotted, and becomes most pronounced approximately at $v_d = 3.9$ which is close to the position of the first minimum of the Airy pattern in the detector space. The asymmetry implies that the phase transfer function, defined by Streibl[4.27] as the odd part of the 3-D WOTF, is not zero and thus that the axial phase structure of the object can be imaged in the transmission confocal system using a finite-sized detector. Transverse cross-sections

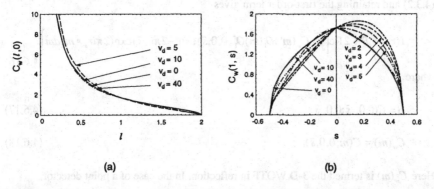

(a)   (b)

Fig. 4.6.2   Transverse (a) and axial (b) cross-sections of the 3-D WOTF for different values of the detector radius, v_d.

of the 3-D WOTF at $s = 0$, which describe imaging of a thick object without axial variations, are displayed in Fig. 4.6.2a for different sizes of the detector. Note that the 3-D WOTF does not demonstrate an appreciable change in the transverse direction on increasing v_d and is always positive regardless of the size of the detector.

4. 6. 2 *Weak-Object Transfer Function in Reflection*

The choice of an appropriate form for the weak object in a reflection confocal system should be carefully considered. Generally, the unscattered light is in the forward direction and is not detectable in reflection. Therefore, there is no undiffracted wave relative to which the scattered light can be considered to be small. However, if we consider a class of objects in which the spectral-power density decreases monotonically with spatial frequency,[4.10] the object can be assumed to be weakly scattering with a strong component at the lowest true axial spatial frequency $\tilde{s}$ ($\tilde{s} = 1 - s0$) because the reflection-mode microscope has a dark-field nature caused by the axial offset of the transfer function according to the discussion in Section 3.2. In this case the spatial spectrum of the object can be expressed as[4.1, 4.10]

$$O(m) = \delta(m)\delta(n)\delta(s - \tilde{s}) + O_1(m). \tag{4.6.14}$$

The corresponding reflectivity of the object is given by the Fourier transform of Eq. (4.6.14):

$$o(r) = \exp(2\pi i z \tilde{s}) + o_1(r). \tag{4.6.15}$$

The first term describes a strong component at spatial frequency $\tilde{s}$ in reflection and the second is the weak object function. The substitution of Eq. (4.6.14) for $O(m)$ in Eq. (4.3.2) and retaining the first-order term gives

$$I(r_s) = \tilde{B} + 2\mathrm{Re}\left[\int_{-\infty}^{\infty} C_w'(m')O_1(m)O_1^*(0,0,\tilde{s})\exp(-2\pi i z_s \tilde{s})\exp(2\pi i r_s \bullet m)dm\right] \tag{4.6.16}$$

where

$$\tilde{B} = C(0,0,\ \tilde{s};0,0,\ \tilde{s}), \tag{4.6.17}$$

$$C_w'(m) = C(m;0,0,\tilde{s}). \tag{4.6.18}$$

Here $C_w'(m)$ is termed the 3-D WOTF in reflection. In the case of a point detector,

$$C_w'(m) = c(m,\ 0), \tag{4.6.19}$$

which is the 3-D CTF in a reflection-mode confocal microscope without the presence of detector offset.

In general, the 3-D WOTF can be derived, for equal circular objective and collector lenses, as

$$C'_w(l,s) = K_r \int_0^{v_a} f_3(l,s) J_0 \left[v_2 (\bar{s} + s0)^{1/2} \right] v_2 dv_2,$$ (4.6.20)

where

$$f_3(l,s) =$$

$$\begin{cases} J_0\left(\dfrac{lv_2}{2}\right) J_0\left[v_2 \sqrt{\bar{s} - \dfrac{l^2}{4}} \right], & \dfrac{l^2}{4} \le \bar{s} \le 1 - l\left(1 - \dfrac{l}{2}\right), \\ \dfrac{1}{\pi^2} \int_0^\pi \left\{ \int_{-\pi/2}^{\pi/2} \cos\left[\dfrac{lv_2}{2}\cos\phi' - v_2\sqrt{\bar{s} - \dfrac{l^2}{4}}\cos(\theta + \phi') \right] d\theta \right\} d\phi', & 1 - l\left(1 - \dfrac{l}{2}\right) \le \bar{s} \le 1, \\ 0, & otherwise. \end{cases}$$

(4.6.21)

It is noticed that $C'_w(l, s)$ is also real and that the region where the values of $C'_w(l, s)$ are non-zero is the same as the 3-D CTF in confocal reflection (see Eq. (3.2.20)). $C'_w(l, s)$ for $v_d = 0$ has an identical form to the 3-D CTF (see Fig. 3.2.3 and Eq. (3.2.20)). In this case, $C'_w(l, s)$ is positive. The 3-D WOTF is displayed in Fig. 4.6.3 for different values of the detector radius v_d and for $\bar{s} = 1 - s0$, where $C'_w(l, s)$ is normalized to unity when $l = 0$ and $s + s0 = 1$. As v_d increases, the value of $C'_w(l, s)$ in the region of $l^2/4 \le s + s0 \le 1 - l(1 - l/2)$ increases and then decreases with an oscillating behaviour along the axial direction. As a result, $C'_w(l, s)$ exhibits negative components (see Fig. 4.6.3b). The negative and oscillating behaviour is clearly illustrated in Fig. 4.6.4 in which the axial cross-sections of $C'_w(l, s)$ at $l = 0$ are plotted. The significance of the negative component in the 3-D WOTF is that the contrast of an image at certain spatial frequencies may be reversed. Another significance of Fig. 4.6.4 is that the modulus squared of its Fourier transform gives the strength of optical sectioning for weak objects. Therefore, the optical sectioning property of a weakly-scattering object is degraded on increasing the detector size.

Eventually, as v_d approaches infinity, $C'_w(l = 0, s) = \delta(s+s0-1)$ and $C'_w(l, s)$ in $l^2/4 < s + s0 \le 1 - l(1 - l/2)$ drops to zero. On the other hand, $C'_w(l, s)$ in the region of $1 - l(1 - l/2) < s + s0 \le 1$ is always positive and eventually approaches a limiting form given by

$$C'_w(l,s) = \frac{2}{l} \frac{1}{\text{Re}\left[\sqrt{\bar{s} + s0 - \left(\frac{\bar{s}-s}{l} + \frac{l}{2}\right)^2}\right]}.$$ (4.6.22)

By recalling Eq. (3.5.31), it is clear that when $\bar{s} = -s0+1$, which is the lowest true axial spatial frequency, Eq. (4.6.22) is also the 3-D CTF for a reflection confocal system with

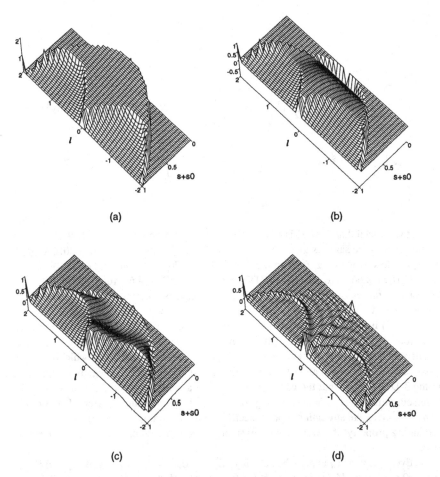

(a)   (b)

(c)   (d)

Fig. 4.6.3   3-D weak-object transfer function in reflection for different values of the detector radius v_d: (a) $v_d = 2$; (b) $v_d = 5$; (c) $v_d = 10$; (d) $v_d = 40$.

one thin annular pupil and one circular pupil. Eq. (4.6.22) shows that the 3-D WOTF has a singularity at $l = 0$ and a missing cone of spatial frequencies around it (see Fig. 3.5.6)). This phenomenon indicates that no optical sectioning effect exists for a weak object if the detector size is large. It should be pointed out that $C_w'(l, s)$ in Eq. (4.6.22) has non-zero values when $s + s0 = 1$, which contrasts with that for a finite value of the detector size (see Fig. 4.6.3). This transverse cross-section at $s + s0 = 1$ is shown as a solid curve in Fig. 4.6.5a, while an axial cross-section at $l = 1$ is placed in Fig. 4.6.5b. The dashed curves in Figs. 4.6.5a and 4.6.5b are the projections of $C_w'(l, s)$ in the focal plane and on the axis, respectively, which accordingly correspond to the 2-D in-focus WOTF for a

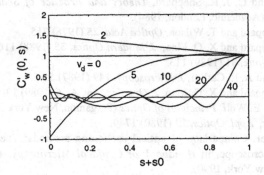

Fig. 4.6.4   Axial cross-section of the 3-D WOTF in reflection at $l = 0$.

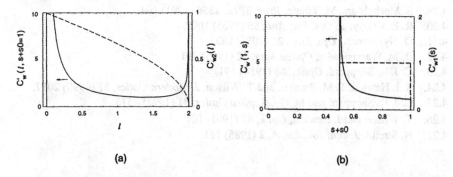

(a)   (b)

Fig. 4.6.5   (a) Transverse cross-section of the 3-D WOTF (solid curve) and the 2-D in-focus WOTF (dashed curve) in reflection for $v_d \rightarrow \infty$; (b) Axial cross-section of the 3-D WOTF (solid curve) and the 1-D on-axis WOTF (dashed curve) in reflection for $v_d \rightarrow \infty$.

thin weak object, given by Eq. (3.5.32), and to the 1-D on-axis WOTF for a line weak object. The significance of the 2-D in-focus and 1-D on-axis WOTF will be described in Chapter 10.

References

4.1.　M. Gu and C. J. R. Sheppard, *J. Modern, Optics*, **41** (1994) 1701.

4.2.　C. J. R. Sheppard and M. Gu, *J. Opt. Soc. Am. A*, **10** (1993) 533.

4.3.　V. Drazic, *J. Opt. Soc. Am. A*, **9** (1992) 725.

4.4.　V. Drazic, *J. Modern Optics*, **39** (1992) 1777.

4.5.　T. Wilson and C. J. R. Sheppard, *Theory and Practice of Scanning Optical Microscopy* (Academic, London, 1984).

4.6.　C. J. R. Sheppard and T. Wilson, *Optica Acta*, **25** (1978) 315.

4.7.　C. J. R. Sheppard and X. Q. Mao, *J. Modern Optics*, **35** (1988) 1169.

4.8.　T. Wilson, *Optik*, **81** (1989) 113.

4.9.　T. Wilson and A. R. Carlini, *J. Microscopy*, **149** (1987) 51.

4.10.　C. J. R. Sheppard and X. Q. Mao, *J. Opt. Soc. Am. A*, **6** (1989) 1260.

4.11.　M. Born and E. Wolf, *Principles of Optics* (Pergamon, New York, 1980).

4.12.　C. J. Koester, *Appl. Optics*, **19** (1980) 1749.

4.13.　C. J. Koester, Comparison of optical sectioning methods: The scanning slit confocal microscope, in *Handbook of Confocal Microscopy*, ed. J. Pawley (Plenum, New York, 1990).

4.14.　M. Gu, C. J. R. Sheppard, and H. Zhou, *Optik*, **93** (1993) 87.

4.15.　C. J. R. Sheppard and M. Gu, *Optik*, **86** (1991) 169.

4.16.　C. J. R. Sheppard and M. Gu, *Opt. Commun.*, **84** (1991) 7.

4.17.　J. T. Lindow, S. D. Bennet, and I. R. Smith, *Proc. SPIE*, **565** (1985) 81.

4.18.　Th. Zapf and R. W. Wijnaendts-van- Resandt, *Microelectron. Eng*, **5** (1986) 573.

4.19.　S. Mechels and M. Young, *Proc. SPIE*, **1556** (1991) 164.

4.20.　R. E. Kinzly, *J. Opt. Soc. Am.*, **55** (1965) 1002.

4.21.　D. Nyyssonen, *Opt. Eng.*, **21** (1982) 882.

4.22.　B. M. Watrasiewitz, *Optica Acta*, **12** (1965) 391.

4.23.　C. J. R. Sheppard, *Optik*, **66** (1984) 191.

4.24.　S. J. Hewlett, S. M. Barnett, and T. Wilson, *J. Modern Optics*, **37** (1990) 2017.

4.25.　C. J. R. Sheppard and M. Gu, *Applied Optics*, **31** (1992) 4575.

4.26.　T. Wilson and J. Hewlett, *Optik*, **87** (1991) 109.

4.27.　N. Streibl, *J. Opt. Soc. Am. A*, **2** (1985) 121.

Chapter 5

CONFOCAL SINGLE-PHOTON
FLUORESCENCE MICROSCOPY

Because of the three-dimensional (3-D) imaging property in confocal scanning microscopy, confocal fluorescence microscopy was achieved in the same period as confocal bright-field microscopy.[5.1] Under the illumination of intermediate power, one incident photon can be absorbed in the sample under inspection to excite the electron transition from the ground state to an excited state. The excited electron returns to the ground state by radiating fluorescence light. The energy of the radiated fluorescence photon is slightly less than the incident one due to the non-radiation relaxation during the downward transition,[5.2] and the corresponding microscopy in which the fluorescence light is imaged is termed single-photon (1-p) fluorescence microscopy. Its 3-D imaging performance differs from two-photon fluorescence microscopy, which will be discussed in the next chapter.

This chapter is focused on the 3-D imaging performance in confocal 1-p fluorescence microscopy. Section 5.1 demonstrates the 3-D incoherent image formation in confocal 1-p fluorescence microscopy. As a result, the 3-D optical transfer function (OTF), which gives the description of an incoherent imaging process, is introduced in Section 5.2. Section 5.3 gives numerical examples of the 3-D OTF by considering the effect of annular pupils as well as of a finite-sized detector. In Section 5.4, signal level, one of the important issues in confocal 1-p fluorescence microscopy when a finite-sized detector is used, is explored. The combined effect of annular lenses and a finite-sized detector on axial resolution is investigated in Section 5.5, giving a method for improving the axial resolution in practice. An analysis on the influence of a finite-sized source on 3-D confocal 1-p fluorescence imaging is described in Section 5.6.

5. 1 Incoherent Image Formation

Fig. 3.1.1a can be used as the schematic diagram of a confocal 1-p fluorescence microscope. Assume that the light from the source has a wavelength λ. We first consider a point source, and the effect of a finite-sized source will be described in Section 5.6. Therefore, the light field in the object space is given by Eq. (3.1.4). Thus the intensity distribution at a point r_1 in the object space may be expressed as

$$I_o(r_1) = \left| \int_{-\infty}^{\infty} \delta(r_0) \exp[ik(z_0 - z_1)] h_1(r_0 + M_1 r_1) dr_0 \right|^2 . \tag{5.1.1}$$

Here the parameter M_1 is defined by Eq. (3.1.5). r_0 and r_1 are position vectors in source and object spaces, respectively. $h_1(r)$ is the 3-D amplitude point spread function for the objective lens at an incident wavelength λ, given in Eq. (3.1.1).

Consider a thick object labelled with fluorescent materials. It can emit fluorescence light with a strength $o_f(r)$. The fluorescence light from a given point at r_1 is proportional to the light intensity impinging on the point[5.1] and has a wavelength λ_f which is slightly larger than λ due to the difference of energy between the fluorescence and incident photons. The total fluorescence strength emitted from the point r_1 in the object under the illumination $I_o(r_1)$ can be expressed as

$$I_1(r_1) = |h_1(M_1 r_1)|^2 o_f(r_s - r_1), \tag{5.1.2}$$

where a scan position r_s has been introduced to include the scanning movement of the object, so that Eq. (5.1. 2) is also a function of r_s.

For a single fluorescent point object at r_1, the intensity at r_2 in the detector space is given, according to Eq. (2.5.1), by

$$I_2(r_1, r_2) = |h_2(r_1 + M_2 r_2)|^2 , \tag{5.1.3}$$

where $h_2(r)$ is the 3-D amplitude point spread function for the collector lens at wavelength λ_f (see Eq. (3.1.1)). M_2 is defined in Eq. (3.1.8).

Because the 1-p fluorescence radiation is a spontaneous process,[5.2] the fluorescence light propagates in all directions and the phase of the light is not correlated, so that the total intensity at r_2 from the 3-D fluorescent object with the fluorescence strength $I_1(r_1)$ is the superposition of the intensity contributed from all points in the object, i. e., the integration of the product of $I_1(r_1)$ and $I_2(r_1, r_2)$ with respect to r_1:

$$I(r_2, r_s) = \int_{-\infty}^{\infty} |h_1(M_1 r_1)|^2 o_f(r_s - r_1) |h_2(r_1 + M_2 r_2)|^2 dr_1, \tag{5.1.4}$$

which has been explicitly expressed as a function of the scan position.

In confocal fluorescence microscopy, a detector of a certain size is usually selected because the signal strength is not so strong. Consider a finite-sized detector with an intensity sensitivity $D(r_2)$. The final measured intensity from the sample is the summation of Eq. (5.1.4) over the detector aperture:

$$I(r_s) = \int_{-\infty}^{\infty} \left[\int_{-\infty}^{\infty} |h_1(M_1r_1)|^2 o_f(r_s - r_1)|h_2(r_1 + M_2r_2)|^2 dr_1 \right] D(r_2)dr_2,$$ (5.1.5)

which can be rewritten as a form of the 3-D convolution operation:

$$I(r_s) = h_i(r_s) \otimes_3 o_f(r_s),$$ (5.1.6)

where $\otimes_3$ denotes the 3-D convolution operation and $h_i(r)$ is given by

$$h_i(r) = |h_1(M_1r)|^2 \int_{-\infty}^{\infty} |h_2(r + M_2r_2)|^2 D(r_2)dr_2.$$ (5.1.7)

Obviously, Eq. (5.1.7) is the image of a single fluorescent point and is thus the 3-D effective intensity point spread function (IPSF) for confocal 1-p fluorescence microscopy. Since Eq. (5.1.6) indicates the superposition of the intensity, it represents an incoherent imaging process. According to Eq. (5.1.7), the 3-D IPSFs for reflection and transmission confocal fluorescence microscopes are identical. Consequently, there is no difference in imaging performance between the two cases for monochromatic, i. e., continuous wave (CW) illumination. This conclusion does not hold any more if a pulsed beam is used, which will be discussed in Chapter 8.

According to the discussion in Section 2.3, a 3-D IPSF for a single lens becomes broader when a longer wavelength is used. The 3-D IPSF for the collector lens is therefore broader than that for the objective lens because the fluorescence wavelength is larger than the incident one. Eventually, when $\lambda_f \rightarrow \infty$, the former 3-D IPSF becomes a constant and the role of the collector lens is completely degraded. The confocal fluorescence system reduces to a conventional incoherent imaging system which has been discussed in Section 2.5 and has no optical sectioning property.[5.3]

By recalling Eq. (4.1.5), Eq. (5.1.7) is identical to Eq. (4.1.5) if the fluorescence wavelength is the same as the incident wavelength. Therefore, the effect of a finite-sized circular detector on the 3-D IPSF for a confocal 1-p fluorescence microscope of two equal circular lenses can be found from Fig. 4.2.1. As expected, the use of a large detector results in a degradation of the 3-D IPSF towards the case of a single lens.

5. 2 Optical Transfer Function

In order to derive the 3-D transfer function for confocal 1-p fluorescence microscopy, we resolve the 3-D object function $o_f(r)$ into a series of periodic components using a Fourier transform:

$$o_f(m) = \int_{-\infty}^{\infty} O_f(m) \exp(2\pi i r \bullet m) dm, \tag{5.2.1}$$

where m represents a spatial frequency vector as we defined before and $O_f(m)$ is the 3-D intensity spectrum of the fluorescent object, given by the inverse Fourier transform of $o_f(r)$. Making use of Eq. (5.2.1) in Eq. (5.1.6) yields

$$I(r_s) = \int_{-\infty}^{\infty} C(m) O_f(m) \exp(2\pi i r_s \bullet m) dm. \tag{5.2.2}$$

Here

$$C(m) = \int_{-\infty}^{\infty} h_i(r) \exp(-2\pi i r \bullet m) dr. \tag{5.2.3}$$

Clearly, $C(m)$ is the efficiency with which each periodic component in the fluorescent object is imaged and is thus called the 3-D optical transfer function (OTF) as it describes the incoherent imaging process. It is given by the 3-D inverse Fourier transform of the 3-D effective intensity point spread function as predicted by Frieden.[5.4] We can rewrite Eq. (5.2.3) as

$$C(m) = F_3 \left\{ |h_1(M_1 r)|^2 \left[|h_2(r)|^2 \otimes_3 D(r / M_2) \right] \right\}, \tag{5.2.4}$$

where F_3 denotes the 3-D inverse Fourier transform defined by Eq. (5.2.3). In terms of the convolution theorem,[5.5] Eq. (5.2.4) can be expressed as

$$C(m) = C_1(m) \otimes_3 C_2(m), \tag{5.2.5}$$

where

$$C_1(m) = F_3 \left[|h_1(M_1 r)|^2 \right] \tag{5.2.6}$$

and

$$C_2(m) = F_3\left[\left|h_2(r)\right|^2\right]F_3\left[D(r\,/\,M_2)\right].$$
(5.2.7)

In Eqs. (5.2.6) and (5.2.7), $F_3[\left|h_1(M_1r)\right|^2]$ and $F_3[\left|h_2(r)\right|^2]$ are the 3-D OTFs for the objective and collector lenses, respectively, and $F_3[D(r/M_2)]$ represents the 3-D inverse Fourier transform of the intensity sensitivity of the detector. Therefore, Eq. (5.2.7) can be called the 3-D effective OTF for a collector lens in the presence of a finite-sized detector. Eqs. (5.2.4)-(5.2.7) are applicable for a general system even with aberration and a finite-sized detector. They imply again that there is no difference of imaging performance between reflection and transmission systems.

For a point detector, the 3-D OTF for confocal 1-p fluorescence microscopy is simply given by the convolution of the 3-D OTF for the objective lens with the 3-D OTF for the collector lens because $F_3[D(r/M_2)]$ is a constant. In this case, Eq. (5.2.4) can be reduced to

$$C(m) = F_3\left[\left|h_1(M_1r)h_2(r)\right|^2\right],$$
(5.2.8)

which becomes

$$C(m) = c(m) \otimes_3 c^*(-m),$$
(5.2.9)

where * denotes the complex conjugate operation and

$$c(m) = F_3[h_1(M_1r)\,h_2(r)].$$
(5.2.10)

Clearly, Eq. (5.2.10) is, in terms of Eq. (3.1.13), the 3-D coherent transfer function (CTF) for transmission confocal bright-field microscopy consisting of a point detector without detector offset. In general, $c(m)$ in Eq. (5.2.9) can be understood to be the 3-D CTF for either transmission or reflection because a linear phase term can be introduced in Eq. (5.2.8) without affecting the conclusion. However, when a pulsed beam is used for illumination, fluorescence imaging performance may differ between reflection and transmission when a time-resolved imaging technique is used (Chapter 8).

The method presented here is different from that by Kimura et al.[5.3, 5.6, 5.7] in which Stockseth's approximation[5.8] was introduced to derive the defocused OTF and then the 3-D OTF was obtained by performing the 1-D inverse Fourier transform with respect to the defocus distance. It has been shown that the 3-D OTF for confocal 1-p fluorescence microscopy based on the Stockseth's approximation has a considerable difference from the exact form of the 3-D OTF presented here.[5.9]

5. 2. 1   Circular Detector

In the case of a confocal system of circular lenses and a circular detector, Eq. (5.2.4) reduces, by making use of Eq. (4.2.3), to

$$C(l,s) = F_3\left[|h_1(v,u)|^2\right] \otimes_3 \left\{ F_3\left[|h_2(v/g,u/g)|^2\right] F_2[D(v/g)] \right\}. \qquad (5.2.11)$$

The variables l and s in $C(l, s)$ are, respectively, the radial and axial spatial frequencies, and have been normalized by Eqs. (2.4.8) and (2.4.9). F_2 denotes the two-dimensional (2-D) Hankel Fourier transform with respect to v. $D(v)$ represents the intensity sensitivity of a circular detector with a normalized radius of v_d and is therefore defined in Eq. (4.2.3), where v_d is given by

$$v_d = \frac{2\pi}{\lambda_f} r_d \sin \alpha_d. \qquad (5.2.12)$$

$h_1(v, u)$ and $h_2(v/g, u/g)$ are given by Eqs. (4.2.5) and (4.2.6), where $g = \lambda_f/\lambda$ is the ratio of the fluorescence wavelength to the incident wavelength. The definitions of v and u are the same as those in Eq. (3.1.17) at the incident wavelength λ.

As we have seen in Chapter 4, use of annular pupils may result in an improvement in axial resolution when they are combined with a finite-sized detector. Therefore let us consider that the objective and collector lenses have annular pupils of the same outer radius a but different inner radii $\varepsilon_1 a$ and $\varepsilon_2 a$. According to Eq. (2.5.16), the 3-D OTF for the objective and collector lenses may be expressed as

$$F_3\left[|h_1(v,u)|^2\right] = \frac{2}{l}\left\{ \mathrm{Re}\left[\sqrt{1-\left(\frac{|s|}{l}+\frac{l}{2}\right)^2}\right] - \mathrm{Re}\left[\sqrt{\varepsilon_1^2-\left(\frac{|s|}{l}-\frac{l}{2}\right)^2}\right]\right\} \qquad (5.2.13)$$

with $|s| \leq (1 - \varepsilon_1^2)/2$ and $l \leq 2$, and

$$F_3\left[|h_2(v/g,u/g)|^2\right] = \frac{2}{gl}\left\{ \mathrm{Re}\left[\sqrt{1-\left(\frac{|s|}{l}+\frac{gl}{2}\right)^2}\right] - \mathrm{Re}\left[\sqrt{\varepsilon_2^2-\left(\frac{|s|}{l}-\frac{gl}{2}\right)^2}\right]\right\} \qquad (5.2.14)$$

with $|s| \leq (1 - \varepsilon_2^2)/2g$ and $l \leq 2/g$. Re[ ] represents the real part of its argument. As expected, when $g \neq 1$, the cut-off spatial frequencies for the collector lens are effectively reduced by a factor of g. Since the 2-D inverse Fourier transform of the sensitivity $D(v)$, i. e., the Hankel transform in the circularly symmetric case, can be derived as

$$F_2[D(v/g)] = v_d \left[\frac{J_1(glv_d)}{l} \right],$$ (5.2.15)

where J_1 is a Bessel function of the first kind of order one, then Eq. (5.2.11) can be finally expressed, after considering the cylindrical symmetry of the microscope system, as[5.10]

$$C(l,s) =$$

$$\iiint\limits_V \frac{K_f v_d}{l_1^2 l_2^2} \left\{ \text{Re}\left[\sqrt{l_1^2 - \left(\left| s' - \frac{s}{2} \right| + \frac{l_1^2}{2} \right)^2} \right] - \text{Re}\left[\sqrt{l_1^2 \varepsilon_1^2 - \left(\left| s' - \frac{s}{2} \right| - \frac{l_1^2}{2} \right)^2} \right] \right\}$$

$$\left\{ \text{Re}\left[\sqrt{l_2^2 - \left(\left| s' + \frac{s}{2} \right| + \frac{g l_2^2}{2} \right)^2} \right] - \text{Re}\left[\sqrt{l_2^2 \varepsilon_2^2 - \left(\left| s' + \frac{s}{2} \right| - \frac{g l_2^2}{2} \right)^2} \right] \right\}$$ (5.2.16)

$$\frac{J_1(g l_2 v_d)}{l_2} dm'\, dn'\, ds'.$$

Here K_f is a constant of normalization in fluorescence. V denotes a volume overlapped by the two non-zero functions within brackets { } in the integrand and

$$l_1 = \sqrt{(m' - l/2)^2 + n'^2},$$ (5.2.17)

$$l_2 = \sqrt{(m' + l/2)^2 + n'^2},$$ (5.2.18)

where variables m' and n' represent the spatial frequencies along two orthogonal transverse directions.

According to Eq. (5.2.2), the Fourier transform of $C(0, s)$ provides the intensity of the axial response to a thin fluorescent sheet which has an object function $\delta(z)$. The axial response is a measure of the strength of optical sectioning, i. e., the axial resolution in 3-D fluorescence imaging, and can be expressed, if a constant factor is neglected, as

$$I(u) = K_f \int_{-\infty}^{\infty} C(l = 0, s) \exp(ius) ds.$$ (5.2.19)

It is thus the 1-D Fourier transform of the axial cross-section of the 3-D OTF. The integration can be evaluated within the axial cut-off spatial frequency. When $l = 0$ and g

$= 1$, $C(0, s)$, the axial cross-section of the 3-D OTF, is a symmetric function as exchanging ε_1 and ε_2, meaning that the axial imaging property is the same when exchanging the objective and the collector. But, in general, there is no such symmetry due to a finite-sized detector.

5. 2. 2   Slit Detector

Similarly, for a slit detector of finite width, the 3-D OTF for confocal 1-p fluorescence can be expressed, if the slit is placed to be parallel to the x axis and has a width $2x_d$ (see Eq. (4.1.4)), as

$$C(l,s) =$$

$$\iiint\limits_{V} \frac{K_f v_{dx}}{l_1^2 l_2^2} \left\{ \mathrm{Re}\left[\sqrt{l_1^2 - \left(\left| s' - \frac{s}{2} \right| + \frac{l_1^2}{2} \right)^2} \right] - \mathrm{Re}\left[\sqrt{l_1^2 \varepsilon_1^2 - \left(\left| s' - \frac{s}{2} \right| - \frac{l_1^2}{2} \right)^2} \right] \right\}$$

$$\left\{ \mathrm{Re}\left[\sqrt{l_2^2 - \left(\left| s' + \frac{s}{2} \right| + \frac{g l_2^2}{2} \right)^2} \right] - \mathrm{Re}\left[\sqrt{l_2^2 \varepsilon_2^2 - \left(\left| s' + \frac{s}{2} \right| - \frac{g l_2^2}{2} \right)^2} \right] \right\}$$

(5.2.20)

$$\frac{\sin[g(m'+l/2)v_{dx}]}{(m'+l/2)} dm' \, dn' \, ds',$$

where v_{dx} is defined as

$$v_{dx} = \frac{2\pi}{\lambda_f} x_d \sin \alpha_d.$$

(5.2.21)

As expected, the 3-D OTF in Eq. (5.2.21) does not have circular symmetry[5.7] Eq. (5.2.19) also holds in the case of a slit detector. In the following discussion, we discuss only the effect of a circular detector on 3-D imaging performance. The method and the associated conclusions can be easily generalized to a slit detector.

5. 3   Effect of Annular Pupils on Optical Transfer Functions

This section provides various numerical examples of Eq. (5.2.16) in order to reveal the dependence of the 3-D OTF on the size of the central obstruction of the annular lens and on the size of the detector.

5. 3. 1   Point Detector

For convenience, we consider a system with a point detector and equal fluorescence and incident wavelengths ($g = 1$). As pointed out in Section 5.1, the longer fluorescence wavelength may result in a further degradation of the 3-D OTF.[5.3]

First, assume that both the objective and the collector have the same radius of central obstruction, so that $\varepsilon_1 = \varepsilon_2$. The 3-D OTF, normalized to unity at the origin, is shown in Fig. 5.3.1 for different values of ε_1 (or ε_2). When $\varepsilon_1 = \varepsilon_2 = 0$, corresponding to

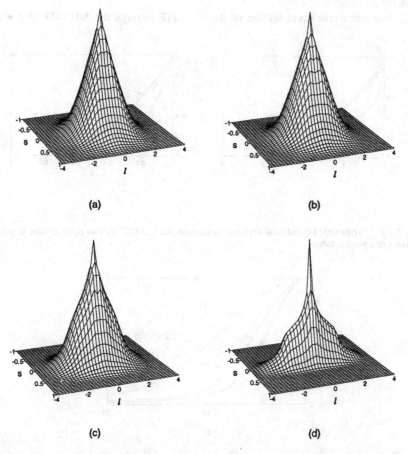

(a)   (b)

(c)   (d)

Fig. 5.3.1   3-D optical transfer function for two equal annular lenses in the case of a point detector:
(a) $\varepsilon_1 = \varepsilon_2 = 0$; (b) $\varepsilon_1 = \varepsilon_2 = 0.25$; (c) $\varepsilon_1 = \varepsilon_2 = 0.5$; (d) $\varepsilon_1 = \varepsilon_2 = 0.75$.

a system of circular pupils, the axial and transverse cut-off spatial frequencies are, respectively, 1 and 4, which are twice as large as those of confocal transmission microscopy. As may be expected from the normal effect of a central obstruction, when the radius of the central obstruction increases, the region in which the 3-D OTF is appreciable becomes narrower in the s direction. But, it becomes broader in the l direction. This broadening of the 3-D OTF accordingly results in a superior response at high spatial frequencies (Fig. 5.3.2a), which implies that when annular lenses are utilized more energy is distributed at higher angles of diffraction than when circular lenses are used. The transverse resolution in the annular system is therefore improved, in agreement with the 2-D analyses.[5.11-5.14]

The transverse cross-section of the 3-D OTF through the 3-D OTF at $s = 0$,

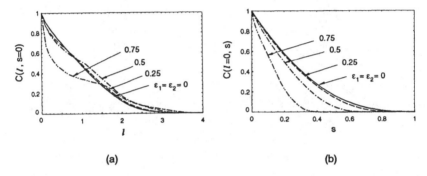

(a)   (b)

Fig. 5.3.2   Transverse (a) and axial (b) cross-sections of the 3-D OTF for two equal annular lenses in the case of a point detector.

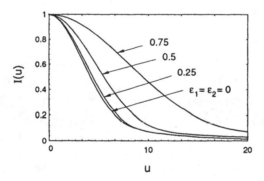

Fig. 5.3.3   Intensity of the axial response to a thin fluorescent sheet for two equal annular lenses in the case of a point detector.

shown in Fig. 5.3.2a, corresponds to imaging of a thick object structure with no variations in fluorescent strength in the axial direction. Fig. 5.3.2b shows the axial cross-section through the 3-D OTF at $l = 0$, and thus corresponds to imaging of planar structures such that there is no variations in fluorescence strength in the transverse directions. In contrast to the transverse behaviour it is seen that the axial cut-off spatial frequency is decreased by $(\varepsilon_1^2 + \varepsilon_2^2)/2$ on increasing radii of the central obstruction. Consequently, the strength of optical sectioning is strongly affected in terms of Eq. (5.2.19). It is confirmed in Fig. 5.3.3 that the strength of optical sectioning is decreased as the radii of the central obstruction are increased. In other words, using annular lenses results in a larger focal depth compared with using two circular pupils ($\varepsilon_1 = \varepsilon_2 = 0$).

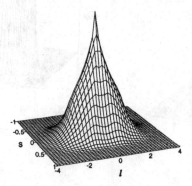

Fig. 5.3.4   3-D optical transfer function for an annular objective and a circular collector in the case of a point detector ($\varepsilon_1 = 0.75$ and $\varepsilon_2 = 0$).

If the objective is an annular lens ($\varepsilon_1 \neq 0$) but the collector is still a circular lens ($\varepsilon_2 = 0$), the higher transverse spatial frequencies can be imaged more strongly with increasing radius of the central obstruction of the objective, while the axial cut-off spatial frequency is decreased by $\varepsilon_1^2/2$. For $\varepsilon_1 = 0.75$ and $\varepsilon_2 = 0$, the 3-D OTF is shown in Fig. 5.3.4. Compared with Fig. 5.3.1 it is most similar to Fig. 5.3.1c except that a weak missing cone of spatial frequencies appears along the axial direction. According to the optical reciprocity and Eq. (5.2.16), Fig. 5.2.4 is still applicable if the objective is of circular aperture ($\varepsilon_1 = 0$) and the collector is of annular aperture ($\varepsilon_2 = 0.75$). Therefore, for a point detector, the 3-D imaging property remains unchanged when exchanging the objective and the collector.

5. 3. 2   Circular Detector

In practice, the detector is not a single point but has a finite size. Usually, a finite-sized circular pinhole is placed immediately in front of the detector, so that the effective size of the detector can be adjusted by the pinhole. The assumption of $g = 1$ is still used in the following discussion.

Let us first consider a confocal 1-p fluorescence system with two equal circular lenses ($\varepsilon_1 = \varepsilon_2 = 0$). When v_d is approximately less than 2, the 3-D OTF is similar to Fig.

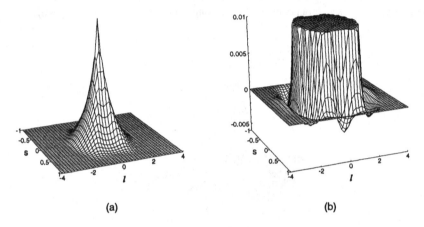

(a)   (b)

Fig. 5.3.5   3-D optical transfer function for two equal circular lenses ($\varepsilon_1 = \varepsilon_2 = 0$) when $v_d = 5$:
(a) normal view; (b) magnified view showing the negative tails.

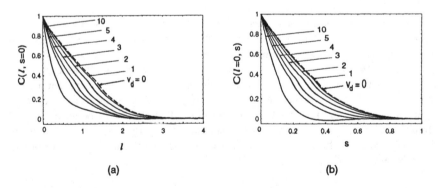

(a)   (b)

Fig. 5.3.6   Transverse (a) and axial (b) cross-sections of the 3-D OTF for two equal circular lenses
($\varepsilon_1 = \varepsilon_2 = 0$) for different values of v_d.

5.3.1a for the confocal 1-p fluorescence microscope with a point detector.[5.12] The transverse and axial cut-off spatial frequencies are nearly 4 and 1, respectively. As v_d increases further, the region where the 3-D OTF has appreciable values is gradually reduced in size (Fig. 5.3.5). Eventually as v_d becomes infinity the 3-D OTF becomes the same as the 3-D OTF for one-lens imaging, showing a missing cone of spatial frequencies (Fig. 2.5.3). This phenomenon can be understood from Eq. (5.2.16), in which the Hankel transform of the detector sensitivity with infinitely-large size is a delta function, so that the 3-D OTF is only determined by the objective.

The transverse and axial cross-sections of the 3-D OTF, $C(l, s = 0)$ and $C(l = 0, s)$, normalized by $C(l = 0, s = 0)$, are shown in Fig. 5.3.6, revealing their dependence on the detector radius for two identical circular lenes ($\varepsilon_1 = \varepsilon_2 = 0$). When $v_d = 0$, $C(l, s = 0)$ and $C(l = 0, s)$ are cut off at the spatial frequencies of 4 and 1, respectively. As v_d increases, $C(l, s = 0)$ and $C(l = 0, s)$ gradually become narrow. The former cut-off approaches 2 while the latter becomes zero. One interesting feature is that the tails of $C(l, s = 0)$ and $C(l = 0, s)$ go negative and are oscillating across the l and s axes, respectively, as v_d increases. Numerical calculations show that when v_d is approximately larger than 2 and 4.5, the negative parts of $C(l, s = 0)$ and $C(l = 0, s)$ emerge, respectively. This phenomenon totally results from the introduction of the finite-sized detector. In terms of Eq. (5.2.16), the Hankel transform of the circular sensitivity $D(v)$ is an oscillating function showing negative values. This accordingly leads to negative components in the 3-D OTF. These negative values in the 3-D OTF (see Fig. 5.3.5b) can result in imaging artefacts as image contrast for some spatial frequency components is reversed. Eventually, as $v_d \rightarrow \infty$, $F_2[D(v)] = \delta(l)$, so that $C(0, 0)$, the value at the origin, approaches infinity, showing that the optical sectioning property is completely degraded.[5.14] The strength of optical sectioning as a function of v_d is in Fig. 5.3.7.

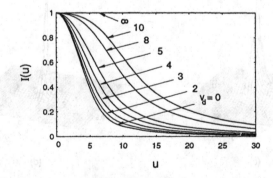

Fig. 5.3.7   Intensity of the axial response to a thin fluorescent sheet for two equal circular lenses for different values of v_d.

In Chapter 4, we have found that the degradation of optical sectioning in confocal bright-field microscopy with a finite-sized detector may be reduced if annular lenses are used. The dependence of the 3-D OTF for two equal annular lenses on the size of the central obstruction is displayed in Fig. 5.3.8 for $v_d = 10$, which is an intermediate size of pinhole approximately equal to the third dark ring of the Airy pattern in the pinhole plane. One can see that the cut-off spatial frequencies are effectively decreased, compared with those for $v_d = 0$ (see Fig. 5.3.1) and a missing cone of spatial frequencies around the origin appears for $\varepsilon_1 = \varepsilon_2 = 0$. Interestingly, when the radius of the central

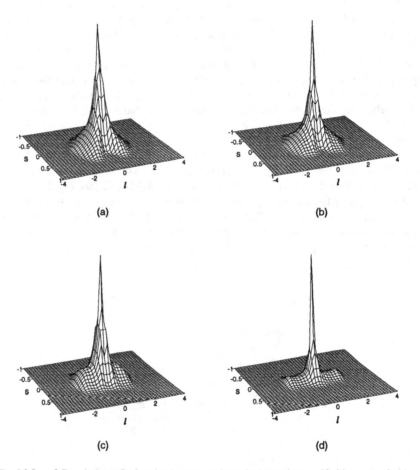

(a)   (b)

(c)   (d)

Fig. 5.3.8   3-D optical transfer function for two equal annular lenses for $v_d = 10$: (a) $\varepsilon_1 = \varepsilon_2 = 0$; (b) $\varepsilon_1 = \varepsilon_2 = 0.25$; (c) $\varepsilon_1 = \varepsilon_2 = 0.5$; (d) $\varepsilon_1 = \varepsilon_2 = 0.75$.

obstruction increases this cone gradually disappears. It should be pointed out that the 3-D OTFs in Fig. 5.3.8 have negative values near the tail of the function, which results from the introduction of the finite-sized circular detector.

The negative tails of the OTF are further observed in the transverse and axial cross-sections of the 3-D OTF (see Fig. 5.3.9). As a result of increasing the radius of the central obstruction, the magnitude of the negative component is reduced, the transverse resolution is slightly improved (see Fig. 5.3.9a) and the axial cut-off spatial frequency is increased slightly (see 5.3.9b). The latter property implies an improved optical sectioning effect.

Fig. 5.3.10 shows the variation of the axial response to the infinitely-thin

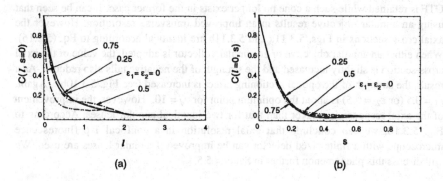

Fig. 5.3.9   Transverse (a) and axial (b) cross-sections of the 3-D OTF for two equal annular lenses when $v_d = 10$.

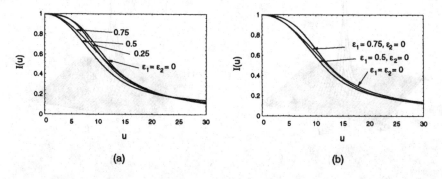

Fig. 5.3.10   Intensity of the axial response to a thin fluorescent sheet for $v_d = 10$: (a) for two equal annular lenses; (b) one annular objective and one circular collector.

fluorescent sheet with the radius of the central obstruction for $v_d = 10$. As the radius of the central obstruction is increased, the width of the axial response $I(u)$ is decreased until a minimum value of the width appears, and then it increases. Eventually, the width becomes infinity when $\varepsilon_1 = \varepsilon_2 = 1$. It is seen that the best case of the optical sectioning effect is approximately at $\varepsilon_1 = \varepsilon_2 = 0.5$. This property implies that for a given value of the detector radius, there is an optimum value $\varepsilon_1 = \varepsilon_2 = \varepsilon_o$.[5.13] When ε_1 (or ε_2) is equal to ε_o the effect of optical sectioning is improved compared with that in a system with circular pupils ($\varepsilon_1 = \varepsilon_2 = 0$).

We now consider an alternative case where either the objective or the collector is an annular lens. Fig. 5.3.11a corresponds to the 3-D OTF when an annular objective is used, while Fig. 5.3.11b to that when an annular collector is used. The behaviour of both OTFs is quite different. In the latter case, a missing cone of spatial frequencies in the 3-D OTF is retained while such a cone no longer exists in the former case. It can be seen that using an annular objective results in an improved transverse resolution. However the axial cross-sections in Figs. 5.3.11a and 5.3.11b are identical according to Eq. (5.2.16). When either an annular objective or an annular collector is adopted, the value of the axial cross-section is slightly increased and the strength of the negative tails also reduced. As a result, the strength of the optical sectioning effect is increased (see Fig. 5.3.10b). Again, $\varepsilon_1 = 0.5$ (or $\varepsilon_2 = 0.5$) is about the optimum point for $v_d = 10$. However, the improvement of the axial response is weaker than that for two identical annular lenses. According to Fig. 5.3.10, we can conclude that axial resolution in a confocal 1-p fluorescence microscope with a finite-sized detector can be improved if annular lenses are used. We will discuss this phenomenon further in Section 5.5.

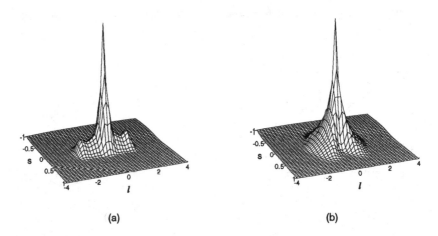

(a)   (b)

Fig. 5.3.11   3-D optical transfer function for one annular lens and one circular lens when $v_d = 10$: (a) $\varepsilon_1 = 0.75$ and $\varepsilon_2 = 0$; (b) $\varepsilon_1 = 0$ and $\varepsilon_2 = 0.75$.

The improvement in the optical sectioning effect also shows that the depth of the focus with an annular objective in confocal microscopy is not always longer than that with a circular objective ($\varepsilon_1 = 0$).[5.15-5.17]

Although the optical sectioning property in confocal 1-p fluorescence microscopy retains unchanged when exchanging the objective and collection lenses, such exchanging does affect the signal strength and accordingly the signal level when annular lenses is used. This issue will be addressed in the next section.

5. 4 Signal Level

Eq. (5.2.16) indicates that the strength of the 3-D OTF is a function of the detector size. This feature is lost when the 3-D OTF is normalized. The dependence of the OTF on the detector size provides information concerning the signal strength for different spatial frequencies.[5.13] To characterize the signal strength of the confocal fluorescence system, we need to assume some simple objects.

First consider a featureless volume object, so that the detected intensity can be expressed, in terms of Eq. (5.2.16), as

$$I(v_d) = C(l = 0, s = 0),$$
(5.4.1)

which can become, for a circularly symmetric system,

$$I(v_d) = \iiint_V \frac{v_d}{l^4} \left\{ \text{Re}\left[\sqrt{l^2 - \left(|s| + \frac{l^2}{2}\right)^2} \right] - \text{Re}\left[\sqrt{l^2 \varepsilon_1^2 - \left(|s| - \frac{l^2}{2}\right)^2} \right] \right\}$$

(5.4.2)

$$\left\{ \text{Re}\left[\sqrt{l^2 - \left(|s| + \frac{l^2}{2}\right)^2} \right] - \text{Re}\left[\sqrt{l^2 \varepsilon_2^2 - \left(|s| - \frac{l^2}{2}\right)^2} \right] \right\} J_1(lv_d) dl ds.$$

Eq. (5.4.2) represents the detected intensity from a given illumination source. Actually, the input energy from a given source is lost in two steps: first, the electric field is obstructed by the annular objective; second, this centrally-obstructed field illuminates the object, leading to a further loss of energy. For a microscopic system, since the source power can be made arbitrarily large, only the signal level η_2, i. e., the energy transformation in the second step, for a given system is of importance. It is easy to derive the fraction of the energy lost by the annular objective as $(1 - \varepsilon_1^2)$. The signal level for a featureless volume object, η_{2v}, we define here, is thus given by $I(v_d)$ in Eq. (5.4.2) divided by $(1 - \varepsilon_1^2)$. It should be noted that Eq. (5.4.2) is also a function of the central obstructions.

The signal level η_{2v} as a function of the detector radius and the radii of the central obstruction is given in Fig. 5.4.1a. The signal level approaches infinity when v_d approaches infinity. This is understandable as we have assumed an infinitely-thick object, and have neglected absorption of the primary radiation. In practice, of course, there is absorption and the object must also have some finite thickness. For a given value of v_d, the signal level decreases with increasing the radius of the central obstruction. For a microscope with just an annular objective the signal level is reduced only slightly compared with a system with two circular pupils. For example, when $v_d = 10$, the signal level for an annular objective only is just 10% lower than that for circular pupils. This property together with Fig. 5.3.10b implies that it is possible to maintain the signal strength at a high level by using an annular objective along while the axial resolution is improved compared with that for circular pupils.[5.10]

For a confocal fluorescence microscope with a point detector, the detector sensitivity function is a delta function, so that the signal level η_{2v} is only a function of the central obstruction size (see Fig. 5.4.1b). Note that the signal level drops to zero for a system with very thin annular lenses. Interestingly, the signal level for a system with an annular objective only ($\varepsilon_1 = \varepsilon$, $\varepsilon_2 = 0$) is always larger than that in a system with either two equal annular lenses ($\varepsilon_1 = \varepsilon_2 = \varepsilon$) or an annular collector only ($\varepsilon_1 = 0$, $\varepsilon_2 = \varepsilon$). At $\varepsilon = 0.5$, the improvements in the signal level are about 18% and 24%, respectively. This conclusion and Fig. 5.3.1 suggest that in practice only an annular objective should be used, so that a higher signal level can be obtained while the transverse resolution can be improved when the pinhole size is kept very small.

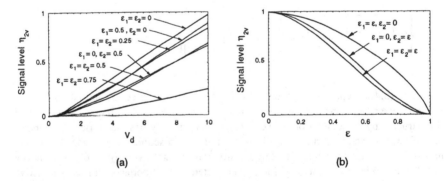

(a)   (b)

Fig. 5.4.1   Signal level for a featureless volume object (a) as a function of detector radius v_d and (b) for a point detector.

For 2-D in-focus imaging in the case of a thin object, the signal level can be derived by considering the object to be a thin uniform fluorescent sheet in the focal plane. The detected intensity is thus given by

$$I(v_d) = \int_0^1 C(l = 0, s)ds. \tag{5.4.3}$$

According to the argument described after Eq. (5.4.2), the signal level for a uniform planar object, η_{2p}, in this case is given by Eq. (5.4.3) divided by $(1 - \varepsilon_1^2)$.

Fig. 5.4.2a represents the signal level η_{2p} as a function of the detector size and the central obstruction size, where η_{2p} has been normalized to unity as v_d approaches infinity when $\varepsilon_1 = \varepsilon_2 = 0$. It is seen that for a given value of the detector size, η_{2p} is increased on increasing the detector size, and decreased on increasing the central obstruction radii. The signal level for using an annular objective only ($\varepsilon_1 = 0.5$, $\varepsilon_2 = 0$) is larger than that for using an annular collector only ($\varepsilon_1 = 0$, $\varepsilon_2 = 0.5$). Further, after v_d is larger than about 6.2, the signal level for $\varepsilon_1 = 0.5$ and $\varepsilon_2 = 0$ is even larger than that for $\varepsilon_1 = \varepsilon_2 = 0.25$.

When a point detector is used, the signal level η_{2p} for a planar object as a function of the central obstruction size is shown in Fig. 5.4.2b, which is similar to Fig. 5.4.1b. Again, a higher signal level can be obtained by using only an annular objective. The signal levels at $\varepsilon = 0.5$ for $\varepsilon_1 = \varepsilon$ and $\varepsilon_2 = 0$ are about 16% and 30% higher than those for $\varepsilon_1 = 0$ and $\varepsilon_2 = \varepsilon$ corresponding to using only an annular collector and for $\varepsilon_1 = \varepsilon_2 = \varepsilon$ corresponding to two equal annular lenses, respectively.

Finally, let us discuss the signal level for a single point object. Because a single point includes all spatial frequencies, the detected intensity is given, if it is placed at the

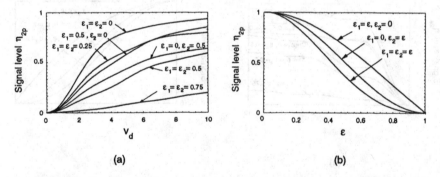

(a)   (b)

Fig. 5.4.2   Signal level for a uniform planar object (a) as a function of detector radius v_d and (b) for a point detector.

center of the focal plane, by

$$I(v_d) = \int_0^1 \int_0^4 C(l,s) l\, dl\, ds.$$ (5.4.4)

Therefore, the signal level η_{2s} may be derived, after including the factor of $(1 - \varepsilon_1^2)$, as

$$\eta_{2s} = \frac{1 - \varepsilon_1^2}{2} \int_0^{v_d} \left[\frac{2J_1(v)}{v} - \varepsilon_2^2 \frac{2J_1(v)}{\varepsilon_2 v} \right]^2 v\, dv,$$ (5.4.5)

which has been normalized to unity as v_d approaches infinity when $\varepsilon_1 = \varepsilon_2 = 0$. Two limiting forms of Eq. (5.4.5) can be derived analytically:

$$\eta_{2s} = (1 - \varepsilon_1^2)(1 - \varepsilon_2^2)^2$$ (5.4.6)

for a point detector, and

$$\eta_{2s} = (1 - \varepsilon_1^2)(1 - \varepsilon_2^2)$$ (5.4.7)

for a very large detector.

Fig. 5.4.3a shows the signal level η_{2s} for a single point object as a function of the pinhole radius for different annular lenses. Note that using either an annular objective only ($\varepsilon_1 = 0.5$, $\varepsilon_2 = 0$) or an annular collector only ($\varepsilon_1 = 0$, $\varepsilon_2 = 0.5$) results in an

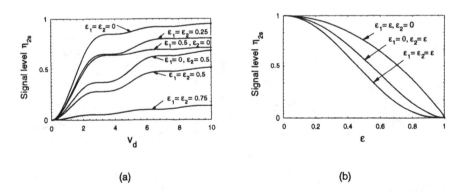

(a)   (b)

Fig. 5.4.3   Signal level for a single point object (a) as a function of detector radius v_d and (b) for a point detector.

improved signal level compared with that for two equal annular lenses ($\varepsilon_1 = \varepsilon_2 = 0.5$). Fig. 5.4.3b gives the signal level for a point detector, showing that using an annular objective only with a central obstruction of 0.5 can result in improvement in signal level of 18% and 33% higher than those for using an annular collector only and for using two equal annular lenses.

Eqs. (5.4.1), (5.4.3) and (5.4.4) are also applicable to a confocal fluorescence microscope with a slit detector of finite width. As may be expected, the signal level increases more quickly than that for a circular pinhole.[5.18]

5. 5 Improvement in Axial Resolution

As pointed in Section 5.3, the axial response of an infinitely-thin fluorescent layer (sheet) in confocal 1-p fluorescence microscopy can become broader than the theoretical limiting value because of a finite-sized pinhole placed in front of the detector. To overcome this practical difficulty, we have proposed to use annular lenses combined with a finite-sized pinhole. For a given size of the detector, the axial response can be made better, by altering the central obstruction of the annular lens to an optimum size, than that for using a circular lens. The measurement of the axial response of a thin fluorescent layer is one of the methods for characterizing axial resolution in confocal fluorescence microscopy.[5.14] This response gives the strength of optical sectioning: the narrower the axial response, the less the cross-talk between two adjacent sections of the images and therefore the higher the axial resolution.

An alternative method for the determination of the axial resolution is to image a thick uniform fluorescent layer scanned in the axial direction. In this case, a sharper axial response of the layer corresponds to a higher axial resolution. This method is more useful as it is easy to prepare a thick fluorescent layer in practice, and further as the derivative of the axial response of the thick layer directly gives the strength of optical sectioning.

To demonstrate this conclusion, let us consider an infinitely-thick fluorescent layer. Its fluorescence strength $o(x, y, z)$ may be expressed by an axial step function, i. e.,

$$o_f(x,y,z) = o_f(z) = \begin{cases} 1, & z \geq 0, \\ \\ 0, & z < 0. \end{cases} \tag{5.5.1}$$

According to Eq. (5.2.2), the fluorescence image can be calculated by the 3-D Fourier transform of the 3-D OTF multiplied by the spatial spectrum of the object, i. e., the 3-D inverse Fourier transform of the fluorescence strength of the object. When the layer is scanned in the axial direction, the measured intensity as a function of the defocus distance z of the surface, $I(z)$, i. e., the axial image, can be derived, for a circularly symmetric system, as

$$I(u) = \int_{-\infty}^{\infty} O_f(s)C(l = 0,s)\exp(ius)ds. \tag{5.5.2}$$

Here $O_f(s)$ is the 1-D inverse Fourier transform of Eq. (5.5.1):

$$O_f(s) = \frac{\delta(s)}{2} + \frac{1}{2\pi is}, \tag{5.5.3}$$

and $C(l = 0, s)$ is the axial cross-section of the normalized 3-D OTF.

Substituting Eq. (5.5.3) into Eq. (5.5.2) yields

$$I(u) = \frac{1}{2} + \frac{1}{\pi}\int_0^{s_c} \frac{C(l = 0,s)\sin(us)}{s}ds, \tag{5.5.4}$$

where s_c is the axial cut-off spatial frequency. It is seen that the image intensity on the surface of the layer ($u = 0$) is one half of its value far away from the surface on the bright side, as may be expected for incoherent imaging. The gradient of the intensity can be easily written as

$$I'(u) = \frac{1}{\pi}\int_0^{s_c} C(l = 0,s)\cos(us)ds. \tag{5.5.5}$$

By recalling Eq. (5.2.19) and the symmetry property of $C(l = 0, s)$ with respect to s, it is now clear that $I'(u)$ in Eq. (5.5.5) is directly proportional to the axial response of an infinitely-thin fluorescent sheet, thus giving the strength of optical sectioning in confocal 1-p fluorescence microscopy. In particular, on the surface of the thick layer, i. e., when $u = 0$, the gradient is completely determined by the area under the axial cross-section of the 3-D OTF, namely,

$$I'(u = 0) = \frac{1}{\pi}\int_0^{s_c} C(l = 0,s)ds, \tag{5.5.6}$$

which is denoted by γ hereafter. In fact, γ defines the sensitivity with which one can locate the surface of the fluorescent layer: the larger the gradient γ the higher the axial resolution in confocal fluorescence microscopy.

To understand how the annular lenses affect the axial resolution, let us take as an example a fluorescent layer of Nile blue dye, which was used for characterizing the axial resolution in a confocal 1-p fluorescence microscope.[5.19, 5.20] If it is excited by a He-Ne laser with a wavelength of 633 nm, the fluorescence signal has a peak at the wavelength of about 725 nm, so that the ratio of the fluorescence wavelength λ_f to the illumination wavelength λ is approximately 1.145, i. e., $g = 1.145$.

Since it has been found in Sections 5.3 and 5.4 that use of an annular objective lens can lead to the improvement in both signal level and the axial response of the thin fluorescent layer, we consider a system either with two circular lenses ($\varepsilon_1 = \varepsilon_2 = 0$) or with an annular objective lens ($\varepsilon_1 = 0.5$) and a circular collector lens ($\varepsilon_2 = 0$). The

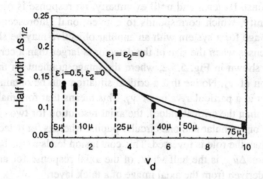

Fig. 5.5.1   Half width of the axial cross-section, $\Delta s_{1/2}$, of the 3-D OTF for Nile blue dye as a function of the detector radius v_d. The dark spots and squares are experimental results with an objective of numerical aperture 0.75. The vertical lines denote the positions of the detector radii in the experiment.

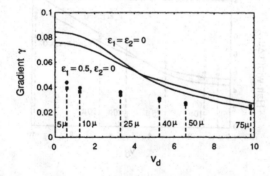

Fig. 5.5.2   Gradient γ of the image of a thick Nile blue fluorescent layer as a function of the detector radius v_d. The dark spots and squares are experimental results. The vertical lines denote the positions of the detector radii in the experiment.

axial cross-section of the 3-D CTF, $C(l = 0, s)$, can be calculated from Eq. (5.2.16) by assuming $g = 1.145$, giving a behaviour similar to Fig. 5.3.6b and revealing negative tails which can be reduced by using an annular objective lens. As a result, the half width at half maximum of $C(l = 0, s)$, $\Delta s_{1/2}$, for a system with an annular objective lens is larger than that with two circular lenses, if the detector size exceeds a certain value (see Fig. 5.5.1). This result implies the possibility of improving axial resolution by using an annular objective lens.

The image intensity of the thick layer can be calculated using $C(l = 0, s)$ and Eq. (5.5.3). For $v_d = 0$ (i. e., a point detector), the response is sharp.[5.19] But as v_d increases, the response flattens, and until eventually, no response is obtained when v_d tends towards infinity, which corresponds to conventional fluorescence microscopy. Importantly, the image for a system with an annular objective may be sharper than that with two circular lenses when the size of the detector is larger than a certain value. This property is clearly shown in Fig. 5.5.2, where the image gradient on the surface, γ, is plotted as a function of v_d. Notice that a critical situation can be obtained, namely, the two curves intersect at a particular value of v_d. This means that for small detector sizes (i. e., v_d is smaller than the critical value), the axial resolution for two circular pupils is higher than that for one annular and one circular pupils, whereas for detector sizes bigger than the critical value, the role is reversed. This conclusion leads to the behaviour shown in Fig. 5.5.3, where $\Delta u_{1/2}$ is the half width of the axial response for an infinitely-thin fluorescent sheet, derived from the axial image of a thick layer.

The comparison of the experimentally measured and theoretical values of γ is shown in Fig. 5.5.2.[5.19] The half width of the axial response in Fig. 5.5.3, $\Delta u_{1/2}$, is derived from the axial image of the thick layer. Notice that for a detector radius smaller

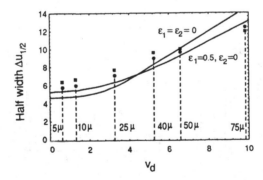

Fig. 5.5.3   Half width of the axial response, $\Delta u_{1/2}$, to a thick Nile blue fluorescent layer as a function of the detector radius v_d. The dark spots and squares are experimental results. The vertical lines denote the positions of the detector radii in the experiment.

than or equal to $50\mu m$, $\Delta u_{1/2}$ for circular pupils is bigger than that for the annular pupil, but for a $75\mu m$-radius detector, $\Delta u_{1/2}$ for the annular pupil becomes bigger than that for the circular pupil.

A further comparison of the theoretical prediction and experimental results may be derived from the Fourier transform of the measured axial image of the thick layer, which delivers the axial cross-section of the 3-D OTF, $C(l = 0, s)$, according to Eq. (5.5.3). Because of aberration presented in the confocal system,[5.19] the derived $C(l = 0, s)$ has real and imaginary parts. The measured half width $\Delta s_{1/2}$ of $\mathrm{Re}[C(l=0, s)]$ as a function of the detector size, shown in Fig. 5.5.1, gives a similar tendency to the theoretical prediction: when the detector is large, the value of $\Delta s_{1/2}$ is larger for annular lenses than that for circular lenses. This explains why the improved resolution shown in Figs. 5.5.2 and 5.5.3 can be obtained.

The values of γ, $\Delta u_{1/2}$ and $\Delta s_{1/2}$, measured in the experiments differ from the theoretical expectations when the detector size is small. This difference is possibly due to the presence of the spherical aberration caused by the mismatch of the refractive indices between the dye solution and the cover glass used for mounting the solution.[5.19] It may also result from the calibration of the piezo-translator or from the fact that the practical fluorescent layer has a finite thickness so that the signals from two surfaces of the layer are partly overlapped each other.

5. 6   Effect of Finite-Sized Source

In practice, a source in a confocal microscope also has finite size as a result of using an illumination pinhole mask. If a laser beam is focused onto a pinhole, the source is a finite-sized coherent source, so that the intensity in the object space should be expressed as

$$I_o(r_1) = \left| \int_{-\infty}^{\infty} \delta(z_0) S(x_0, y_0) \exp[ik(z_0 - z_1)] h_1(r_0 + M_1 r_1) dr_0 \right|^2. \tag{5.6.1}$$

Here $S(x_0, y_0)$ represents the field distribution of light from the coherent source and $\delta(z_0)$ implies the source is a planar one. Accordingly, the 3-D OTF for a confocal 1-p fluorescence microscope including a finite-sized detector becomes

$$C(l,s) = \left\{ F_3[h_1(v,u)] F_2[S(v)] \right\} \otimes_3 \left\{ F_3[h_1^*(v,u)] F_2[S^*(v)] \right\}$$

$$\tag{5.6.2}$$

$$\otimes_3 \left\{ F_3\left[|h_2(v/g, u/g)|^2 \right] F_2[D(v/g)] \right\},$$

if a system is of circular symmetry. Here $F_3[h_1(v, u)]$ (or $F_3[h_1^*(v, u)]$) is the 3-D CTF for a thin lens discussed in Eq. (2.4.10) and $F_2[S(v)]$ (or $F_2[S^*(v)]$) represents the 2-D inverse Fourier transform of the source field. The rest of the terms are the same as those in Eq. (5.2.11). Since the illumination beam from a laser has a TEM_{00} Gaussian mode. The source distribution is a Gaussian function truncated by the illumination pinhole.[5.21]

On the other hand, if an incoherent source is used,[5.21] Eq. (5.6.1) can be generalized to

$$I_o(r_1) = \int_{-\infty}^{\infty} \delta(z_0) S(x_0, y_0) |h_1(r_0 + M_1 r_1)|^2 \, dr_0, \tag{5.6.3}$$

where $S(x_0, y_0)$ is the intensity distribution of the incoherent source. Therefore the 3-D OTF for a circularly symmetric system is

$$C(l,s) = \left\{ F_3 \left[|h_1(v,u)|^2 \right] F_2[S(v)] \right\} \otimes_3 \left\{ F_3 \left[|h_2(v/g, u/g)|^2 \right] F_2[D(v/g)] \right\}, \tag{5.6.4}$$

where $F_2[S(v)]$ denotes the 2-D inverse Fourier transform of the intensity distribution of the source. The other terms are given by Eqs. (5.2.13), (5.2.14) and (5.2.15). If $g = 1$ and the source and the detector have an identical distribution, the 3-D imaging property remains unchanged when the objective and the collector exchange. The effect of a finite-sized source is the same as that of a finite-sized detector if they are the same in size.

The 3-D OTF in Eqs. (5.6.2) and (5.6.4) suffers from a further degradation when a finite-sized source is used because $F_2[S(v)]$ is not a constant due to the finite size of the source. If the function of the source distribution is the same in Eqs. (5.6.2) and (5.6.4), the 3-D OTF in the former case is poorer than that in the latter.[5.21, 5.22] In practice, the source $S(v)$ is coherent and has a truncated Gaussian function. The negative tails in the 3-D OTF can be made less strong if the illumination pinhole is opened up to a certain size so that the truncation becomes less pronounced.[5.21]

References

5.1.   A. F. Slomba, D. F. Wasserman, G. I. Gaufman, J. F. Nester, *J. Assoc. Adv. Med. Instrum.*, **6** (1972) 230.

5.2.   A. Yariv, *Quantum Electronics* (John Wiley & Son, New York, 1975).

5.3.   S. Kimura and C, Munakata, *J. Opt. Soc. Am. A*, **6** (1989) 1015.

5.4.   B. R. Frieden, *J. Opt. Soc. Am.*, **57** (1967) 56.

5.5.   B. R. Bracewell, *The Fourier Transform and Its Applications* (McGraw-Hill, New York, 1965).

5.6.   S. Kimura and C, Munakata, *Applied Optics*, **29** (1990) 1004.

5.7.   S. Kimura and C, Munakata, *Applied Optics*, **29** (1990) 3007.

5.8.   P. A. Stockseth, *J. Opt. Soc. Am.*, **59** (1969)1314.

5.9.   M. Gu and C. J. R. Sheppard, *J. Opt. Soc. Am. A*, **9** (1992)151.

5.10. M. Gu and C. J. R. Sheppard, *J. Modern Optics*, **38** (1991) 2247.

5.11. T. Wilson and C. J. R. Sheppard, *Theory and Practice of Scanning Optical Microscopy* (Academic, London, 1984).

5.12. M. Gu and C. J. R. Sheppard, *Optik*, **89** (1991) 65.

5.13. M. Gu and C. J. R. Sheppard, *J. Modern Optics*, **39** (1992) 1883.

5.14. T. Wilson, *J. Microscopy*, **154** (1989) 143.

5.15. O. Nakamura and S. Kawata, *J. Opt. Soc. Am. A*, **7** (1990) 522.

5.16. S. Kawata, R. Arimoto and O. Nakamura, *J. Opt. Soc. Am. A*, **8** (1991) 171.

5.17. R. Arimoto and S. Kawata, *Optik*, **86** (1990) 7.

5.18. C. J. R. Sheppard, C. J. Cogswell, and M. Gu, *Scanning*, **13** (1991) 233.

5.19. M. Gu, T. Tannous, and C. J. R. Sheppard, *Opt. Commun.*, **110**, (1994) 533.

5.20. S. Hell and E. H. K. Stelzer, *J. Opt. Soc. Am. A*, **9** (1992) 2159.

5.21. C. J. R. Sheppard and M. Gu, *J. Modern Optics*, **41** (1994) 1521.

5.22. V. Drazic, *J. Modern Optics*, **40** (1993) 879.

Chapter 6

CONFOCAL TWO-PHOTON
FLUORESCENCE MICROSCOPY

The original idea of two-photon (2-p) fluorescence scanning microscopy was proposed by Sheppard et al. along with other nonlinear scanning microscope modes[6.1, 6.2] and was experimentally demonstrated by Denk et al.[6.3] Physically, in 2-p fluorescence microscopy, two incident photons are simultaneously absorbed by the sample under inspection to excite the electron transition from the ground state to an excited state. The excited electron then jumps back to the ground state by radiating a photon having energy approximately twice as large as that of the incident photon. One of the advantages of this technique is that because of cooperative 2-p excitation, the photo-bleaching associated with single-photon (1-p) fluorescence imaging is only confined to the vicinity of the focal region. In addition, the 2-p technique offers access to ultra-violet (UV) photon excitation without using a UV source. These advantages promise some interesting applications involving 3-D optical storage,[6.4] photochemical processes[6.3] and calcium-ion activity.[6.5]

The fact that the 2-p excitation is confined to the focal region leads to an optical sectioning property for 2-p fluorescence microscopy without the necessity for a confocal pinhole mask,[6.3, 6.6] which contrasts with single-photon (1-p) fluorescence microscopy where point detection must be used in order to achieve the optical sectioning effect. However the strength of the optical sectioning effect for 2-p fluorescence microscopy does depend on the size of the pinhole.[6.7-6.10] Furthermore, using the confocal geometry has a unique advantage in 4Pi confocal 2-p fluorescence microscopy in the sense that axial resolution can be increased without the appearance of the pronounced axial side lobes.[6.11-6.13]

The three-dimensional (3-D) imaging properties in confocal 2-p fluorescence microscopy are investigated in this chapter. Discussions regarding 4Pi confocal 2-p fluorescence microscopy will be given in Chapter 9 after introducing the imaging theory for a high-aperture lens. In Section 6.1, similar procedures used in Chapter 5 are adopted to discuss 3-D incoherent image formation in 2-p fluorescence microscopy. The effect of the circular pinhole size on the 3-D intensity point spread function (IPSF) is given in Section 6.2. In Section 6.3, the corresponding 3-D optical transfer function (OTF) is studied. In Section 6.4, resolution in 2-p and 1-p fluorescence microscopy is compared in terms of the modelled images of three objects, a planar layer, a sharp edge and a grating. Finally, in Section 6.5, the detected intensity as a function of the detector radius, i. e., signal level, in 2-p fluorescence microscopy is investigated in comparison with that for 1-p fluorescence imaging.

2-p fluorescence imaging is usually performed in practice by using an ultrashort pulsed beam, so that nonlinear 2-p fluorescence radiation can be efficiently generated. The description in the present chapter is based on continuous wave (CW) illumination. A more accurate theory will be given in Chapter 8 after the effect of a pulsed beam is considered. The difference of the prediction between the two descriptions is not pronounced when the temporal width of the pulse is larger than 10 femtoseconds.

6. 1 Incoherent Image Formation

To analyse 3-D image formation in confocal 2-p fluorescence microscopy, we still use Fig. 3.1.1a as our schematic diagram for the system, so that a point source is considered in this chapter. We assume that incident and fluorescence wavelengths are λ and λ_f, respectively.

According to the discussion in the last chapter, the intensity of the illumination light in object space, $I_o(r_1)$, is given by Eq. (5.1.1). A two-photon absorption process is a nonlinear process, so that the strength of fluorescence radiation is proportional to the square of the intensity of the light illuminated on the object.[6.14] Thus the total strength of the 2-p fluorescence radiation from the object which has an object function $o_f(r)$ is

$$I_1(r_1) = \left| h_1(M_1 r_1) \right|^4 o_f(r_s - r_1), \tag{6.1.1}$$

where $h_1(r)$ is the 3-D amplitude point spread function for the objective lens given by Eq. (3.1.1) for the incident wavelength λ.

Like the 1-p fluorescence radiation, the 2-p fluorescence process is also a spontaneous process.[6.14] The image intensity at r_2 in the detector space is the contribution of the intensity from all points in the 3-D fluorescent object with the fluorescent strength $I_1(r_1)$ because the 2-p fluorescence light propagates in all directions and the phase of the light is not correlated. This process is accordingly an incoherent imaging process. The image intensity at r_2 of a single point object is given by Eq. (5.1.3) and consequently the image intensity from the 3-D object with a 2-p fluorescent strength $I_1(r_1)$ in Eq. (6.1.1) can be expressed as

$$I(r_2, r_s) = \int_{-\infty}^{\infty} |h_1(M_1 r_1)|^4 o_f(r_s - r_1)|h_2(r_1 + M_2 r_2)|^2 dr_1, \tag{6.1.2}$$

where $h_2(r)$ is the 3-D amplitude point spread function for the collector lens given by Eq. (3.1.1) for the fluorescence wavelength λ_f. If a finite-sized detector with an intensity sensitivity $D(r_2)$ is used, the final measured intensity from the sample is the integration of Eq. (6.1.2) over the detector aperture:

$$I(r_s) = \int_{-\infty}^{\infty} \left[\int_{-\infty}^{\infty} |h_1(M_1 r_1)|^4 o_f(r_s - r_1)|h_2(r_1 + M_2 r_2)|^2 dr_1 \right] D(r_2) dr_2, \tag{6.1.3}$$

or

$$I(r_s) = h_i(r_s) \otimes_3 o_f(r_s), \tag{6.1.4}$$

where $\otimes_3$ denotes the 3-D convolution operation. Here $h_i(r)$ is the image of a single fluorescent point under illumination of a point source and is thus the 3-D effective intensity point spread function (IPSF) for confocal 2-p fluorescence microscopy:

$$h_i(r) = |h_1(M_1 r)|^4 \int_{-\infty}^{\infty} |h_2(r + M_2 r_2)|^2 D(r_2) dr_2. \tag{6.1.5}$$

The above discussion clearly shows that there is no difference in the imaging performance between the reflection-mode and transmission-mode 2-p fluorescence microscopes under CW illumination, as expected. In fact, this conclusion is true for any fluorescence microscopy containing a CW source.

6. 2 Intensity Point Spread Function

We also consider a finite-sized circular pinhole of a radius r_d placed in front of the detector. According to Eq. (6.1.5) and the discussion in Section 4.2, the 3-D IPSF for confocal 2-p fluorescence microscopy of two circular lenses becomes

$$h_i(v, u) = |h_1(v / g', u / g')|^4 \left[|h_2(v, u)|^2 \otimes_2 D(v) \right]. \tag{6.2.1}$$

Here $g' = \lambda / \lambda_f$ is the ratio of the incident wavelength to the fluorescence wavelength. The amplitude spread functions h_1 and h_2 are given by Eqs. (4.2.4) and (4.2.6) but the radial and axial optical coordinates v and u are normalized by the fluorescence wavelength:

$$v = \frac{2\pi}{\lambda_f} r \sin \alpha_o, \tag{6.2.2}$$

$$u = \frac{8\pi}{\lambda_f} z \sin^2 (\alpha_o / 2), \tag{6.2.3}$$

where $\sin \alpha_o$ is the numerical aperture of the objective in the object space. $D(v)$ is the intensity sensitivity of the detector, defined as

$$D(v) = \begin{cases} 1 & , \quad v < v_d, \\ \\ 0 & , \quad otherwise, \end{cases} \tag{6.2.4}$$

where

$$v_d = \frac{2\pi}{\lambda_f} r_d \sin \alpha_d \tag{6.2.5}$$

is the normalized detector radius and $\sin \alpha_d$ denotes the numerical aperture of the collector lens in the detection space. It should be mentioned that $g' = 1/g$, where g is defined in Section 5.2. If g is introduced in Eq. (6.2.1), v and u should be defined at the incident wavelength λ.

For a point detector, Eq. (6.2.1) may be reduced to

$$h_i(v, u) = \left| h_1(v / g', u / g') \right|^4 \left| h_2(v, u) \right|^2 \tag{6.2.6}$$

whereas it approaches

$$h_i(v, u) = \left| h_1(v / g', u / g') \right|^4 \tag{6.2.7}$$

for an infinitely-large detector. Eq. (6.2.7) implies that the role of the collector lens is completely degraded. However, the 3-D IPSF is proportional to the square of the intensity of the incident light, showing that system has an optical sectioning property even for a large detector. This can be further seen from Eq. (4.2.7) for two identical imaging lenses. In this case, Eq. (6.2.7) has the same form as Eq. (4.2.7) for confocal 1-p fluorescence microscopy but the scale of Eq. (6.2.7) is reduced by a factor of g'. Therefore, 2-p fluorescence microscopy with a large detector, termed conventional 2-p fluorescence microscopy in this book, does have an optical sectioning property but its strength is decreased by a factor of g' in comparison with that in confocal 1-p fluorescence microscopy for the assumption of the equal fluorescence wavelength.

To understand the property of the 3-D IPSF for 2-p fluorescence, let us consider that the incident wavelength is twice as large as the 2-p fluorescence wavelength, which is equivalent to neglecting the non-radiation transition. This assumption also implies that any following comparison of 2-p and 1-p fluorescence imaging is based on the equal fluorescence wavelength. Fig. 6.2.1 shows contours of the 3-D IPSF of Eq. (6.2.1) in the v-u plane for two identical lenses for different values of v_d. It is seen that most of light is

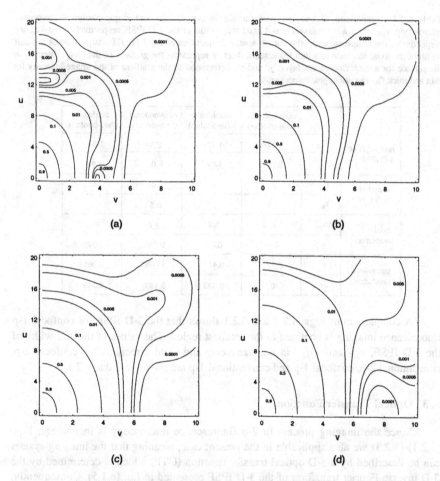

(a)

(b)

(c)

(d)

Fig. 6.2.1   Contours of the 3-D intensity point spread function (IPSF) for a confocal two-photon microscope comprising two identical circular lenses and a circular detector. The effect of the detector size is given by the normalized radius v_d: (a) $v_d = 0$; (b) $v_d = 2$; (c) $v_d = 5$; (d) $v_d \to \infty$.

confined to a small region for the point-detector case. As the detector size increases, the region of the 3-D IPSF is gradually expanded. This phenomenon implies that confocal 2-p fluorescence microscopy should give higher resolution than conventional 2-p fluorescence microscopy, and that the behaviour of the signal level for 2-p fluorescence may be different from that for the 1-p fluorescence case, which will be discussed in Section 6.5.

Table 6.2.1 Comparison of imaging performance in two-photon and single-photon fluorescence microscopy. $v_{1/2}$ and $u_{1/2}$ are transverse and axial half widths of the 3-D IPSF, respectively. l_c and s_c are, respectively, the transverse and axial cut-off spatial frequencies of the 3-D OTF. $\Delta u_{1/2}$ denotes the half width of the axial response to a thin fluorescent sheet. γ represents the gradient of the image intensity at the surface for a thick fluorescent layer. γ' and γ'' correspond to the gradient of the image intensity for thin and thick fluorescent edges, respectively.

		conventional (one-photon)	confocal (one-photon)	conventional (two-photon)	confocal (two-photon)
half width of 3-D IPSF	$v_{1/2}$	1.62	1.17	2.34	1.34
	$u_{1/2}$	5.56	4.01	8.02	4.62
cutoff of 3-D OTF	l_c	2	4	2	4
	s_c	0.5	1	0.5	1
axial resolution	$\Delta u_{1/2}$	∞	4.3	8.6	5.1
	γ	0	0.09	0.045	0.093
transverse resolution	γ'	0.27	0.417	0.208	0.361
	γ''	0	0.333	0.167	0.340

A comparison of Figs. 4.2.1 and 6.2.1 shows that the 3-D IPSF for confocal 1-p fluorescence imaging is confined to the smallest region. The values of the half width of the 3-D IPSF, $v_{1/2}$ and $u_{1/2}$, in the transverse and axial directions for confocal 2-p, conventional 2-p, confocal 1-p and conventional 1-p are shown in Table 6.2.1.

6.3 Optical Transfer Function

Since the imaging process in 2-p fluorescence microscopy is incoherent, Eqs. (5.2.1)-(5.2.3) are also applicable in the present case, meaning that the imaging system can be described by a 3-D optical transfer function (OTF) which is determined by the 3-D inverse Fourier transform of the 3-D IPSF presented in Eq. (6.1.5). Consequently, the 3-D OTF, $C(m)$, for confocal 2-p fluorescence microscopy may be expressed, in terms of Eqs. (6.1.5) and (5.2.3), as

$$C(m) = F_3\left\{|h_1(M_1r)|^4\left[|h_2(r)|^2 \otimes_3 D(r/M_2)\right]\right\},\tag{6.3.1}$$

where F_3 denotes the 3-D inverse Fourier transform defined by Eq. (5.2.3). Eq. (6.3.1) can be rewritten as

$$C(m) = C'_1(m) \otimes_3 C_2(m).\tag{6.3.2}$$

Here $C_2(m)$ is given by Eq. (5.2.7) but $C'_1(m)$ by

$$C'_1(m) = F_3\left[|h_1(M_1r)|^4\right],\tag{6.3.3}$$

which is also the 3-D OTF for a confocal 1-p fluorescence microscopy of two identical lenses at wavelength $g'\lambda_f$, in terms of Eq. (5.2.8).

For a confocal system of circular symmetry, Eq. (6.3.1) reduces, by making use of Eq. (6.2.1), to

$$C(l,s) = F_3\left[|h_1(v/g',u/g')|^4\right] \otimes_3 \left\{F_3\left[|h_2(v,u)|^2\right]F_2[D(v)]\right\}.\tag{6.3.4}$$

Here l and s in $C(l, s)$ are, respectively, the radial and axial spatial frequencies. In the present case, they are normalized by

$$\sin\alpha_o/\lambda_f\tag{6.3.5}$$

and

$$4\sin^2(\alpha_o/2)/\lambda_f,\tag{6.3.6}$$

respectively. Eq. (6.3.4) can be rewritten as

$$C(l,s) = \left\{F_3\left[|h_1(v/g',u/g')|^2\right] \otimes_3 F_3\left[|h_1(v/g',u/g')|^2\right]\right\}$$

$$\otimes_3\left\{F_3\left[|h_2(v,u)|^2\right]F_2[D(v)]\right\}.\tag{6.3.7}$$

Consider that the objective and collector lenses have annular pupils of the same outer radius a but different inner radii $\varepsilon_1 a$ and $\varepsilon_2 a$. According to Eq. (2.5.16), we have

$$F_3\left[\left|h_2(v,u)\right|^2\right] = \frac{2}{l}\left\{\mathrm{Re}\left[\sqrt{1-\left(\frac{|s|}{l}+\frac{l}{2}\right)^2}\right] - \mathrm{Re}\left[\sqrt{\varepsilon_2^2-\left(\frac{|s|}{l}-\frac{l}{2}\right)^2}\right]\right\} \qquad (6.3.8)$$

with $|s| \leq (1 - \varepsilon_2^2)/2$ and $l \leq 2$. $F_3\left[\left|h_1(v/g',u/g')\right|^2\right]$ denotes the 3-D OTF for the objective lens and is also the 3-D OTF for a conventional 1-p fluorescence microscope of wavelength $g'\lambda_f$. According to the Fourier transform theory[6.15] and Eq. (2.5.16), it has the following form:

$$F_3\left[\left|h_1(v/g',u/g')\right|^2\right] = \frac{2}{g'l}\left\{\mathrm{Re}\left[\sqrt{1-\left(\frac{|s|}{l}+\frac{g'l}{2}\right)^2}\right] - \mathrm{Re}\left[\sqrt{\varepsilon_1^2-\left(\frac{|s|}{l}-\frac{g'l}{2}\right)^2}\right]\right\}$$
$$(6.3.9)$$

with $|s| \leq (1 - \varepsilon_1^2)/2g'$ and $l < 2/g'$. The 2-D inverse Fourier transform of the intensity sensitivity $D(v)$, i. e., the Hankel transform in the circularly symmetric case, is given by Eq. (5.2.15).

Therefore the 3-D OTF for confocal 2-p fluorescence microscopy can be calculated using Eqs. (6.3.7)-(6.3.9) and Eq. (5.2.15). For an infinitely-large detector $F_2[D(v)]$ is a delta function and thus Eq. (6.3.7) is given by the factor in the first braces which represents the 3-D OTF for conventional 2-p fluorescence microscopy and is identical to the 3-D OTF for a confocal 1-p fluorescence microscope at wavelength λ. On the other hand, $F_2[D(v)]$ becomes constant for the true confocal case with a point detector.

In order to carry out numerical calculations, we assume $g' = 2$ and $\varepsilon_1 = \varepsilon_2 = 0$ in Eq. (6.3.7) hereafter. The passbands of the 3-D OTF, $C(l, s)$, for confocal ($v_d = 0$) and conventional ($v_d \to \infty$) 2-p fluorescence microscopy are shown in Figs. 6.3.1a and 6.3.1b, whereas Figs. 6.3.1c and 6.3.1d represent those in Eqs. (6.3.8) and (6.3.9) for conventional 1-p fluorescence microscopy with wavelengths λ_f and $2\lambda_f$. Notice that the passband in Fig. 6.3.1a is the same as that for confocal 1-p fluorescence microscopy,[6.16] which can be analytically expressed as

$$|s| = \begin{cases} 1 & , \quad 0 \leq l \leq 2, \\[2mm] l - \dfrac{l^2}{4} & , \quad 2 < l \leq 4. \end{cases} \qquad (6.3.10)$$

It is seen that Fig. 6.3.1b has the same shape as Eq. (6.3.10) but its scale is decreased by a factor of two, which implies that the cut-off spatial frequencies for confocal 2-p fluorescence microscopy are twice as large as those in the conventional case, and thus that more information can be imaged in the confocal 2-p case. In both cases, no missing cone of spatial frequencies exists. This property promises the capability of 3-D image formation in a conventional 2-p fluorescence microscope. In this respect, it contrasts with the conventional 1-p fluorescence microscope which has a missing cone of spatial frequencies as shown in Fig. 6.3.1c, the passband of which is given by $| s | = l(1 - l/2)$. The 3-D OTFs within the areas of Figs. 6.3.1c and 6.3.1d have a singularity at $s = 0$ and $l = 0$, as shown in Fig. 2.5.3.

The 3-D OTFs, normalized to unity by $C(0, 0)$, for different radii v_d are shown in Fig. 6.3.2. For the true confocal case ($v_d = 0$), the 3-D OTF has cut-off spatial frequencies of 4 and 1 in the transverse and axial directions, respectively, which are the same as those for confocal 1-p fluorescence microscopy (see Table 6.2.1). For finite values of detector size (Figs. 6.3.2b and 6.3.2c), the 3-D OTFs give negative values. A particular example is shown in Fig. 6.3.3 for $v_d = 5$. The negative OTF may reverse the contrast of images at particular spatial frequencies. As the detector size becomes large, the effective cut-off spatial frequencies in the OTF are reduced (Figs. 6.3.2b and 6.3.2c). Eventually as v_d tends to infinity, the 3-D OTF for a conventional 2-p fluorescence microscope is obtained, shown in Fig. 6.3.2d. The shape of Fig. 6.3.2d is the same as that

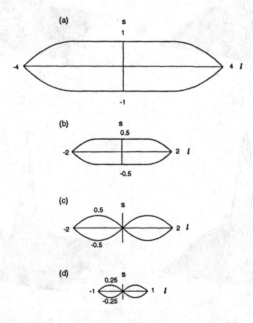

Fig. 6.3.1   Passband of 3-D optical transfer functions: (a) confocal 1-p and 2-p fluorescence microscopy at wavelength λ_f, (b) conventional 2-p fluorescence microscopy at wavelength λ_f, (c) conventional 1-p fluorescence microscopy at wavelength λ_f, (d) conventional 1-p fluorescence microscopy at wavelength $2\lambda_f$.

for confocal single-photon fluorescence microscopy (Fig. 5.3.1a) but its scale is shrunk by a factor of two. Comparing Fig. 6.3.2a with Fig. 6.3.2d, we can conclude that confocal 2-p fluorescence microscopy gives super-resolutions in both axial and transverse directions as the cut-off spatial frequencies in the confocal 2-p case are twice as large as those for the conventional case.

Comparing Fig. 6.3.2a for the confocal 2-p case with Fig. 5.3.1a for confocal 1-p fluorescence microscopy shows that the 3-D OTF for confocal 2-p fluorescence microscopy gives stronger responses at low spatial frequencies and reduced responses at

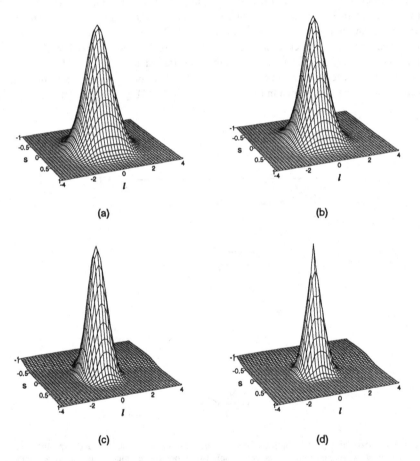

(a)   (b)

(c)   (d)

Fig. 6.3.2   3-D optical transfer function for two equal circular lenses at different values of v_d: (a) $v_d = 0$; (b) $v_d = 2$; (c) $v_d = 5$; (d) $v_d \to \infty$.

high spatial frequencies. This phenomenon results from the nonlinear 2-p excitation process as described by Eq. (6.1.1), and can be further observed in the cross-sections of the 3-D OTF, displayed in Figs. 6.3.4a and 6.3.4b.

Figs. 6.3.4a and 6.3.4b represent the transverse and axial cross-sections through the 3-D OTF at $s = 0$ and $l = 0$, respectively. In addition to the negative values in the OTF caused by the finite-sized pinhole, it is noted that the effects of the detector size on the transverse and axial cross-sections are different. When the detector size is small, the axial cross-section (Fig. 6.3.4b) does not change appreciably, while the transverse cross-section (Fig. 6.2.3a) does. This is caused by the detector sensitivity: its Fourier transform (Eq. (5.2.15)) is only a function of the transverse spatial frequency.

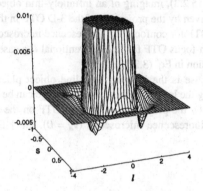

Fig. 6.3.3   Magnified view of the 3-D optical transfer function for two equal circular lenses for $v_d = 5$.

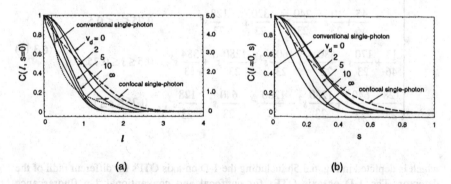

Fig. 6.3.4   Transverse (a) and axial (b) cross-sections of the 3-D OTF for two equal circular lenses for different values of v_d.

For comparison, the cross-sections of the 3-D OTFs for confocal and conventional 1-p fluorescence microscopy are depicted in Fig. 6.3.4. Note that enhanced responses at low spatial frequencies are obtained in the confocal 2-p fluorescence microscope, so that the areas under the curves for the confocal 2-p case may be slightly larger than those for the confocal 1-p case. The significance of this result is that the image of a thick edge or a thick layer object in the confocal 2-p fluorescence microscope may be expected to be sharper than that in the confocal single-photon fluorescence microscope as the gradient of the edge or layer image is determined by the areas under the functions of $C(l, s = 0)$ or $C(l = 0, s)$. This feature will be further demonstrated in Section 6.4. As a result of the enhanced OTF at low spatial frequencies, the slope of the 3-D OTF for confocal 2-p fluorescence is zero at the origin, while that for the conventional 2-p case is not.

According to Eq. (5.2.3), imaging of an infinitely-thin object can be described by the 2-D in-focus OTF given by the projection of the 3-D OTF in the focal plane (Chapter 10). The 2-D in-focus OTF for confocal 2-p fluorescence microscopy is displayed in Fig. 6.3.5a, where the 2-D in-focus OTF for the conventional 1-p case is included according to the analytical expression in Eq. (3.2.25).

Another special case is the imaging of a line object placed along the axis and varying in strength along the length. In this case, the image can be determined by the 1-D on-axis OTF produced by the projection of the 3-D OTF on the axis (Chapter 10). For the true confocal 2-p fluorescence microscope ($v_d = 0$), the 1-D on-axis OTF can be analytically derived as

$$C_1(s) = \begin{cases} 1 - \dfrac{160}{23}s^2 + \dfrac{640}{23}s^4 - \dfrac{640}{23}s^5, & 0 \le s \le 0.25, \\[2mm] \dfrac{45}{46} + \dfrac{10}{23}s - \dfrac{240}{23}s^2 + \dfrac{320}{23}s^3 - \dfrac{128}{23}s^5, & 0.25 \le s \le 0.5, \\[2mm] \dfrac{13}{46} + \dfrac{170}{23}s - \dfrac{880}{23}s^2 + \dfrac{1600}{23}s^3 - \dfrac{1280}{23}s^4 + \dfrac{384}{23}s^5, & 0.5 \le s \le 0.75, \\[2mm] \dfrac{128}{23} - \dfrac{640}{23}s + \dfrac{1280}{23}s^2 - \dfrac{1280}{23}s^3 + \dfrac{640}{23}s^4 - \dfrac{128}{23}s^5, & 0.75 \le s \le 1, \end{cases}$$

(6.3.11)

which is depicted in Fig. 6.3.5b including the 1-D on-axis OTFs for different radii of the detector. The 1-D on-axis OTFs for confocal and conventional 1-p fluorescence microscopy given by

$$C_1(s) = \begin{cases} 1 - 6s^2 + 6s^3 & , \quad 0 \leq |s| < 0.5, \\ 2 - 6s + 6s^2 - 2s^3 & , \quad 0.5 \leq |s| \leq 1, \end{cases} \tag{6.3.12}$$

and

$$C_1(s) = 1 - 2|s| \quad , \quad 0 \leq |s| \leq 0.5, \tag{6.3.13}$$

are also included in Fig. 6.3.5. Note that Eq. (6.3.13) is also the 1-D on-axis CTF for confocal transmission microscopy with a point detector (Chapter 10). It is again noticed that there are negative values in the OTF and that the effect of the detector size in the transverse direction is stronger than that in the axial direction. Although the cut-off spatial frequencies of the OTF for confocal 2-p fluorescence microscopy are the same as those for the confocal 1-p case, the strength of the 2-D in-focus OTFs in the former case is weaker than that in the latter. This conclusion is meant that the 2-D and 1-D imaging performance in confocal 1-p fluorescence microscopy is better than that in the confocal 2-p fluorescence case.

In confocal fluorescence microscopy, axial resolution can be evaluated by considering the axial response to a thin uniform fluorescent plane scanned in the axial direction (see Section 5.5). According to Eqs. (5.2.19) and (5.5.5), the strength of the optical sectioning property can be calculated by the Fourier transform of the axial cross-section of the 3-D OTF with respect to s. The calculated axial responses normalized to unity at the focal plane are shown in Fig. 6.3.6 for different detector radii. As expected, the axial response for 2-p fluorescence becomes broader when v_d increases, meaning that

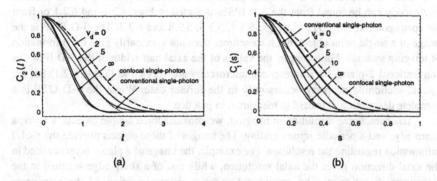

Fig. 6.3.5   2-D in-focus (a) and 1-D on-axis (b) OTFs for two equal circular lenses for different values of v_d.

the axial resolution is degraded when a finite-sized detector is used. This effect results from the fact that the effective cut-off spatial frequency in the axial direction is decreased with detector size (Fig. 6.3.4b). It is confirmed that conventional 2-p fluorescence ($v_d \rightarrow \infty$) gives an optical sectioning effect, although the half width of the axial response for conventional 2-p fluorescence is increased approximately by 68%, compared with that for the true confocal case ($v_d = 0$).

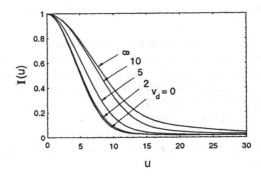

Fig. 6.3.6   Intensity of the axial response to a thin 2-p fluorescent sheet for two equal circular lenses for different values of v_d.

6. 4 Comparison of Resolution in Two-Photon and Single-Photon Fluorescence Microscopy

The difference of 3-D imaging properties between 2-p and 1-p fluorescence microscopy can be found from the 3-D IPSFs depicted in Figs. 4.2.1 and 6.2.1 or from the corresponding 3-D OTFs in Figs. 5.3.1, 5.3.5, 5.3.8 and 6.3.2. The 3-D IPSF is the image of a single point object, but it sometimes does not necessarily give the information of imaging systems. For example, the values of the axial half width of the 3-D IPSF for conventional 2-p and 1-p fluorescence microscopy are very close (Table 6.2.1) but the optical sectioning property occurs only in the former case. While the 3-D OTF is a complete description, it is hard to measure it in practice.

To characterize resolution in imaging, we consider three simple objects, a layer, a sharp edge and a periodic square grating. The images of these objects provide the useful information regarding the resolution. For example, the image of a planar layer scanned in the axial direction gives the axial resolution, while that of a sharp edge scanned in the transverse direction can be considered to be a characterization of the transverse resolution.[6.17] The image of a thick periodic grating is an approximate model of biological striated muscle fibres.[6.18] The following comparison is given only for confocal 2-p ($v_d = 0$), conventional 2-p ($v_d \to \infty$), confocal 1-p ($v_d = 0$) and conventional 1-p ($v_d \to \infty$) fluorescence microscopy.

6. 4. 1   Image of a Planar Layer

We first consider an infinitely-thin fluorescent sheet scanned in the axial direction. As pointed out in Section 5.5, the axial response of the sheet can be used as an evaluation of axial resolution in confocal fluorescence microscopy. This result is the so-called optical sectioning property. The narrower the axial response, the high the axial resolution and therefore the stronger the optical sectioning effect.

As the thin planar layer does not include transverse structures, the transverse spatial frequency is zero, i. e. $l = 0$. Therefore, the axial image of this layer can be calculated by the Fourier transform of the axial cross-section of the corresponding 3-D OTF with respect to s, as shown in Eq. (5.2.19).

The calculated axial responses normalized to unity at the focal plane are shown in Fig. 6.4.1. The values of the half width at half maximum, $\Delta u_{1/2}$, of the axial response for confocal 1-p, conventional 1-p, confocal 2-p, and conventional 2-p fluorescence imaging are given in Table 6.2.1. As expected, the axial response for conventional 2-p fluorescence is broader than that for confocal 2-p fluorescence. The half width of the axial response for confocal 2-p fluorescence is decreased approximately by 41%, compared with that for the conventional 2-p case. This property demonstrates an advantage over the conventional 1-p fluorescence imaging method which does not give rise to the optical sectioning property.

A comparison of the axial response for confocal 2-p fluorescence with that for

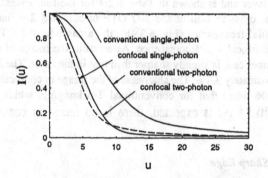

Fig. 6.4.1   Intensity of the axial response to a thin fluorescent sheet for two equal circular lenses for confocal 1-p, conventional 1-p, confocal 2-p, and conventional 2-p fluorescence imaging.

confocal 1-p fluorescence reveals that as a result of the enhanced OTF at low axial spatial frequencies, the axial response for confocal 2-p fluorescence is slightly broader than that for single-photon fluorescence when the defocus distance u small, and is narrower when u is large.

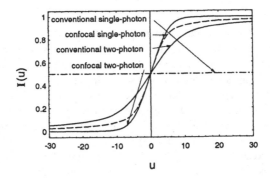

Fig. 6.4.2   Intensity of the axial response to a thick fluorescent layer for two equal circular lenses for confocal 1-p, conventional 1-p, confocal 2-p, and conventional 2-p fluorescence imaging.

When the thickness of the fluorescent layer becomes much larger than the focal depth of the objective, the object can be considered to be a thick object, the image of which can be calculated by Eq. (5.5.3). The gradient, $\gamma = I'(u = 0)$, of the image at the surface of the layer is simply determined by the area under the axial cross-section of the 3-D OTF divided by π. It actually defines the sensitivity with which one can locate the surface of the thick layer and is shown in Table 6.2.1 for the four cases. Because the response of the axial cross-section of the 3-D OTF in confocal 2-p fluorescence is enhanced at low spatial frequencies (Fig. 6.3.4b), the area under the OTF is slightly larger than that for confocal 1-p fluorescence. As a result, the image of a thick layer (Fig. 6.4.2) in the former case is slightly sharper than that in the latter. The improvement in sharpness is approximately 3.4%. The sharpness of the image in confocal 2-p imaging is approximately 2.06 times that for conventional 2-p imaging, which agrees with experimental results.[6.8] As is expected, there is no image in conventional 1-p microscopy.

6.4.2 Image of a Sharp Edge

For a sharp fluorescent edge scanned in the focal plane, the corresponding object function is

$$o_f(x,y,z) = \delta(z) \begin{cases} 1 & , \quad x \geq 0, \\ \\ 0 & , \quad x < 0, \end{cases} \tag{6.4.1}$$

for a thin edge, and

$$o_f(x,y,z) = \begin{cases} 1 & , \quad x \geq 0, \\ \\ 0 & , \quad x < 0, \end{cases} \tag{6.4.2}$$

for a thick edge. The respective images can be expressed as

$$I(v_x) = \frac{1}{2} + \frac{1}{\pi} \int_0^{l_c} C_2(m=l,n=0) \frac{\sin(v_x m)}{m} dm \tag{6.4.3}$$

and

$$I(v_x) = \frac{1}{2} + \frac{1}{\pi} \int_0^{l_c} C(m=l,n=0,s=0) \frac{\sin(v_x m)}{m} dm. \tag{6.4.4}$$

Here $C_2(m=l, n=0)$ and $C(m=l, n=0, s=0)$ are the 2-D in-focus OTF shown in Fig. 6.3.5a and the transverse cross-section of the 3-D OTF shown in Fig. 6.3.4a, respectively. l_c is the transverse cut-off spatial frequency. The gradient of the image at the edge is determined by the area under $C_2(l)$ for a thin edge and under $C(l, 0)$

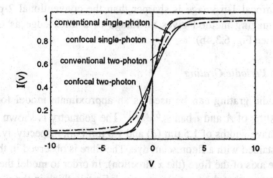

Fig. 6.4.3 Intensity of the image of a thin fluorescent edge for two equal circular lenses for confocal 1-p, conventional 1-p, confocal 2-p, and conventional 2-p fluorescence imaging.

for a thick edge and gives the transverse resolution.

Fig. 6.4.3 displays the image of the thin edge for confocal 1-p, conventional 1-p, confocal 2-p, and conventional 2-p fluorescence imaging and the corresponding gradient γ at the edge is shown in Table 6.2.1. In both confocal and conventional imaging methods, the 1-p fluorescence method results in sharper images compared with those using 2-p excitation.

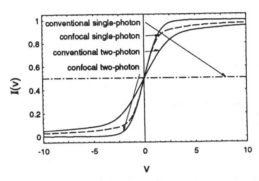

Fig. 6.4.4   Intensity of the image of a thick fluorescent edge for two equal circular lenses for confocal 1-p, conventional 1-p, confocal 2-p, and conventional 2-p fluorescence imaging.

Table 6.2.1 also gives the gradient γ' of the image at the edge for a thick edge for confocal 1-p, conventional 1-p, confocal 2-p, and conventional 2-p fluorescence imaging. As a result of the enhanced response of $C(l, 0)$ at low spatial frequencies, confocal 2-p imaging (Fig. 6.4.4) gives the sharpest image of the thick edge among the four cases. The confocal 1-p image is sharper than the conventional 2-p image. The conventional 1-p imaging method gives no image of a thick edge as $C(l, 0)$ has a singularity at $l = 0$ (see Fig. 6.3.4b).

6. 4. 3   Image of a Periodic Grating

A thick periodic grating can be used as an approximate model for the striated muscle fibre consisting of A and I bands.[6.18, 6.19] The geometry is shown in Fig. 6.4.5. The A and I bands have widths of 1.5 μm (d) and 1.0 μm (x_0), respectively. The I bands are assumed to be stained with a fluorescent dye. The fibre is observed in the z direction, perpendicular to the axis of the fibre (the x direction). In order to model the image of the muscle fibre, we assume that the fibre structure is infinitely thick in the y direction. It is also assumed that the fibre is imaged using a water-immersion objective of numerical aperture 1.0 and that the fluorescence wavelength λ_f is 488 nm.

The object function shown in Fig. 6.4.5 and its Fourier spectrum can be easily expressed as[6.18]

$$o_f(x,y,z) = \zeta(z)\left[\frac{x_0}{T} + \frac{2}{\pi}\sum_{j=1}^{\infty}\frac{1}{j}\sin\left(\frac{j\pi x_0}{T}\right)\cos\left(\frac{2j\pi x}{T}\right)\right]$$

(6.4.5)

and

$$O_f(m,n,z) =$$

(6.4.6)

$$\left\{\frac{x_0}{T}\delta(m) + \frac{1}{\pi}\sum_{j=1}^{\infty}\frac{1}{j}\sin\left(\frac{j\pi x_0}{T}\right)\left[\delta\left(m-\frac{j}{T}\right) + \delta\left(m+\frac{j}{T}\right)\right]\right\}\delta(n)\frac{\sin(\pi z_0 s)}{\pi s},$$

where $\zeta(z)$ is defined as

$$\zeta(z) = \begin{cases} 1 & , \quad z \le |z_0|/2, \\ \\ 0 & , \quad otherwise, \end{cases}$$

(6.4.7)

with z_0 as the thickness of the object, and $T = d + x_0$. The image of the thin ($z_0 = 0$) and thick ($z_0 \to \infty$) fibres can calculated using Eq. (5.2.2) and the 2-D in-focus OTF (Fig. 6.3.5a) and the transverse cross-section of the 3-D OTF (Fig. 6.3.4a). Figs. 6.4.6 and 6.4.7 give a comparison of 2-p fluorescence and 1-p fluorescence microscopy for thin and thick objects, respectively. In Fig. 6.4.6, the image of the thin grating in confocal 2-p

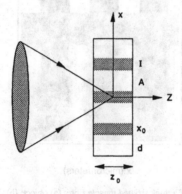

Fig. 6.4.5   Geometry of imaging a striated muscle fibre consisting of A and I bands.

fluorescence is slightly worse than that for confocal 1-p fluorescence as the 2-D in-focus OTF in the former is slightly weaker than that in the latter (Fig. 6.3.5a). But the situation is reversed for conventional cases, as expected from the fact that the 2-D in-focus OTF for conventional 1-p fluorescence gives a weaker response at low spatial frequencies than that for conventional 2-p fluorescence (Fig. 6.3.5a). In the case of the thick grating (Fig. 6.4.7), confocal and conventional 2-p fluorescence images are both better than those for 1-p fluorescence. One can expect this result from the transverse cross-section of

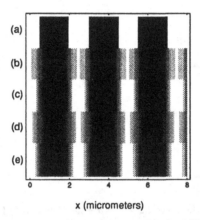

Fig. 6.4.6   Image intensity of a thin striated muscle fibre: (a) object; (b) conventional 1-p fluorescence; (c) confocal 1-p fluorescence; (d) conventional 2-p fluorescence; (e) confocal 2-p fluorescence.

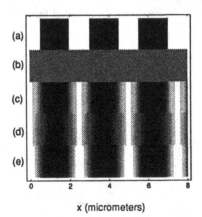

Fig. 6.4.7   Image intensity of a thick striated muscle fibre: (a) object; (b) conventional 1-p fluorescence; (c) confocal 2-p fluorescence; (d) conventional 2-p fluorescence; (e) confocal 2-p fluorescence.

3-D OTF in Fig. 6.3.4a. The images in Fig. 6.4.7 just give another example demonstrating the advantage of the enhanced 3-D OTF for 2-p fluorescence at low spatial frequencies.

The method used here for calculating the image of the muscle fibres requires that multiple scattering of light is weak. However, this is not true in practice. Therefore, a more accurate model based on the electromagnetic wave theory should be introduced.[6.20]

6. 5 Signal Level

The 2-p excitation occurs efficiently near the focal region of the objective owing to the cooperation interaction. This accordingly affects the signal detection in 2-p fluorescence microscopy for different dimensional objects under investigation. In this section, we study the detected intensity as a function of the detector size, which we call, after normalization for a large detector, the signal level, η_2.

Three 2-p fluorescent objects, a point, a uniform thin plane, and a uniform volume specimen, are considered. In the point object case, the signal level is the same as that for 1-p fluorescence if the fluorescence wavelengths are assumed to be the same in both cases:

$$\eta_{2s} = 1 - J_0^2(v_d) - J_1^2(v_d). \tag{6.5.1}$$

where J_0 and J_1 are Bessel functions of the first kind of order zero and order unity, respectively.

In terms of the investigation in Section 5.4, the signal level for thin and volume objects are proportional to the 2-D in-focus OTF at $l = 0$ and the 3-D OTF at $l = 0$ and $s = 0$, respectively. Thus we have

$$\eta_{2p} \propto C_2(l = 0) \tag{6.5.2}$$

and

$$\eta_{2v} \propto C(l = 0, s = 0) \tag{6.5.3}$$

for planar and volume objects, respectively, where $C_2(l = 0)$ and $C(l = 0, s = 0)$ are the functions of the detector radius.

The signal level for 1-p fluorescence microscopy has been, in detail, discussed in Section 5.4. However, in the case of two equal circular pupils, alternative expressions of the signal level for planar and volume objects can be derived. For a thin planar object which has uniform fluorescence and is located on the focal plane, the signal level in this case is also proportional to the integration of the intensity over the detector aperture, which can be mathematically expressed as[6.21]

$$\eta_{2p} \propto \int_0^{v_d} \left\{ \int_0^2 \left[\cos^{-1}\left(\frac{l}{2}\right) - \frac{l}{2}\sqrt{1-\frac{l^2}{4}} \right]^2 J_0(lv)l\,dl \right\} v\,dv.$$ (6.5.5)

In the case of a featureless volume object, the detected signal strength is now proportional to $C(l = 0, s = 0)$, so that the signal level can be reduced, according to Eq. (5.2.16), to

$$\eta_{2v} = v_d \iint_\sigma \frac{1}{l^4}\left[l^2 - \left(|s| + \frac{l^2}{2}\right)^2 \right] J_1(lv_d)\,dl\,ds,$$ (6.5.6)

where σ denotes the area enclosed by $|s| < l(1 - l/2)$.

The calculated results are shown in Fig. 6.5.1 for 2-p and 1-p fluorescence microscopy, where all curves except the one for the 1-p volume object are normalized to unity by the values for an infinitely-large detector, It is note that the signal level for the 2-p volume object is lower than that for the 2-p planar object, while the latter signal level is lower than that for the point object. This interesting phenomenon, a result of the normalization, demonstrates that the 2-p excitation is only confined to a small focal region. As the object is slightly extended away from the focal plane, the 2-p excitation becomes less efficient.

Another important feature in Fig. 6.5.1 is that the signal level for the 2-p volume object approaches unity for a large detector, which contrasts with that for 1-p fluorescence microscopy where the signal level becomes infinity as the detector size is infinity. This difference is also caused by the finite confinement of the 2-p excitation.

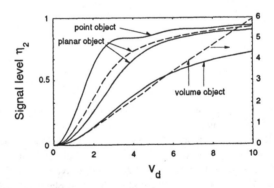

Fig. 6.5.1   Comparison of signal level of 2-p (solid curves) and 1-p (dashed curves) fluorescence microscopy for point, planar and volume objects.

This feature also results in the slightly lower signal level for a planar object for 2-p fluorescence microscopy in comparison with that for 1-p fluorescence microscopy.

References

6.1. C. J. R. Sheppard, R. Kompfner, J. Gannaway, and D. Walsh, *IEEE J. Quantum Electron.* **QE-13** (1977) 100D.
6.2. C. J. R. Sheppard and R. Kompfner, *Applied Optics*, **17** (1978) 2879.
6.3. W. Denk, J. H. Strickler, and W. W. Webb, *Science*, **248** (1990) 73.
6.4. J. H. Strickler and W. W. Webb, *Opt. Lett.*, **15** (1991) 1780.
6.5. D. W. Piston, M. S. Kirby, H. Cheng, W. J. Lederer, and W. W. Webb, *Applied Optics*, **33** (1994) 662.
6.6. J. N. Gannaway and C. J. R. Sheppard, *Opt. Quantum Elect.*, **10** (1978) 435.
6.7. M. Gu and C. J. R. Sheppard, *J. Modern Optics*, **40** (1993) 2009.
6.8. E. N. K. Stelzer, S. Hell, S. Lindek, R. Stricker, R. Pick, C. Storz, G. Ritter, and N. Salmon, *Opt. Commun.*, **104** (1994) 223.
6.9. C. J. R. Sheppard and M. Gu, *Optik*, **86** (1990) 104.
6.10. O. Nakamura, *Optik*, **93** (1993) 39.
6.11. S. Hell and E. H. K. Stelzer, *J. Opt. Soc. Am. A*, **9** (1992) 2159.
6.12. S. Hell and E. H. K. Stelzer, *Opt. Commun.*, **93**, (1992) 277.
6.13. S. Hell, S. Lindek, and E. H. K. Stelzer, *J. Modern Optics*, **41** (1994) 675.
6.14. W. Kaiser and C. G. B. Garrett, *Phys. Rev. Lett.*, **7** (1961) 229.
6.15. B. R. Bracewell, *The Fourier Transform and Its Applications* (McGraw-Hill, New York, 1965).
6.16. C. J. R. Sheppard, *Optik*, **74** (1986) 128.
6.17. M. Gu and C. J. R. Sheppard, *J. Microscopy*, **177** (1994) 128.
6.18. C. J. R. Sheppard and M. Gu, *J. Microscopy*, **169** (1993) 339.
6.19. A. F. Huxley, *Proc. R. Soc. Lond. B*, **241** (1990) 65.
6.20 J. T. Sheridan and T. O. Körner, *J. Microscopy*, **177** (1994) 95.
6.21. M. Gu and C. J. R. Sheppard, *Optik*, **89** (1991) 65.

Chapter 7

FIBRE-OPTICAL CONFOCAL MICROSCOPY

This chapter is devoted to a complete study of three-dimensional (3-D) image formation in confocal scanning microscopy consisting of single-mode optical fibres for illumination and collection. In Section 7.1, the recent development of a fibre-optical confocal scanning microscope (FOCSM) and its main features are described. We then derive, in Section 7.2, the purely-coherent imaging feature in a bright-field FOCSM in which the sample is non fluorescent. Sections 7.3 and 7.4 discuss the images of two special objects, a perfect reflector and a sharp edge, in the bright-field FOCSM. The results give rise to resolution in the axial and transverse directions in the FOCSM, respectively. Signal level in the FOCSM is of particular importance and is properly explored in Section 7.5. Section 7.6 demonstrates the principle of fibre-optical confocal interference microscopy. The last section, Section 7.7, deals with 3-D imaging properties in fibre-optical confocal fluorescence microscopy.

7.1 Fibre-Optical Confocal Scanning Microscopy

Currently, most of commercial confocal scanning microscopes have been constructed using conventional bulk optical components. An alternative novel implementation can be developed using fibre-optical components such as optical fibres, optical fibre couplers, graded-index (GRIN) rod lenses, fibre gratings and so on.[7.1-7.8] One of the advantages of these arrangements is that a simple and compact microscope system can be constructed. They may also prove to have significant cost advantage.

When a single-mode optical fibre is used, the light from a laser (or a combination of lasers) can be delivered into the confocal microscope system (Fig. 7.1.1). This arrangement does not require the laser to be suspended near the input port of the microscope, as is frequently done in commercial confocal systems, and in addition can help significantly with isolation of vibrations from the laser cooling fan. The single-mode fibre gives out a beam which has approximately Gaussian cross-section.[7.9] As a result, the geometry can be designed to more or less fill the objective pupil (see Section 7.2.4). If the pupil is overfilled the maximum resolution is obtained but at the expense of loss of light, whereas if the pupil is underfilled resolution is degraded (see Sections 7.3 and 7.4). A single-mode optical fibre can also be used in place of the physical pinhole to collect the signal in a confocal microscope system (Fig. 7.1.1).

If both the illumination and the collection fibres have the same geometry, this

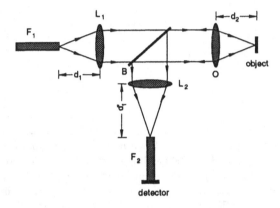

Fig. 7.1.1   A fibre-optical confocal scanning microscope comprising two lengths of the single-mode fibre as an illumination source and a collection device.

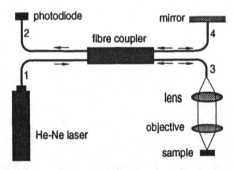

Fig. 7.1.2   A fibre-optical confocal scanning microscope comprising a 4-port fibre coupler.

arrangement suggests that a single fibre can be used for illumination and detection, and further that the beamsplitter be replaced by a fibre-optical coupler, as shown in Fig. 7.1.2. For normal confocal microscopy, the light from port 4 is terminated, whereas it provides a reference beam in confocal interference microscopy. The arrangement in Fig. 7.1.2 is similar to bulk optical implementations using a single pinhole for both illumination and detection. It has the advantage that the system is automatically aligned, but care has to be taken to avoid reflections from the end of the optical fibre. An image of an integrated circuit in a microscope using a 4-port coupler is displayed in Fig. 7.1.3.

The implementations described above can be used with beam scanning, mechanical object scanning or lens scanning, as with the analogous bulk optical system. A further possibility with optical fibres is to scan by mechanically moving the fibre itself. As the fibre is very light it can be scanned quite fast, and this is the case for axial as well as transverse scanning.

With regard to imaging, since the diameter of the tip of an optical fibre is about few micrometers, one would think that optical fibres could play the same role as the confocal pinholes.[7.4-7.7] However, rigorous theoretical analyses[7.1, 7.2] have shown that image formation for a non-fluorescent object in a confocal system using optical fibres is fundamentally different from that using pinhole masks in that imaging in the former case is purely coherent even for a finite value of the fibre spot size,[7.1, 7.2] while imaging in the latter is partially coherent due to using a finite-sized pinhole,[7.10-7.11] as has been discussed in Chapter 4.

The physical reason for this important difference is that a finite-sized detector incoherently responds to the light *intensity* impinging upon the detector, whereas an optical fibre is a waveguide and has a set of eigenfield modes[7.9] which can coherently

Fig. 7.1.3   Image of an integrated circuit in a FOCSM based on a 4-port coupler. The scale is 100 x 100 μm².

respond to the *amplitude and phase* of light.[7.1, 7.2] Because of this basic difference, it has been found that the effect of the fibre spot size on confocal images differs from that of the detector of finite size.[7.1-7.6, 7.12-7.16]

The purely-coherent nature of the fibre-optical confocal scanning microscope (FOCSM) is advantageous for allowing the imaging performance to be determined in terms of a 3-D coherent transfer function.[7.13] This has obvious advantages for image restoration. The purely-coherent feature of imaging in the FOCSM may also prove advantageous when it is used for spectroscopic imaging such as coherent anti-Stokes Raman spectroscopy (CARS), for quantitative microscopy in confocal metrology[7.14] (Section 7.4), and in particular for performance of confocal interference microscopy[7.12] (Section 7.6).

Confocal interference microscopy[7.17] is a useful method for measuring both the phase and amplitude information, with which one can measure aberration of the optical system,[7.18, 7.19] and for improving axial resolution of confocal imaging.[7.20] As is well-known, the degree of coherence determines the visibility of interference fringes. In this respect, the FOCSM gives better results than that using a pinhole as imaging with fibre detection is purely coherent (Section 7.6). Various forms of interference microscope can be envisaged based on the use of optical fibre interferometer techniques.[7.21-7.23] The fibre optical system also has the advantage that air currents are avoided.

The coherent imaging property also results in an improvement in signal strength in the FOCSM because a good matching of the fibre mode with the incoming diffraction pattern can be achieved[7.15] (Section 7.5). When a FOCSM is used in combination with a point-source illumination, the signal level in the FOSCM can be optimized to a value of 81%.

If optical fibres are employed with a fluorescent object, the system now of course behaves as an incoherent microscope, and thus can be described by a 3-D optical transfer function (OTF).[7.24, 7.25] The important property resulting from using fibres in confocal fluorescence microscopy is that there is no missing cone of spatial frequencies in the 3-D OTF (Section 7.7) as occurs with a pinhole detector (Chapter 5). This feature thus causes an improvement in the 3-D fluorescence imaging performance.

A multi-mode fibre can be also used in confocal scanning microscopes.[7.26-7.31] The purely-coherent nature still exists in non-fluorescence imaging, which can be used for confocal differential microscopy.[7.26, 7.27] The theory of 3-D image formation using multi-mode fibres is not within the scope of this chapter and readers who are interested in work using multi-mode fibres can generalize the methods described in this chapter or refer to Wilson's publication[7.26-7.29] for two-mode fibres and to other publications for multi-mode fibres.[7.30, 7.31]

7. 2 Bright-Field Imaging

A schematic diagram of the reflection-mode fibre-optical confocal scanning microscope is shown in Fig. 7.2.1, where two lengths of the single-mode optical fibres F_1 and F_2 are used as an illumination source and a collection device. The latter delivers the signal to a large area detector D. We use P_1 and P_2 to denote the aberration-free objective and collector lenses, respectively. The 3-D amplitude point spread functions for them are given by Eq. (3.1.1) and the defocused pupil functions by Eq. (3.1.3). Because of the aberration-free assumption, the corresponding pupil functions are real. In terms of Fig. 3.1.1b, a transmission-mode FOCSM can be considered to be an unfolded reflection FOCSM.

7. 2. 1 Purely-Coherent Imaging Nature

Assume that $f_1(x, y)$ and $f_2(x, y)$ are the two-dimensional (2-D) amplitude mode profiles of the fibres F_1 and F_2. If the coordinates systems shown in Fig. 7.2.1 are used, the light field at r_1 in object space can be expressed, according to Eq. (2.3.9), by

$$\int_{-\infty}^{\infty} f_1(x_0, y_0)\delta(z_0)\exp[ik(z_0 - z_1)]h_1(r_0 + M_1 r_1)dr_0, \qquad (7.2.1)$$

where the delta function means that the output end of the illumination fibre is placed at

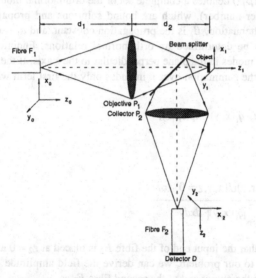

Fig. 7.2.1   Schematic diagram of a fibre-optical confocal scanning microscope in reflection.

$z_0 = 0$. In a similar way presented in Section 3.1, the light field at r_2 in detection space from a scan point at r_s can be derived as

$$U(r_2, r_s) = \int_{-\infty}^{\infty} \left[\int_{-\infty}^{\infty} f_1(x_0, y_0) \delta(z_0) \exp[ik(z_0 - z_1)] h_1(r_0 + M_1 r_1) dr_0 \right]$$

(7.2.2)

$$o(r_s - r_1) \exp[ik(\pm z_1 - z_2)] h_2(r_1 + M_2 r_2) dr_1.$$

Here $o(r)$ denotes the 3-D amplitude reflectivity of the thick object in a reflection-mode FOCSM or the 3-D amplitude transmittance of the object in a transmission-mode FOCSM. In Eqs. (7.2.1) and (7.2.2), the parameters M_1 and M_2 are defined in Eqs. (3.1.5) and (3.1.8), respectively.

In a FOCSM, a second fibre F_2 is employed to collect the light field and deliver it to a detector. In order to obtain the field amplitude on the output end of the fibre F_2, we now consider the propagation of the field $U(r_2, r_s)$ through the optical fibre F_2. In terms of Snyder et al.,[7.9] an arbitrary light field $U(x, y, z)$ can be expanded as a superposition of a complete set of orthonormal modes in an optical fibre, viz.,

$$U(x, y, z) = \sum_{j=1}^{\infty} a_j \tilde{q}_j(x, y) \exp(i \tilde{\beta}_j z),$$

(7.2.3)

where $\tilde{q}_j(x, y) \exp(i \tilde{\beta}_j z)$ denotes a complete set of the orthonormal modes in a particular fibre (j is an integer number), which are bound solutions and propagate along the z direction without attenuation, $\tilde{\beta}_j$ is the propagation constant, and a_j is called the modal amplitude and can be determined by orthonormal relations. Functions $\tilde{q}_j(x, y)$ are profiles of the fibre modes in the plane perpendicular to the z axis. If the optical fibre is a single-mode fibre, the summation over j includes only the first term with $j = 1$ and thus Eq. (7.2.3) becomes

$$U(x, y, z) = a_1 \tilde{q}_1(x, y) \exp(i \tilde{\beta}_1 z),$$

(7.2.4)

where

$$a_1 = \frac{\int \int_{-\infty}^{\infty} \tilde{q}_1^*(x, y) U(x, y, z = 0) dx dy}{\int \int_{-\infty}^{\infty} |\tilde{q}_1(x, y)|^2 dx dy}.$$

(7.2.5)

Considering that the input end of the fibre F_2 is placed at $z_2 = 0$ and applying Eqs. (7.2.4) and (7.2.5) to our problem, we can derive the field amplitude $U(x_3, y_3, r_s)$ at a point (x_3 and y_3) on the output end of the second fibre F_2 as

$$U(x_3,y_3,r_s) = \frac{f_2(x_3,y_3)\exp(i\bar{\beta}_1 L_f)\iint_{-\infty}^{\infty} f_2^*(x_2,y_2)U(x_2,y_2,0,r_s)dx_2dy_2}{\iint_{-\infty}^{\infty}|f_2(x_2,y_2)|^2 dx_2dy_2}, \qquad (7.2.6)$$

where L_f is the length of the optical fibre F_2. Therefore the intensity measured by the large area detector D, which has a uniform intensity sensitivity, is

$$I(r_s) = \iint_{-\infty}^{\infty}|U(x_3,y_3,r_s)|^2 dx_3dy_3,$$

or

$$I(r_s) = \left|\iint_{-\infty}^{\infty} f_2^*(x_2,y_2)U(x_2,y_2,0,r_s)dx_2dy_2\right|^2, \qquad (7.2.7)$$

where constant factors have been neglected. Compared with Eq. (4.1.1), Eq. (7.2.7) differs from it, giving rise to the superposition of the amplitude, while Eq. (4.1.1) implies the superposition of the intensity. Substituting Eq. (7.2.6) into Es. (7.2.7), one can finally find the detected image intensity from a scan point $r_s(x_s, y_s, z_s)$:

$$I(r_s) = \iiint\iiint_{-\infty}^{\infty} f_1(x_1,y_1)\delta(z_0)f_2^*(x_2,y_2)\delta(z_2)h_1(r_0 + M_1r_1)o(r_s - r_1)h_2(r_1 + M_2r_2)$$

$$f_1^*(x_0',y_0')\delta(z_0')f_2(x_2',y_2')\delta(z_2')h_1^*(r_0' + M_1r_1')o^*(r_s - r_1')h_2^*(r_1' + M_2r_2')$$

$$\exp[ik(z_0 - z_1 \pm z_1 - z_2)]\exp[-ik(z_0' - z_1' \pm z_0' - z_2)]$$

$$dr_0dr_0'dr_1dr_1'dr_2dr_2',$$

$$(7.2.8)$$

where the bold letters r_j' ($j = 0, 1, 2$) have the same definition as those in Eq. (3.1.10) and the symbol * corresponds to the conjugate operation,

In terms of the definition of the convolution operation,[7.32] Eq. (7.2.8) can be expressed as a form of the 3-D convolution:

$$I(r_s) = |h_a(r_s) \otimes_3 o(r_s)|^2. \qquad (7.2.9)$$

It can be recognized that the microscope described by Eq. (7.2.9) behaves as a purely-coherent microscope because the superposition of amplitude is applicable, and that $h_a(r_s)$ is the 3-D effective amplitude point spread function given by

$$h_a(r) = \exp[ik(-z \pm z)][f_1(M_1x, M_1y) \otimes_2 h_1(M_1r)][f_2^*(M_1x, M_1y) \otimes_2 h_2(r_2)], \quad (7.2.10)$$

where $\otimes_2$ represents the 2-D convolution operation in the transverse plane.

Eq. (7.2.10) indicates that a 3-D coherent transfer function can be introduced to describe the 3-D image formation in a FOSCM, as we have done in Chapter 3. Using Eqs. (3.2.1) and (3.2.2) in Eq. (7.2.9) gives

$$I(r_s) = \left| \int_{-\infty}^{\infty} c(m)O(m)\exp(2\pi i r_s \bullet m)dm \right|^2. \quad (7.2.11)$$

Here $c(m)$ is called the 3-D coherent transfer function (CTF) with a spatial frequency vector m and is given by

$$c(m) = \int_{-\infty}^{\infty} h_a(r)\exp(-2\pi i r \bullet m)dr, \quad (7.2.12)$$

which is the 3-D inverse Fourier transform of the 3-D effective point spread function $h_a(r)$ given in Eq. (7.2.10).

Considering a practical case, we can assume that the objective, the collector and the fibres are all of circular geometry. By substituting Eq. (7.2.10) into Eq. (7.2.12) and taking the cylindrical symmetry into account, the 3-D CTF, $c(m)$, therefore reduces to

$$c(l,s) = \int_{-\infty}^{\infty} c(l,z)\exp(-2\pi i z s)dz, \quad (7.2.13)$$

where $l = (m^2 + n^2)^{1/2}$ denotes the radial spatial frequency. $c(l, z)$ is the 2-D defocused CTF given by

$$c(l,z) = \exp[ik(-z \pm z)]\left[\tilde{f}_1\left(\frac{l}{M_1}\right)P_1(\lambda d_2 l, -M_1^2 z) \right] \otimes_2 \left[\tilde{f}_2\left(\frac{l}{M_1}\right)P_2(\lambda d_2 l, z) \right], \quad (7.2.14)$$

where $\tilde{f}_1$ and $\tilde{f}_2$ are the Fourier transforms of fibre mode profiles f_1 and f_2, respectively. They also represent the weighting functions of the pupil functions for the objective and collector lenses, respectively. As expected, the 3-D CTF is given by the one-dimensional (1-D) inverse Fourier transform of the 2-D defocused coherent transfer function with respect to z, which is in fact the 3-D convolution of two caps of the paraboloids with weighting functions $\tilde{f}_1$ and $\tilde{f}_2$. If the pupil function for the objective is normalized by its radius a, one has

$$c(l,s) = \int_{-\infty}^{\infty} c(l,u)\exp(-ius)du, \quad (7.2.15)$$

and

$$c(l,u) = \exp[is_0(-u \pm u)]\left[\tilde{f}_1\left(\frac{la}{d_1\lambda}\right)P_1(l,u)\right] \otimes_2 \left[\tilde{f}_2\left(\frac{la}{d_1\lambda}\right)P_2(l,u)\right], \qquad (7.2.16)$$

where l and s have been normalized by Eqs. (2.4.8) and (2.4.9). s_0 is defined in Eq. (2.4.12)

The above results are applicable to both reflection and transmission FOCSMs with two different pupil functions. In fact, they also hold even for a system suffering from aberration. We will discuss only the aberration-free reflection and transmission FOCSM systems in the rest of this chapter. Readers can cope with the effect of aberration on the FOCSM in accordance with the discussion presented in Section 3.3.

7. 2. 2   Coherent Transfer Function in Reflection

For a reflection-mode fibre-optical confocal scanning microscope (see Fig. 7.2.1), the defocused pupil functions for identical objective and collector lenses are given by Eqs. (3.2.12) and (3.2.13) and a negative sign in the pre-phase term in Eq. (7.2.16) should be taken. Eq. (7.2.16) is thus

$$c_r(l,u) = K_r \exp(-2is_0u)\left[\tilde{f}_1\left(\frac{la}{d_1\lambda}\right)P_1(l,u)\right] \otimes_2 \left[\tilde{f}_2\left(\frac{la}{d_1\lambda}\right)P_2(l,u)\right], \qquad (7.2.17)$$

where K_r is a constant of normalization.

In many practical arrangements, we can assume that both illumination and collection fibres have the same mode profiles, which can be considered to be a case in Fig. 7.1.2. Under the Gaussian approximation[7.9] for circular fibres, the amplitude profile of a single-mode fibre with a spot size r_0 is

$$f_1(r) = \exp\left[-\frac{1}{2}\left(\frac{r}{r_0}\right)^2\right], \qquad (7.2.18)$$

so that the corresponding Fourier transform is

$$\tilde{f}_1(l) = 2\pi r_0^2 \exp\left[-\frac{1}{2}(2\pi r_0 l)^2\right]. \qquad (7.2.19)$$

By substituting Eqs. (7.2.19), (3.2.12) and (3.2.13) into Eq. (7.2.17), the 2-D convolution operation in Eq. (7.2.17) can be evaluated, by the same method presented in Section 3.2, as

$$c_r(l,u) = K_r \exp(-2is_0 u) \exp\left[-\left(A - \frac{iul^2}{4}\right)\right] \int_0^{\pi/2} \int_0^{\rho_0} \exp\left[-\left(A - iu\rho'^2\right)\right]\rho' \, d\rho' \, d\theta',$$

(7.2.20)

where

$$A = \left(\frac{2\pi a r_0}{\lambda d_1}\right)^2$$

(7.2.21)

and ρ_0 is given by Eq. (3.2.18). The parameter A incorporates the effect of the fibre spot size r_0, the pupil radius a, and the distance d_1.

Substituting Eq. (7.2.20) for $c(l, u)$ in Eq. (7.2.15) yields the 3-D CTF for the reflection-mode FOCSM:

$$c_r(l,s) = K_r \exp\left(-\frac{Al^2}{4}\right) \int_0^{\pi/2} \int_0^{\rho_0} \exp\left(-A\rho'^2\right) \delta\left(s + s0 - \rho'^2 - \frac{l^2}{4}\right)\rho' \, d\rho' \, d\theta',$$ (7.2.22)

where $s0$ is equal to $2s_0$. Like the case in Eq. (3.2.19), the δ-function implies that the above integral can be evaluated along a curve on the ρ'–θ' plane. The curve is a circle or a part of a circle as shown in Fig. 3.2.2, depending on the value of ρ_0, but is now weighted by a function $\exp(-A\rho'^2)$. The 3-D CTF $c_r(l, s)$ normalized by the value of $c_r(l = 0, s = -s0)$ can be analytically expressed as

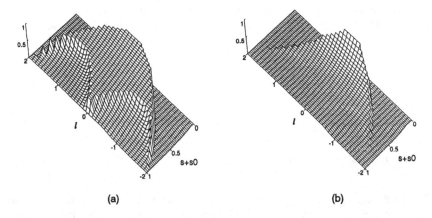

(a)   (b)

Fig. 7.2.2   3-D CTF in a reflection-mode FOCSM: (a) $A = 1$; (b) $A = 5$.

$$c_r(l, s) = \exp[-A(s + s0)] f_r(l, s), \tag{7.2.23}$$

where $f_r(l, s)$ is given by Eq. (3.2.20) and is the 3-D CTF in the reflection-mode confocal scanning microscope with a point source and a point detector. The 3-D CTF is of course shifted by $s0$ along the axial direction. Interestingly, the effect of the fibre spot size results in only a decrease of the value of $f_r(l, s)$ by a factor of $\exp[-A(s + s0)]$.

Fig. 7.2.2 shows the 3-D plots of the coherent transfer functions, $c_r(l, s)$, in the reflection-mode FOCSM for non-zero values of the parameter A. Because of the constant spatial frequency shift $s0$, the 3-D coherent transfer function $c_r(l, s)$ remains non-zero only in the region of $l^2/4 \le (s + s0) \le 1$, where the value of $c_r(l, s)$ is finite and the non-zero region is divided into three sub-regions with the transverse cut-off spatial frequency of 2. As shown in Fig. 3.2.3 for $A = 0$, $c_r(l, s)$ is a constant in one of the non-zero sub-regions but varies in the other two regions. Therefore, the borders between them are sharp. The constant region decreases quickly when A increases from zero, i. e, when the fibre spot size increases. Eventually, it reduces smoothly towards the two varying regions (see Fig. 7.2.2b). For large values of A, the value of the coherent transfer function near $l = 2$ becomes very small, so that the cut-off spatial frequency is effectively decreased.

Importantly, it should be pointed out that the finite distribution of the cross-section $c_r(l = 0, s)$ shown in Fig. 7.2.3 results in the optical sectioning effect in such a microscope, because the strength of the optical sectioning effect is determined by the modulus squared of the Fourier transform of $c_r(l = 0, s)$, which will be, in detail, studied in Section 7.3.

In general, when fibres F_1 and F_2 in the FOCSM have different structures, the mode profiles can be expressed as

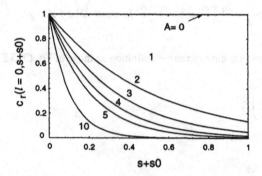

Fig. 7.2.3 Axial cross-section of the 3-D CTF in a reflection-mode FOCSM.

$$f_1(r) = \exp\left[-\frac{1}{2}\left(\frac{r}{r_{01}}\right)^2\right], \tag{7.2.24a}$$

$$f_2(r) = \exp\left[-\frac{1}{2}\left(\frac{r}{r_{02}}\right)^2\right], \tag{7.2.24b}$$

where r_{01} and r_{02} denote the spot sizes of fibre profiles for illumination and collection. The corresponding Fourier transforms are

$$\tilde{f}_1(l) = 2\pi r_{01}^2 \exp\left[-\frac{1}{2}(2\pi r_{01}l)^2\right], \tag{7.2.25a}$$

$$\tilde{f}_2(l) = 2\pi r_{02}^2 \exp\left[-\frac{1}{2}(2\pi r_{02}l)^2\right]. \tag{7.2.25b}$$

Thus the corresponding 3-D CTF for a reflection-mode FOCSM including two unequal annular pupils[7.33] can be generalized, using the methods presented above and in Section 3.5, to

$$c_r(l,s) = K_r \left(\iint_{\sigma_1} + \iint_{\sigma_2} - \iint_{\sigma_3} - \iint_{\sigma_4} \right)$$

$$\exp\left(-\frac{A_1+A_2}{2}\frac{l^2}{4} - \frac{A_1+A_2}{2}\rho'^2 - \frac{A_1-A_2}{2}l\rho'\cos\theta'\right)\delta\left(s+s0-\rho'^2-\frac{l^2}{4}\right)\rho'\,d\rho'\,d\theta', \tag{7.2.26}$$

where σ_1, σ_2, σ_3 and σ_4 have the same definition as those in Eq. (3.5.8), and

$$A_1 = \left(\frac{2\pi a r_{01}}{\lambda d_1}\right)^2 \tag{7.2.27a}$$

and

$$A_2 = \left(\frac{2\pi a r_{02}}{\lambda d_1}\right)^2. \tag{7.2.27b}$$

In general, it is difficult to derive an analytical expression for Eq. (7.2.26) if $A_1 \neq A_2$. However, when $A_1 = A_2 = A$, an analytical solution to Eq. (7.2.26) takes a form of Eq. (7.2.23) where $f_r(l, s)$ is given by the solutions in Section 3.5 for reflection microscopy of a point detector. For example, in the case of two identical annular pupils,[7.34] $f_r(l, s)$ is given by Eqs.(3.5.25)-(3.5.29) and the corresponding 3-D CTF for the FOCSM when $\varepsilon = 0.25$, where ε is the normalized radius of the central obstruction of the annular pupils, is shown in Fig. 7.2.4. Compared with Fig. 3.5.3a, the value of the 3-D CTF in Fig. 7.2.4 is decreased in the axial direction.

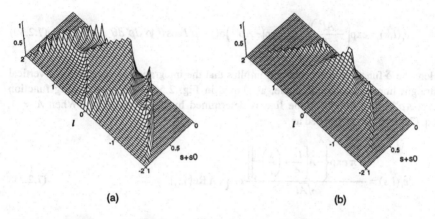

(a)   (b)

Fig. 7.2.4   3-D CTF in a reflection-mode FOCSM consisting of two equal annular pupils ($\varepsilon = 0.25$): (a) $A = 1$; (b) $A = 5$.

7. 2. 3   Coherent Transfer Function in Transmission

In the case of the transmission-mode FOCSM, the relationship between the two defocused pupil functions is given by Eqs. (3.2.12) and (3.2.27) if the corresponding pupil functions are same. The defocused CTF is thus

$$c_t(l,u) = K_t \left[\tilde{f}_1\left(\frac{la}{d_1\lambda}\right) P_1(l,u) \right] \otimes_2 \left[\tilde{f}_2\left(\frac{la}{d_1\lambda}\right) P_2(l,u) \right]. \tag{7.2.28}$$

Here K_t is again a constant of normalization. For the Gaussian approximation,[7.9] Eqs. (7.2.18) and (7.2.19) still hold. By using the same method as in the reflection case, the 2-

D defocused CTF in a transmission FOCSM of two identical single-mode fibres can be derived as

$$c_t(l,u) = K_t \exp\left(-\frac{Al^2}{4}\right) \int_0^\pi \int_0^{\rho_0} \exp\left(-A\rho'^2\right) \exp(i\rho' \, lu\cos\theta') \rho' \, d\rho' \, d\theta'. \qquad (7.2.29)$$

It should be noticed that $c_t(l = 0, u)$ is, in fact, not a function of the defocus distance u. Therefore there is a singularity at the origin in the spatial frequency space, so that the 3-D CTF cannot be normalized to unity at the origin. Performing the Fourier transform of $c_t(l, u)$ with respect to u leads to the 3-D CTF in the transmission-mode FOCSM:

$$c_t(l,s) = \exp\left(-\frac{Al^2}{4}\right) \int_0^\pi \int_0^{\rho_0} \exp\left(-A\rho'^2\right) \delta(s - \rho' \, l\cos\theta') \rho' \, d\rho' \, d\theta'. \qquad (7.2.30)$$

Here the δ-function in Eq. (7.2.30) implies that the integration is taken along a vertical straight line on the $\rho' - \theta'$ plane, as shown in Fig. 2.5.2, with a weighting function $\exp(-A\rho'^2)$. The length of the line is determined by the value of ρ_0. When $A \neq 0$, Eq. (7.2.30) can be derived as

$$c_t(l,s) = \frac{\sqrt{\pi} \exp\left\{-A\left[\frac{l^2}{4} + \left(\frac{s}{l}\right)^2\right]\right\}}{\sqrt{A}l} \operatorname{erf}\left[\sqrt{A} \operatorname{Re}(y_0)\right], \qquad (7.2.31)$$

where Re( ) means the real part of its argument, y_0 is given by Eq. (3.3.11) and erf is the

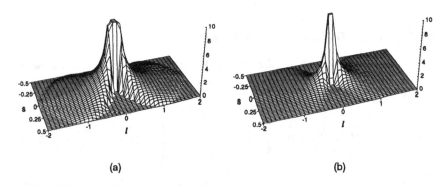

(a)   (b)

Fig. 7.2.5   3-D CTF for a transmission-mode FOCSM: (a) $A = 1$; (b) $A = 5$.

error function[7.35] defined in Eq. (3.3.13). It should be pointed out that Eq. (7.2.31) is also the 3-D OTF for a single lens with a Gaussian pupil function.

When $A = 0$, Eq. (7.2.31) reduces to

$$c_i(l,s) = \frac{2}{l}\text{Re}(y_0).$$

(7.2.32)

This is the 3-D CTF in the transmission-mode confocal scanning microscope with a point source and a point detector, as shown in Fig. 2.5.3.

The 3-D coherent transfer functions for the transmission-mode FOCSM are displayed in Fig. 7.2.5 for non-zero values of A. As mentioned above, the value of $c_f(l, s)$ at $s = 0$ and $l = 0$ is infinite, leading to a missing cone of spatial frequencies. It can be seen that the non-zero region of $c_f(l, s)$ is bounded by $|s| \le l(1 - l/2)$, and that a δ-function is found when $l = 0$, implying that it is impossible to achieve real optical sectioning in the transmission-mode fibre-optical confocal scanning microscope when the distance between the two lenses is fixed.

The 3-D CTF for a FOCSM consisting of two unequal annular pupils and unequal fibre spot sizes[7.34] can be generalized to

$$c_r(l,s) = K_r \left(\iint_{\sigma_1} + \iint_{\sigma_2} - \iint_{\sigma_3} - \iint_{\sigma_4} \right)$$

$$\exp\left(-\frac{A_1 + A_2}{2}\frac{l^2}{4} - \frac{A_1 + A_2}{2}\rho'^2 - \frac{A_1 - A_2}{2}l\rho'\cos\theta' \right)\delta(s - \rho'l\cos\theta')\rho' \, d\rho' \, d\theta'.$$

(7.2.33)

Because of the delta function, we can derive the solution to one of the four terms in Eq. (7.2.33), if we assume ρ_1 and ρ_2 to be the radii of the two circular pupils, as[7.33]

1) $\rho_1 > \rho_2$

1a) $0 < l < \rho_1 - \rho_2$

$$c_i(l,s) = \frac{\sqrt{\pi}\exp\left[-\frac{A_1}{2}\left(\frac{l}{2} + \frac{s}{l}\right)^2 - \frac{A_2}{2}\left(\frac{l}{2} - \frac{s}{l}\right)^2 \right]}{\sqrt{\frac{A_1 + A_2}{2}}l}$$

$$\left\{ \begin{array}{ll} \mathrm{erf}\left\{ \sqrt{\dfrac{A_1+A_2}{2}} \, \mathrm{Re}\left[\sqrt{\rho_2^2 - \left(\dfrac{s}{l}-\dfrac{l}{2}\right)^2} \, \right] \right\}, & -l\left(\rho_2-\dfrac{l}{2}\right) \le s \le l\left(\rho_2+\dfrac{l}{2}\right), \\[4mm] 0, & \textit{otherwise.} \end{array} \right.$$

(7.2.34)

1b) $\rho_1 - \rho_2 < l < \rho_1 + \rho_2$

$$c_t(l,s) = \frac{\sqrt{\pi}\exp\left[-\dfrac{A_1}{2}\left(\dfrac{l}{2}+\dfrac{s}{l}\right)^2 - \dfrac{A_2}{2}\left(\dfrac{l}{2}-\dfrac{s}{l}\right)^2 \right]}{\sqrt{\dfrac{A_1+A_2}{2}}\, l}$$

$$\left\{ \begin{array}{ll} \mathrm{erf}\left\{ \sqrt{\dfrac{A_1+A_2}{2}} \, \mathrm{Re}\left[\sqrt{\rho_1^2 - \left(\dfrac{s}{l}+\dfrac{l}{2}\right)^2} \, \right] \right\}, & \dfrac{\rho_1^2-\rho_2^2}{2} \le s \le l\left(\rho_1-\dfrac{l}{2}\right), \\[5mm] \mathrm{erf}\left\{ \sqrt{\dfrac{A_1+A_2}{2}} \, \mathrm{Re}\left[\sqrt{\rho_2^2 - \left(\dfrac{s}{l}-\dfrac{l}{2}\right)^2} \, \right] \right\}, & -l\left(\rho_2-\dfrac{l}{2}\right) \le s \le \dfrac{\rho_1^2-\rho_2^2}{2}, \\[5mm] 0, & \textit{otherwise.} \end{array} \right.$$

(7.2.35)

When A_1 and A_2 approach zero, Eqs. (7.2.34) and (7.2.35) reduce to Eqs. (3.5.20) and (3.5.21).

2) $\rho_1 < \rho_2$

The corresponding 3-D CTF has a similar form to Eqs. (7.2.34) and (7.2.35) if the same replacements as described in Eqs. (3.5.22)-(3.5.24) are used.

If two equal annular lenses are employed in the transmission-mode FOCSM with two equal fibres, the 3-D CTF is expressed, in terms of Eqs. (7.2.33)-(7.2.35), as

$$c_t(l,s) = \frac{\sqrt{\pi}\exp\left\{ -A\left[\dfrac{l^2}{4}+\left(\dfrac{s}{l}\right)^2\right] \right\}}{\sqrt{A}\, l} \left\{ \mathrm{erf}\left[\sqrt{A}\,\mathrm{Re}(y_0)\right] - \mathrm{erf}\left[\sqrt{A}\,\mathrm{Re}(y_1)\right] \right\},$$

(7.2.36)

where

$$y_1 = \sqrt{\varepsilon^2 - \left(\frac{|s|}{l} - \frac{l}{2}\right)^2}. \qquad (7.2.37)$$

Eq. (7.2.36) is also identical to the 3-D optical transfer function for an annular lens of the central obstruction ε and a Gaussian-shaped pupil function. When $\varepsilon \to 0$, Eq. (7.2.36) reduces to the 3-D CTF in the transmission FOCSM for circular pupils (see Eq. (7.2.31)). On the other hand, if $A \to 0$, Eq. (7.2.36) approaches the 3-D CTF for the transmission confocal scanning microscope with a point source and a point detector (see Eq. (2.5.16)). For $\varepsilon = 0.25$, the 3-D CTFs for a transmission FOCSM with two equal annular lenses are shown in Fig. 7.2.6. It is seen that as expected, the 3-D CTF is cut off at 1- $\varepsilon^2/2$ in the axial direction and has a singularity at $l = 0$. Further, the values of the 3-D CTF in Eq. (7.2.36) are decreased with increasing A, in comparison with Fig. 3.5.4a.

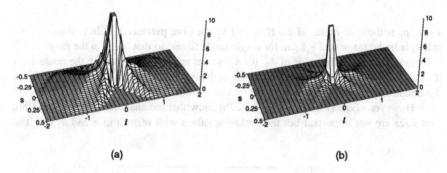

(a)   (b)

Fig. 7.2.6   3-D CTF for a transmission-mode FOCSM consisting of two identical annular lenses ($\varepsilon = 0.25$): (a) $A = 1$; (b) $A = 5$.

According to the calculation in Figs. 7.2.2-7.2.6, we can conclude that using optical fibres in the confocal system tends to reduce the values of the 3-D CTF: the larger the values of A_1 (or A_2), the stronger the reduction. Therefore, one may expect such reduction in the 3-D CTFs for those confocal systems discussed in Section 3.5, if optical fibres are used.

7. 2. 4   Role of Fibre Spot Size

We have shown that since the tip of an optical fibre has finite-sized cross-section, the properties of the image formation in fibre-optical confocal microscopy are accordingly affected. In the case of single-mode fibres with Gaussian profiles, the above analyses have shown that the FOCSM can be described by two system parameters, A_1 and A_2, defined in Eq. (7. 2.27). A_1 and A_2 denote the normalized fibre spot sizes, which play a similar role to the normalized pinhole radius v_d for a finite-sized detector (see Eq. (4.2.4)), and therefore represent the effects of the fibre spot sizes. The values of A_1 and A_2 can be changed by altering the fibre spot sizes.

Let us take as an example a single-mode fibre with a step refractive index profile,[7.9] which is usually a practical case. In this case, the radius of the fibre core and the mode spot size r_0 satisfy the following relationship under the Gaussian approximation (see p. 341 in Ref. 7.9):

$$r_0 = \frac{\rho_c}{\sqrt{2 \ln V_f}},\qquad\qquad(7.2.38)$$

where ρ_c is the core radius of the fibre and V_f the fibre parameter.[7.9] In practice, $V_f \approx 2$ and ρ_c is in the range of $2 \sim 5$ μm for single-mode fibres, so that r_0 is in the range of 1.7 ~ 4.2 μm. Therefore, the value of A_1 (or A_2) varies approximately within the range $1 \sim 7$ if d_1 and a are set as 50 mm and 5 mm, respectively. This means that the range of A_1 (or A_2) obtainable by alteration of the fibre spot size is not large.

However, Eqs. (7.2.27a) and (7.2.27b) show that the absolute values of the fibre spot sizes are not important but their relative values with respect to a and d_1 are. The

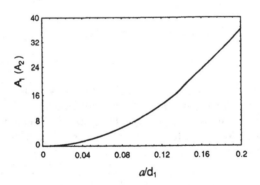

Fig. 7.2.7   Dependence of A_1 (or A_2) on the ratio of a to d_1.

values of A_1 and A_2 may be altered by changing the distance d_1. This can be realized by inserting an additional lens between the fibre and the imaging lens. For example, the reflection FOCSM can be arranged as in Fig. 7.1.1, where the focal length of the lenses L_1 and L_2 can be altered. It can be proved that the image properties shown in Sections 7.2.1-7.2.3 are still applicable in Fig. 7.1.1. In terms of the theory of the Gaussian beam propagation,[7.36] it is known that the focal length of L_1 and L_2 affects the matching condition between the fibre mode profile and the field distribution on the tip of the fibre. As an example, if $\lambda = 0.633$ μm and $\rho_c = 4$ μm, then the variation of A_1 (or A_1) with (a/d_1) is shown in Fig. 7.2.7. For the values of A_1 (or A_2) at the lower end of the curve, the imaging performance in the FOCSM, both for later and axial resolution, is substantially the same as in the true confocal microscope with point source and detection. On the other hand, it has been found that the values of the 3-D CTF in reflection are reduced in the axial direction when A_1 (or A_2) is large (see Fig. 7.2.3). The reduction consequently affects resolution in the FOCSM in the both axial and transverse directions. The dependence of the resolution on the fibre spot size and the central obstruction of annular pupils will be considered in Sections 7. 3 and 7.4.

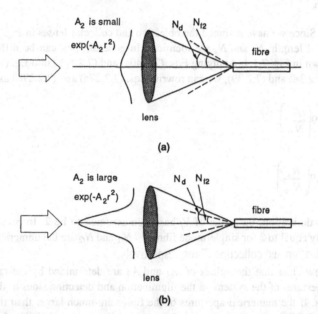

Fig. 7.2.8   Coupling of a Gaussian beam into a single-mode optical fibre: (a) the numerical aperture of the fibre is larger than that of the illumination; (b) the numerical aperture of the fibre is smaller than that of the illumination.

The physical meanings of A_1 and A_2 can be further explained as follows. Under the Gaussian approximation,[7.9] the fibre parameter V_f can be expressed, for a single-mode fibre with a step refractive-index profile, as

$$V_f = \frac{2\pi \rho_c N_f}{\lambda}. \tag{7.2.39}$$

Here N_f is the numerical aperture of the single-mode fibre. In addition, the numerical apertures of the objective on the detection and illumination sides can be represented, under the paraxial approximation, by

$$N_i = \frac{a}{d_1} \tag{7.2.40a}$$

and

$$N_d = \frac{a}{d_1} \tag{7.2.40b}$$

respectively. Since we have assumed the objective and collector lenses in Fig. 7.2.1 have the same focal length, N_i and N_d are identical. In general, they can be different each other, as shown in Fig. 7.1.1. Applying Eqs. (7.2.40a) and (7.2.40b) into Eq. (7.2.27) and using Eqs. (7.2.38) and (7.2.39), we can rewrite Eqs. (7.2.27a) and (7.2.27b) as

$$A_1 \approx 2.9 \left(\frac{N_i}{N_{f1}} \right), \tag{7.2.41a}$$

$$A_2 \approx 2.9 \left(\frac{N_d}{N_{f2}} \right). \tag{7.2.41b}$$

In the above derivation, the value of V_f has been assumed to be 2. In practice, V_f is approximately equal to 2 for single-mode fibres.[7.9] N_{f1} and N_{f2} are the numerical aperture for the illumination and collection fibres, respectively.

It is now clear that the values of A_1 and A_2 are determined by the ratios of the numerical apertures of the system on the illumination and detection sides to those of the optical fibres. If the numerical apertures of the fibres are much larger than those of the system (Fig. 7.2.8a), only a small part of the source light is coupled into the confocal system. In other words, the pupil of the objective (or collector) is overfilled. The tips of the optical fibres can thus be considered to be a point source and a point detector with A_1

= 0 and $A_2 = 0$ and the FOCSM system behaves as an ideal confocal microscope. On the other hand, when the numerical apertures of the fibres are smaller than or comparable to those of the system, i. e., when A_1 and A_2 have finite values, the finite-sized cross-sections of the fibres play an important role in the FOCSM. The pupil of the objective (or collector) is underfilled (Fig. 7.2.8b) and its effective pupil function is now shaded with a function which depends on the fibre profiles, for example with a Gaussian function under the Gaussian approximation. In other words, the FOCSM is now determined by the properties of the optical fibres. This suggests that one can operate the FOCSM in novel modes by changing fibre parameters.

7. 3 Axial Resolution

Axial resolution or optical sectioning in a reflection-mode FOCSM can be studied by considering the axial response to a perfect planar reflector scanned along the axial direction. For a good system, the width of the axial response is needed to be as narrow as possible. In the case of FOCSM, the axial response, i. e., the detected intensity as a function of the defocus distance, can be expressed, according to Eq. (7.2.11), as

$$I(u) = K_r |c_r(l = 0, u)|^2 , \tag{7.3.1}$$

where $c_r(l, u)$ is the so-called defocused coherent transfer function given by Eq. (7.2.16). $c_r(l = 0, u)$ is the Fourier transform of the axial cross-section of the 3-D CTF with respect to s. From Eq. (7.2.26) we can derive $c_r(l = 0, u)$ as

$$c_r(l = 0, u) = \exp(-is0u) \int_0^1 P_1(\rho') P_2(\rho') \exp\left[-\left(\frac{A_1 + A_2}{2} - iu \right) \rho'^2 \right] \rho' d\rho'. \tag{7.3.2}$$

For annular lenses, their pupil functions can be expressed as

$$P_{1,2}(\rho') = \begin{cases} 1, & \varepsilon_{1,2} \le \rho' \le 1, \\ \\ 0, & otherwise, \end{cases} \tag{7.3.3}$$

where $\varepsilon_{1,2}$ is the radius of the central obstruction normalized by the outer radius of the objective lens. We have assumed that the outer radii are the same for the objective and collector lenses. Since the value of $c_r(l = 0, u)$ is determined by the product of the two pupil functions, $I(u)$ remains unchanged for a central obstruction placed in either the objective or the collector. Without losing generality, one can set ε as the larger value of ε_1 and ε_2, so that the axial response $I(u)$ can be solved, by substituting Eqs. (7.3.2) and (7.3.3) into Eq. (7.3.1), as

$$I(u) =$$

$$\frac{(A_1 + A_2)^2 \left\{ 1 + \exp\left[(A_1 + A_2)(\varepsilon^2 - 1) \right] - 2\exp\left[\dfrac{(A_1 + A_2)(\varepsilon^2 - 1)}{2} \right] \cos\left[(1 - \varepsilon^2)u \right] \right\}}{4\left\{ 1 - \exp\left[\dfrac{(A_1 + A_2)(\varepsilon^2 - 1)}{2} \right] \right\}^2 \left[\dfrac{(A_1 + A_2)^2}{4} + u^2 \right]},$$

$$(7.3.4)$$

which has been normalized by $I(u = 0)$. It should be noted that, unlike the system with a finite-sized pinhole detector (Section 4.4), the axial response can be expressed by a simple analytical expression. This feature of the FOCSM results from the fact that the imaging system is purely coherent.

It is noted that the axial response is only determined by the sum of A_1 and A_2. We can set $A = (A_1 + A_2)/2$. The variations of the axial response with A for the FOCSM using circular lenses ($\varepsilon = 0$) and annular pupils ($\varepsilon = 0.5$) are shown Fig. 7.3.1. As we expect, the axial response becomes broader as A increases for a given central obstruction. For $A = 0$, corresponding to a point source and a point detector, the axial response also becomes broader as the central obstruction radius ε increase (solid curves in Fig. 7.3.2). For non-zero values of A, for example, when $A = 10$, the axial response for $\varepsilon = 0.5$ is almost identical to that for $\varepsilon = 0$ (dashed curve in Fig. 7.3.2b).

A comparison of the half width $\Delta u_{1/2}$ of the axial response at half maximum (HWHM) as a function of A for circular pupils ($\varepsilon = 0$) with that for annular pupils ($\varepsilon = 0.55$, which is chosen for comparison with experimental results) is shown in Fig. 7.3.3. It is seen that for large values of A, the value of $\Delta u_{1/2}$ for annular pupils is the same as that

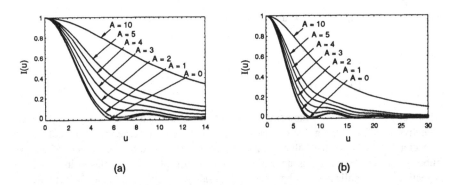

| (a) | (b) |

Fig. 7.3.1   Intensity of the axial response to a perfect reflector for different values of ε in a FOCSM: (a) $\varepsilon = 0$; (b) $\varepsilon = 0.5$.

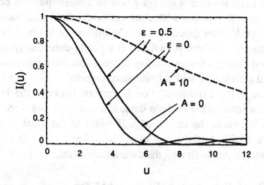

Fig. 7.3.2   Intensity of the axial response to a perfect reflector for a given value of A in FOCSM.

for circular pupils. In other words, for both annular and circular pupils, $\Delta u_{1/2}$ approaches
the same limiting value, which can be derived as

$$\Delta u_{1/2} = A \qquad\qquad\qquad (7.3.5)$$

when A is very large, and is shown in Fig. 7.3.3 (dashed line). This property implies that

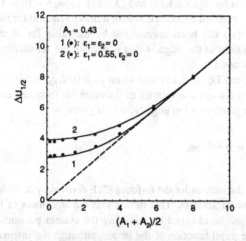

Fig. 7.3.4   Half width of the axial response to a perfect reflector, $\Delta u_{1/2}$, as a function of $A = (A_1 + A_2)/2$.
The dark spots and squares correspond to experimental values when $A_1 = 0.43$. Curve 1 ($\varepsilon_1 = \varepsilon_2 = 0$)
denotes two circular lenses and curve 2 ($\varepsilon_1 = 0.55$, $\varepsilon_2 = 0$) represents an annular objective lens and a
circular collector lens. The dashed line represents the limiting value of $\Delta u_{1/2}$ when A is large.

no improvement in axial resolution occurs when an annular pupil is combined with a finite fibre spot size, and contrasts with confocal bright-field microscopy consisting a finite-sized incoherent detector (Section 4.4). The difference is caused by the degree of coherence of the imaging systems. A FOCSM is a purely-coherent system in which the image is linear in amplitude, whereas imaging becomes partially coherent in the case of a confocal bright-field microscope with a finite-sized detector.

The difference in the axial-resolution behaviour between the FOCSM and the confocal system with finite-sized pinhole detection can be found from Figs. 4.4.1 and 7.3.3, both of which include the measured half width of the axial response.[7.3] In the experiment using a FOCSM, the value of A_1 was fixed to be 0.43. Note that the experimentally measured points fit well the theoretical results.

7. 4 Transverse Resolution - Image of a Straight Edge

Because imaging in a FOCSM is purely coherent even for finite values of fibre spot sizes, the image intensity at the edge is always one quarter of its value far from the edge, as may be expected for coherent imaging, and therefore the edge-setting criterion is independent of the parameters of the optical system. The sharpness of the edge image is, however, controlled by two parameters of the system, the fibre spot size and the central obstruction size when using annular pupils.

To demonstrate this point, let us consider a thin straight edge to be perpendicular to the x axis. According to the discussion in Section 3.3.4, we can express the image intensity and its gradient by Eqs. (3.3.16) and (3.3.17). $c_2(m, n = 0)$ in Eqs. (3.3.16) and (3.3.17) denotes the normalized in-focus coherent transfer function for the FOCSM along the x direction, and $I(v_x)$ has been normalized by its value far from the edge. The importance of the gradient at the edge is that it defines the sensitivity with which the edge position can be located.

It can be seen from Eq. (3.3.16) that when $v_x = 0$, $I(v_x) = 1/4$, so that the image intensity at the edge is always one quarter of its value far from the edge, as expected. Note that the intensity gradient at the edge ($v_x = 0$) is given by

$$I'(v_x) = \frac{1}{\pi} \int_0^\infty c_2(m, n = 0) dm. \tag{7.4.1}$$

Eq. (7.4.1) represents the area under the in-focus CTF divided by π, which is in general a function of the system parameters. This means that the sharpness of the image of the edge in the FOCSM can be changed by adjusting the system parameters, e. g., by the fibre spot size and the pupil function of the lenses, although the intensity at the edge is independent of the system parameters.

7. 4. 1 Effect of Fibre Spot Size

For a FOCSM consisting of two equal objective and collector lenses and $\varepsilon = 0$, $c_2(m, 0)$ is given by the projection of the 3-D CTF at $n = 0$ and can be expressed, from Eq. (7.2.26), as

$$c_2(m, n = 0) = \frac{A_1 + A_2}{\pi\left[1 - \exp\left(-\dfrac{A_1 + A_2}{2}\right)\right]} \exp\left[-\frac{(A_1 + A_2)m^2}{8}\right]$$

$$\hspace{6cm}(7.4.2)$$

$$\int_0^\pi \int_0^{\rho_0} \exp\left(-\frac{A_1 + A_2}{2}\rho'^2 - \frac{A_1 - A_2}{2}m\rho'\cos\theta'\right)\rho'\,d\rho'\,d\theta',$$

which has been normalized to unity at $m = 0$. When both A_1 and A_2 approach zero, the limiting form of $c_2(m, 0)$ corresponds to an analytical coherent transfer function for point illumination and detection, given in Eq. (3.2.25).

First we consider a FOCSM with equal illumination and collection fibres, corresponding to $A_1 = A_2 = A$, in which case $c_2(m, 0)$ as a function of A can be found in Ref. 7.1. Using Eqs. (3.3.16), (3.3.17), (7.4.1) and (7.4.2), we can calculate the image of the edge and its gradient at the edge as a function of A. Fig. 7.4.1a shows the image of the edge for different values of A. It is seen that the image intensity is one quarter of its value far from the edge even for finite values of fibre spot size, meaning that using a FOCSM provides a unique edge-setting criterion regardless of the system configuration. In this respect, this property contrasts with a confocal microscope with finite-sized

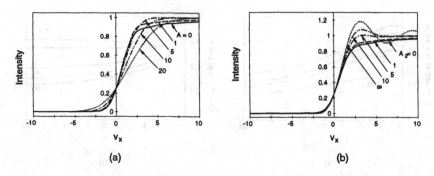

Fig. 7.4.1   Image of a sharp edge for different values of A (a) and for different values of A_2 when $A_1 = 0$ (b).

detectors, in which the image intensity at the edge varies between one quarter and one third, depending on the size of the detector (see Section 4.5). When A becomes large the image becomes spread out. However, the image of the edge when $A = 1$ is about 1.8% sharper than that even for $A = 0$. This can be further observed from the intensity gradient at the edge as a function of A (see the curve in Fig. 7.4.2a when $\varepsilon = 0$). Note that the intensity gradient has a maximum value of 0.2749 at about $A = 1$, while that for $A = 0$, corresponding to a point source and a point detector, is $8/3\pi^2 \approx 0.2701$. This is understandable from the coherent transfer function $c_2(m, 0),$[7.1] which for $A = 1$ gives a stronger response at lower spatial frequencies than that for $A = 0$, while its reduction at higher spatial frequencies is smaller. Therefore, the area under $c_2(m, 0)$ for $A = 1$ is slightly larger than that for $A = 0$, which leads to a larger intensity gradient at the edge in the former case.

As A increases further, the intensity gradient at the edge decreases monotonically. An asymptotic form of the intensity gradient as a function of A can be analytically derived according to Eq. (7.4.2):

$$\Gamma (v_x = 0) = \frac{1}{\sqrt{A\pi}},$$ (7.4.3)

meaning that the image of the edge becomes flat when a larger fibre spot size is used.

Another interesting phenomenon shown in Fig. 7.4.1a is that the image intensity for $A = 5$ approaches the maximum value more quickly than that for $A = 1$, while the intensity gradient at the edge in the former case is decreased slightly, compared with that in the latter case. This conclusion may be useful in practice, because only a small region of the object is needed to be scanned in order to locate the position at which the intensity of the edge is one quarter.

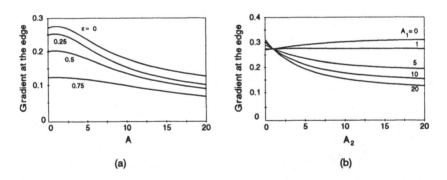

Fig. 7.4.2   Gradient at the edge as a function of A (a) and as a function A_2 for $A_1= 0, 1, 5, 10, 20$ (b).

An alternative FOCSM for $A_1 = 0$ and $A_2 \neq 0$ can also be considered, which corresponds to using a point source and a fibre to collect the signal. The image of the edge in this case is shown in Fig. 7.4.1b. When A_2 is increased, the image becomes sharper. Eventually, when A_2 approaches infinity the image of the edge is identical to that in a conventional coherent microscope. The reason for this is that the role of the collector lens is fully degraded when A_2 is very large, so that the imaging property in the FOCSM is determined only by the objective, meaning that the FOCSM is equivalent to a conventional coherent microscope whose coherent transfer function $c_2(m, 0)$ in the x direction is a square function with a cut-off spatial frequency of unity.[7.15, 7.37] Note that the gradient of the image at the edge in the FOCSM for $A_1 = 0$ is always increased on increasing A_2, as shown in Fig. 7.4.2b. When A_2 approaches infinity, the gradient of the image at the edge has a limiting value of $1/\pi$, meaning that the image of the edge is 17%

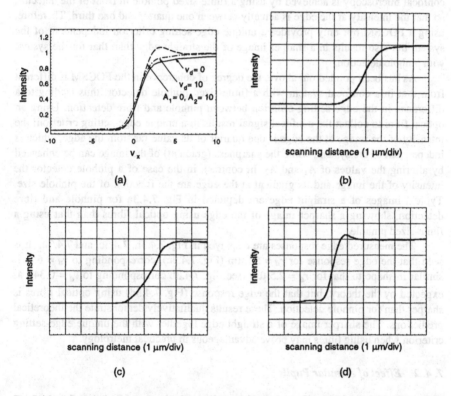

Fig. 7.4.3   Image of a sharp edge: (a) theoretical results; (b) measured result for $v_d = 0.34$; (c) measured result for $v_d = 6.7$; (d) measured result for $A_1 = 0.43$ and $A_2 = 15.8$.

sharper than that for $A_1 = A_2 = 0$. Therefore for a point source, the image of the edge is always made sharper by using fibre detection rather than a point detector.

In general, for a FOCSM with different values of A_1 and A_2, the gradient of the edge image at the edge is displayed in Fig. 7.4.2b. When $A_1 = 0$, the gradient approaches a limiting value of $1/\pi$, as mentioned above. Interestingly, it seems that the gradient hardly changes on increasing A_2 when $A_1 = 1$: the gradient is increased slightly and then decays very slowly as A_2 is enlarged. For values of A_1 larger than 1, the gradient is decreased monotonically on increasing A_2.

Figs. 7.4.2a to 7.4.2b have practical significance. According to the previous estimation in Section 7.2.4, for given values of the fibre spot size, the values of A_1 and A_2 can be changed by adjusting the distance between the collector and the collection fibre, so that A_1 and A_2 can be fixed to those values at which the image of the edge is sharper than that for $A_1 = A_2 = 0$. On the other hand, in practice, point detection in confocal microscopy is achieved by using a finite-sized pinhole in front of the detector, so that the intensity at the edge is actually between one quarter and one third. Therefore, using a FOCSM not only provides a unique edge-setting criterion independent of the system but also results in a sharper image of the straight edge than that for the system with pinhole detection.

As has been pointed out above, the degree of coherence in the FOCSM is different from that in a confocal system with a finite-sized pinhole detector, thus leading to a difference in the edge-setting criterion between pinhole and fibre detection. Using an optical fibre to collect the confocal signal results in a unique edge-setting criterion: the intensity of the image at the edge is one quarter of its value far from the edge, which is independent of the system, while the sharpness (gradient) of the image can be enhanced by altering the values of A_1 and A_2. In contrast, in the case of a pinhole detector the intensity of the image and its gradient at the edge are the function of the pinhole size. Typical images of a straight edge are depicted in Fig. 7.4.3a for pinhole and fibre detection, showing a sharper image of the edge using optical fibres than that using a finite-sized pinhole.

The measured edge responses are displayed in Fig. 7.4.3b, 7.4.3c, and 7.4.3d. It is seen that the edge response for $r_d = 50$ μm (Fig. 7.4.3c), corresponding to $v_d = 6.7$, is similar in shape to that for $r_d = 2.5$ μm (see Fig. 7.4.3b corresponding to $v_d = 0.34$), as expected by the theory, but that the edge response (Fig. 7.4.3d) using optical fibres is sharper than for pinhole detection. These results qualitatively demonstrate the theoretical predictions. The sharper image of a straight edge together with the unique edge-setting criterion when using fibres may prove advantageous in practical metrology.

7.4.2   Effect of Annular Pupils

Next we turn to the effect of annular pupils on the image of the edge. Suppose that both the objective and the collector have the same annular pupil that has an outer radius

a and an inner radius εa, and that the illumination and collection fibres have the same mode profiles. Under those conditions, the 3-D CTF for the transmission FOCSM with two equal annular lenses has been expressed in an analytical form in Eq. (7.2.36). The 2-D in-focus coherent transfer function $c_2(m, 0)$ is therefore the integration of the 3-D CTF with respect to the axial spatial frequency s (Chapter 10). In other words, the in-focus CTF is the projection of the 3-D CTF into the focal plane. Therefore, the in-focus CTF, $c_2(m, 0)$, is given by

$$c_2(m, n = 0) = \int_0^{(1-\varepsilon^2)/2} \frac{\exp\left\{-A\left[\frac{m^2}{4} + \left(\frac{s}{m}\right)^2\right]\right\}}{\sqrt{A}m}$$

(7.4.4)

$$\left\{\mathrm{erf}\left\{\sqrt{A}\,\mathrm{Re}\left[\sqrt{1-\left(\frac{|s|}{m}+\frac{m}{2}\right)^2}\right]\right\} - \mathrm{erf}\left\{\sqrt{A}\,\mathrm{Re}\left[\sqrt{\varepsilon^2-\left(\frac{|s|}{m}-\frac{m}{2}\right)^2}\right]\right\}\right\}ds.$$

Fig. 7.4.4a shows the image of the edge for different values of A for a given central obstruction radius ($\varepsilon = 0.5$). Again, the intensity gradient at the edge for $A = 1$ is slightly larger than that for $A = 0$. From Fig. 7.4.2a, it is seen that the maximum on the curves becomes less pronounced, as the central obstruction size is increased. Eventually, it disappears and the intensity gradient approaches zero as ε approaches one (see Fig. 7.4.4b). Figs. 7.4.5a and 7.4.5b are the images of the edge for different values of the central obstruction radius at $A = 1$ and 5, respectively. The images of the edge for $A = 1$

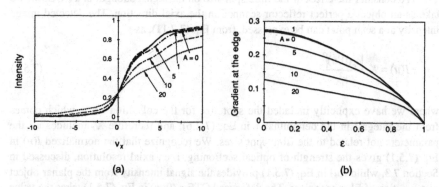

Fig. 7.4.4   (a) Image of a sharp edge when annular pupils are used ($\varepsilon = 0.5$); (b) gradient at the edge.

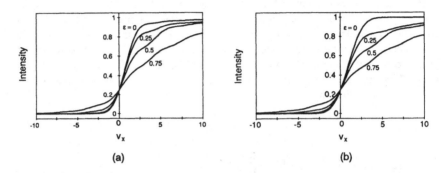

Fig. 7.4.5   Image of a sharp edge when annular pupils are used: (a) $A = 1$; (b) $A = 5$.

are very similar to those for a confocal system consisting of a point source and a point detector,[7.38] because $c_2(m, 0)$ for the FOCSM for small values of A is very close to that for a confocal system with a point source and a point detector.

7.5 Signal Level

Because fibres have finite sizes of cross-section, the imaging properties with fibres differ from those in confocal microscopes with a point source and a point detector, and the detected signal intensity from a scan position of the object can also be affected by the finite-sized cross-section of the fibres. This is reflected in the parameters A_1 and A_2. In practice, the maximum signal can be achieved by matching the fibre-mode profile with the diffraction pattern on the fibre tip.

To consider the effect of the fibre spot size on the signal strength in a FOCSM, we take as an object a perfect reflector scanned in the axial direction. The detected image intensity at a scan point can be expressed, from Eq. (7.2.11), as

$$I(u) = K_r \frac{|c_r(0, u)|^2}{r_{02}^2}, \tag{7.5.1}$$

where we have explicitly included the spot size for the collection fibre, which comes from the integral in the denominator in Eq. (7.2.6), and therefore K_r includes all the parameters not related to the fibre spot sizes. We recognize that the normalized $I(u)$ in Eq. (7.5.1) gives the strength of optical sectioning, i. e., axial resolution, discussed in Section 7.3, while $I(u)$ in Eq. (7.5.1) provides the signal intensity from the planar object as a function of fibre spot sizes. The defocused CTF $c_r(0, u)$ in Eq. (7.5.1) takes the value at the zero transverse spatial frequency and is not normalized.

To solve $c_r(0, u)$ as an explicit function of the fibre sizes, we substitute the Fourier transforms of the fibre profiles given in Eqs. (7.2.25a) and (7.2.25b) into Eq. (7.2.16). For two equal lenses, we can obtain

$$c_r(0,u) = \exp(-is0u) \frac{A_1 A_2 \left[1 - \exp\left(-\frac{A + A - 2iu}{2} \right) \right]}{A_1 + A_2 - 2iu},$$

(7.5.2)

where a constant factor irrelevant to the fibre spot size has been neglected.

The excitation efficiency,[7.9] η, of light power from the input fibre F_1 into the fibre F_2 is given by the detected intensity $I(u)$ divided by the total input power P_i. In our case, the input power from the fibre F_1 can be derived as

$$P_i = K_1 r_{01}^2,$$

(7.5.3)

where K_1 is a parameter irrelevant to the fibre spot size.

By using Eqs. (7.5.1)-(7.5.3), η becomes

$$\eta = \left| \frac{2\sqrt{A_1 A_2} \left[1 - \exp\left(-\frac{A + A - 2iu}{2} \right) \right]}{A_1 + A_2 - 2iu} \right|^2,$$

(7.5.4)

where the normalization factor of 2 is given by the condition that the efficiency η equals unity when there is no defocus ($u = 0$), both fibres have the same spot sizes and the lenses are quite large. When the radius of the lens approaches infinity, Eq. (7.5.4) reduces to Eq. (20-15) given by Snyder et al.[7.9] for the case when a Gaussian field illuminates a single-mode fibre with a Gaussian mode profile.

It should be pointed out that η denotes the total fraction of the power converting from the illumination fibre to the collection fibre in the fibre-optical confocal scanning microscope. Actually, the input power from the fibre F_1 is lost by two steps: first, the electric field is truncated by the finite-sized objective; second, this truncated electric field is focused onto the entrance of the fibre F_2, leading to a further loss of the power. If η_1 and η_2 are used to denote efficiencies of power transformation in these two steps, respectively, we have

$$\eta = \eta_1 \eta_2.$$

(7.5.5)

For a microscopic system, only the efficiency η_2, which is the so-called signal level, is important since the source power can be made arbitrarily large. It is easy to derive η_1, according to the Fresnel diffraction theory in Chapter 2, as

$$\eta_1 = 1 - \exp(-A_1).\tag{7.5.6}$$

Therefore, the signal level, or the efficiency η_2, becomes

$$\eta_2 = \left|\frac{\sqrt{A_1 A_2}\left[1 - \exp\left(-\frac{A + A - 2iu}{2}\right)\right]}{(A_1 + A_2 - 2iu)\sqrt{1 - \exp(-A_1)}}\right|^2.\tag{7.5.7}$$

When $A_1 \rightarrow 0$, e. g., $r_{01} \rightarrow 0$, Eq. (7.5.7) reduces to, for $u = 0$,

$$\eta_2 = \frac{4\left[1 - \exp(-A_2 / 2)\right]^2}{A_2},\tag{7.5.8}$$

which has the same form as Eq. (20-30) by Snyder et al.[7.9] This is understandable that when $A_1 \rightarrow 0$, the light beam after the objective becomes approximately a uniform parallel beam. This is the condition under which Snyder et al. derived Eq. (7.5.8). Eq. (7.5.8) implies that the maximum signal level η_2 is 81% when $A_2 \approx 2.56$.

As a comparison, the signal level for a finite-sized detector can be calculated from Eqs. (4.3.16) and (4.3.21) for circular and slit detectors, respectively. In the former case, it becomes, when the reflector is placed in the focal plane,

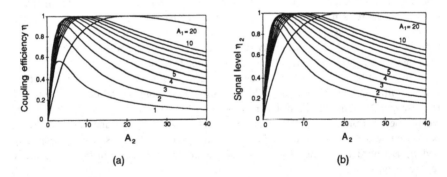

(a)   (b)

Fig. 7.5.1   Signal as a function of A_2: (a) coupling efficiency (i. e., total power transformation) η;
(b) signal level η_2.

$$\eta_2 = 1 - J_0^2(v_d) - J_1^2(v_d), \tag{7.5.9}$$

which approaches unity after normalization when the detector is large.

The in-focus efficiency η of total power transformation as a function of the parameter A_2 for different values of A_1 is shown in Fig. 7.5.1a. For a given value of A_1, each η - A_2 curve shows a maximum. The magnitude of the maximum and the value of A_2 at which the maximum occurs increase with the values of A_1. Fig. 7.5.1b gives the in-focus efficiency η_2, which has been called the signal level of the microscope for a given source. The behaviour of the plots in Fig. 7.5.1b is similar to that in Fig. 7.5.1a. The difference between them is that the scale of the maximum is increased by a factor of [1-exp($-A_1$)] in the latter case. Therefore, η for $A_1 = 0$ approaches zero, while η_2 for the same condition is given by Eq. (7.5.8) and shown as a dashed line in Fig. 7.5.2. The appearance of these maxima in Figs. 7.5.1a and 7.5.1b is due to the matching between the fibre mode profile and the electric field distribution on the entrance face of the fibre F_2. It suggests two points. First, the magnitude of the signal level depends on the proper matching. Second, for arbitrarily given values of A_1 and A_2, one cannot obtain the maximum signal level because of the mismatching between the fibre mode profile and the field distribution on the entrance face of the fibre F_2. However, for a given value of A_1, a proper matching between them may be achieved by choosing the value of A_2.

In order to compare η_2 with the signal levels of the confocal scanning microscope of a pinhole detector or of a slit detector, we set $A_2 = (v_d)^2$, where $v_d = 2\pi a r_{02}/(\lambda d_1)$, denoting the normalized fibre spot size. For a point source, the signal level η_2 is given by Eq. (7.5.8). In this case, a comparison of the signal levels as a function of v_d is displayed in Fig. 7.5.2 for a pinhole detector, a slit detector and a detector with a fibre. It should be pointed out that the definition of v_d in the cases of pinhole and slit detectors is $v_d = 2\pi a r_d/(\lambda d_1)$ and $v_d = 2\pi a x_d/(\lambda d_1)$, where r_d and x_d are the radius of the pinhole and the

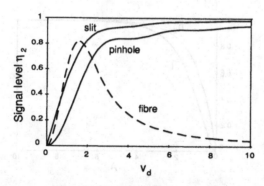

Fig. 7.5.2   Comparison of signal level of fibre, pinhole and slit detection.

half width of the slit, respectively. Although the definitions of the relative magnitudes of v_d for the different cases are arbitrary, we can see some interesting comparisons. The numerical result confirms that only 81% of power is converted to the single-mode fibre when $v_d \approx 1.6$. When $v_d \to \infty$, η_2 for fibre detection approaches zero because of the mode mismatching, while that for pinhole and slit detection becomes unity.

The measured data together with theoretical predictions under the experimental conditions are placed in Fig. 7.5.3. For $A_1 = 0.43$, it is seen that a maximum signal level occurs in fibre detection. The appearance of the maximum signal level for fibre detection gives an advantage when one optimizes the axial resolution.[7.3]

When $A_1 = A_2 = A$, which is also the case if the system employs a single fibre for illumination and collection together with a fibre coupler instead of a beam splitter,[7.39] the

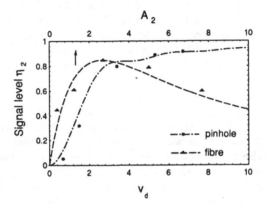

Fig. 7.5.3   Measured signal level of fibre detection and pinhole detection.

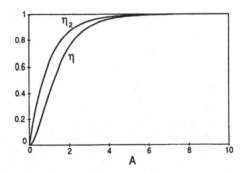

Fig. 7.5.4   Signal level as a function of A when $A_1 = A_2 = A$.

maxima of η and η_2 disappear, and the efficiencies η and η_2 eventually approach unity (see Fig. 7.5.4). This property shows that the input power from the fibre F_1 cannot be collected by the fibre F_2 because the other higher-order modes in the fibre can be excited due to the mismatching between the fibre mode profile and the field distribution on the fibre F_2. Only when both illumination and collection fibres are the same and the radius of the lens is very large, can all power from the input be transformed to the collection fibre.

For a given value of A_1, defocused efficiencies of power transformation as a function of A_2, shown in Fig. 7.5.5, demonstrate that as the axial optical coordinate u, i.e., the defocus distance from the focal plane, increases, the efficiency η and the signal level η_2 become very much lower than those for the in-focus case, and that maxima still

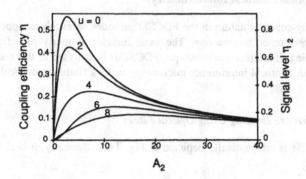

Fig. 7.5.5   Defocused η and η_2 as a function of A_2 when $A_1 = 1$.

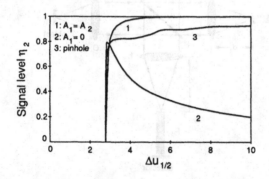

Fig. 7.5.6   Signal level η_2 as a function of the half width $\Delta u_{1/2}$ of the axial response to a perfect reflector.

exist on the curves when the defocus distance u is not very large ($u \leq 8$).

In order to see the practical significance of Fig. 7.5.5, it is helpful to estimate the real defocus distance z for a practical case. When $\lambda = 0.633$ µm and for an objective with a numerical aperture of 0.54, the real defocus distance z in units of µm is about equal to 0.4 times the axial optical distance u. In other words, $z = 4$ µm if $u = 10$.

For a given value of fibre spot size, the signal level is a function of the half width $\Delta u_{1/2}$ of the axial response to a reflector, which is shown in Fig. 7.5.6 for pinhole and fibre detection. Interestingly, using fibre detection only ($A_1 = 0$) results in a slightly higher signal level compared with that for pinhole detection while the axial resolution nearly maintains the same value as using point detection. This phenomenon has been observed in the experiment.[7.3]

7. 6 Fibre-Optical Confocal Interferometry

The purely-coherent nature in the FOCSM provides an excellent opportunity for performing interference microscopy. The axial imaging property of a fibre-optical confocal scanning interference microscope (FOCSIM) is analysed in this section and a comparison with confocal interference microscopy using a finite-sized circular detector is also given.

7. 6. 1 Interference Imaging Using Optical Fibres

A FOCSIM is schematically depicted in Fig. 7.6.1. Basically, it is similar to the

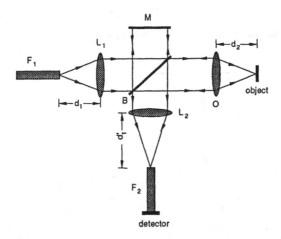

Fig. 7.6.1   Schematic diagram of a fibre-optical confocal scanning interference microscope using two lengths of single-mode fibres.

geometry of the FOCSM. The difference between them is that a reference beam is introduced via a beamsplitter (B) and a mirror (M) in order to perform interference imaging in the FOCSM. Two lengths of single-mode optical fibres (F_1 and F_2) are employed as the illumination source and for signal collection instead of a finite-sized source and a finite-sized detector. The light from the illumination fibre F_1 is collimated via a lens L_1 and then focused with an objective (O) onto the object. The reflected signal beam interfering with the reference beam is collected by the fibre F_2.

As the tip of an optical fibre has finite-sized cross-section, it accordingly affects the properties of the image formation in confocal interference microscopy. Suppose that $U_r(r)$ and $U_s(r, r_s)$ are the amplitudes of the reference and signal beams on the tip of the collection fibre F_2, where r is the radial coordinate on the tip of the fibre F_2 and r_s is the scan position at the object. The measured intensity of the interference image can be expressed, in terms of Eq. (7.2.6) and (7.2.7), as

$$I(r_s) = \left| \int_0^\infty f_2^*(r)[U_r(r) + U_s(r, r_s)]r dr \right|^2 , \tag{7.6.1}$$

where $f_2(r)$ is the mode profile of the collection fibre. Eq. (7.6.1) can be rewritten as

$$I(r_s) = |R + U(r_s)|^2 . \tag{7.6.2}$$

Here

$$R = \int_0^\infty f_2^*(r)U_r(r)r dr \tag{7.6.3}$$

represents the uniform reference beam but the strength of the reference beam R is dependent on the distribution of $U_r(r)$. The second term in Eq. (7.6.2)

$$U(r_s) = \int_0^\infty f_2^*(r)U_s(r, r_s)r dr \tag{7.6.4}$$

is the amplitude of the signal beam. Eq. (7.6.2) implies that the detected intensity is given by the modules squared of the sum of the amplitudes of the reference and signal beams. This conclusion is consistent with the purely-coherent nature of the FOCSM.

For a perfect reflector scanned in the axial direction, the amplitude of the axial response in the FOCSM using single-mode fibres of Gaussian profiles is determined, according to Eq. (7.3.2), by the defocus CTF $c_r(0, u)$, and is denoted by $U(u)$ in this section. For two identical circular lenses, one has

$$U(u) = \exp\left[-\frac{iu}{2\sin^2(\alpha_o/2)}\right] \frac{\frac{A_1+A_2}{2}\left[1-\exp\left(-\frac{A_1+A_2}{2}+iu\right)\right]}{\left[1-\exp\left(-\frac{A_1+A_2}{2}\right)\right]\left(\frac{A_1+A_2}{2}+iu\right)}, \qquad (7.6.5)$$

where the defocused phase factor resulting from $s0$ has been explicitly included because of its importance in interference imaging.

The calculated interference axial responses to a perfect reflector are shown in Fig. 7.6.2 for $R = 1$. Since $U(u)$ is dependent only on the sum of A_1 and A_2, we let $A = (A_1 + A_2)/2$. When $A = 0$, corresponding to point source illumination and point detection, the interference fringes are pronounced. The intensity of the fringes is decreased with the defocus distance due to the optical section property. As the value of A increases, the

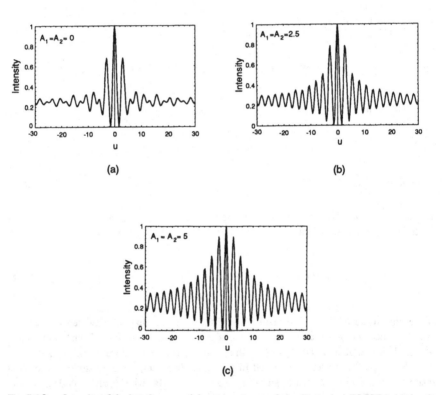

(a)

(b)

(c)

Fig. 7.6.2   Intensity of the interference axial response to a perfect reflector in a FOCSIM: (a) $A = 0$; (b) $A = 2.5$; (c) $A = 5$.

envelope of the interference axial response becomes broad but the interference fringes are still well observed. To describe the interference fringes, we introduce the visibility Δ of the interference fringe, which is defined as

$$\Delta = \frac{I_{max} - I_{min}}{I_{max} + I_{min}}, \qquad (7.6.7)$$

where I_{max} represents the maximum intensity and I_{min} is the first minimum intensity from the central peak. In terms of this definition, the visibility of the interference axial response in the FOCSIM is unity, independent of the system parameter A (see Fig. 7.6.3). This property indicates that optical fibres of finite-sized cross-section do not change the visibility of the confocal interference images. The reason for this phenomenon is that the FOCSM is a purely-coherent imaging system. In this respect, one may expect that the interference axial response behaves differently from that in a confocal system using a finite-sized pinhole, which will be described in Section 7.6.2.

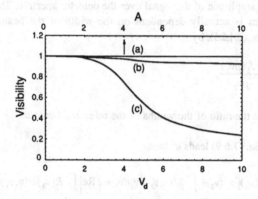

Fig. 7.6.3   Visibility Δ defined in Eq. (7.6.7) as a function of A and v_d: (a) FOCSIM; (b) confocal microscope with a finite-sized detector when $p = 1$; (c) confocal microscope with a finite-sized detector when $p = 0$.

7. 6. 2   Interference Imaging Using a Finite-Sized Circular Detector

For comparisons, let us consider a confocal system consisting of a point source and a finite-sized circular detector. In this case, the interference image is given, from Eq. (4.1.1), by

$$I(r_s) = \int_0^{v_4} |R(v_2) + U_s(v_2, r_s)|^2 v_2 dv_2, \tag{7.6.8}$$

where v_2, defined in Eq. (3.1.18), is the radial coordinate over the detector aperture. $U_s(v_2, r_s)$ is the signal amplitude on the detector. It is seen that the interference image depends on the distribution of the reference beam $R(v_2)$, which contrasts with that in the FOCSM. The interference axial response to a perfect reflector is then given by

$$I(u) = \int_0^{v_4} |R(v_2) + U(u, v_2)|^2 v_2 dv_2, \tag{7.6.9}$$

where $U(u, v_2)$ is the amplitude from the perfect reflector and can be found from Eqs. (4.3.14) and (4.3.15):

$$U(u, v_2) = \exp\left[-\frac{iu}{2\sin^2(\alpha_o/2)}\right] \int_0^1 2J_0(v_2\rho)\exp(iu\rho^2)\rho d\rho, \tag{7.6.10}$$

which denotes the amplitude of the signal over the detector aperture. The distribution of the reference beam is actually dependent on the width of the beam, which can be expressed, from Eq. (2.1.25), by

$$R(v_2) = R_0 \frac{2J_1(pv_2)}{pv_2}, \tag{7.6.11}$$

where p represents the ratio of the widths of the reference beam to the signal beam and R_0 is a constant.

Expanding Eq. (7.6.9) leads to

$$I(u) = \int_0^{v_4} |R(v_2)|^2 v_2 dv_2 + \int_0^{v_4} |U(u, v_2)|^2 v_2 dv_2 + 2\,\mathrm{Re}\left[\int_0^{v_4} R(v_2)U(u, v_2)v_2 dv_2\right]. \tag{7.6.12}$$

These three terms, denoted by I_1, I_2 and I_3, respectively, represent the intensity of the reference beam, the intensity of the signal beam, and the intensity of the interference fringes. It is seen that I_1 is a constant for both a point detector and an infinitely-large detector. The second term I_2 is the normal axial response to a perfect reflector and has been discussed in Section 4.4. It is therefore given by Eq. (3.2.24) for $v_d = 0$ and a constant for $v_d \rightarrow \infty$. The last term I_3 can be expressed, for $v_d = 0$, as

$$I_3 = R_0 \cos\left[\frac{u}{2} - \frac{u}{2\sin^2(\alpha_o/2)}\right]\frac{\sin(u/2)}{u/2}. \tag{7.6.13}$$

On the other hand, when $v_d \to \infty$, I_3 becomes

$$I_3 = 2\,\mathrm{Re}\left\{\exp\left[-\frac{iu}{2\sin^2(\alpha_o/2)}\right]\int_0^\infty\left[R(v_2)\int_0^1 2J_0(v_2\rho)\exp(iu\rho^2)\rho\,d\rho\right]v_2\,dv_2\right\}. \quad (7.6.14)$$

The integration in Eq. (7.6.14) with respect to v_2 is a two-dimensional (2-D) Hankel transform of $R(v_2)$ and is therefore equal to $P(\rho/p)$, according to Eq. (7.6.11). We thus find that Eq. (7.6.14) for $p = 1$ is the same as Eq. (7.6.13) except a constant factor. It is therefore concluded that the interference term is identical for $v_d = 0$ and $v_d \to \infty$.

A confocal microscope with a finite-sized value of v_d is a partially-coherent system (Chapter 4). Thus, the interference axial response may be degraded on increasing the size of the pinhole. In the following, the intensity of the in-focus signal beam is assumed to

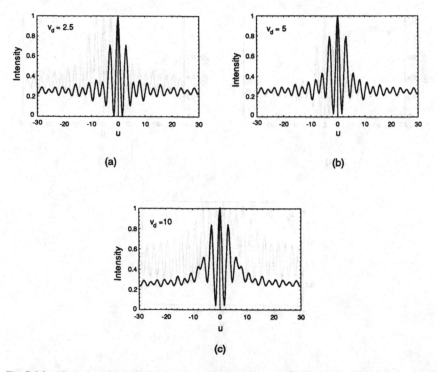

(a)

(b)

(c)

Fig. 7.6.4   Intensity of the interference axial response to a perfect reflector in a confocal microscope with a finite-sized detector when $p = 1$: (a) $v_d = 2.5$; (b) $v_d = 5$; (c) $v_d = 10$.

be the same as that of the reference beam. Fig. 7.6.4 gives the interference axial responses for $v_d = 2.5$, 5 and 10 when $p = 1$ corresponding to the case when the width of the reference beam is the same as that of the signal beam. Note that the minimum intensity of the interference fringes becomes non-zero and slowly increases on increase of v_d. Therefore, the visibility Δ of the interference axial response is, according to the definition in Eq. (7.6.7), decreased with v_d (see curve b in Fig. 7.6.3). Eventually, the visibility approaches a limiting value of about 0.91 for a confocal system of numerical aperture 0.8 with a large detector, in which case imaging is equivalent to that in a conventional microscope with two equal lenses, as pointed out in Section 4.3.

If $p \ll 1$, the width of the reference beam is much smaller than that of the signal beam. In this case, Eq. (7.6.11) decreases very slowly with v_2. In the limiting case, i. e., $p = 0$, the reference beam is independent of v_2 and can be considered to be a constant. The interference axial responses in the case are depicted in Fig. 7.6.5. They are strongly

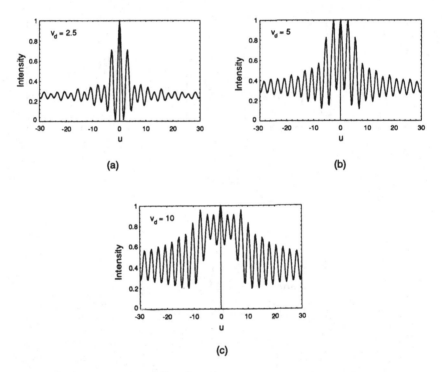

Fig. 7.6.5   Intensity of the interference axial response to a perfect reflector in a confocal microscope with a finite-sized detector when $p = 0$: (a) $v_d = 2.5$; (b) $v_d = 5$; (c) $v_d = 10$.

degraded on increase of the pinhole size: the decrease of the envelopes of the signal is not monotonic and the first minimum intensity increases quickly when the detector size becomes large. Accordingly, the visibility Δ of the interference fringes drops more quickly than that in the previous case (see curve c in Fig. 7.6.3). The reason for the difference between Figs. 7.6.4 and 7.6.5 is that the reference beam in the former case is in phase with the signal beam near the focal plane. For an intermediate value of p ($p = 0.5$), the interference axial response is given in Fig. 7.6.6, showing less degradation.

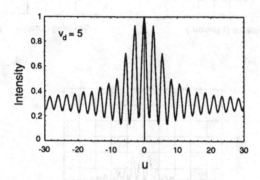

Fig. 7.6.6   Intensity of the interference axial response to a perfect reflector in a confocal microscope with a finite-sized detector when $p = 0.5$ and $v_d = 5$.

The measured interference axial responses in a FOCSM and a confocal microscope with a finite-sized detector for $p = 0.5$ are displayed in Fig. 7.6.7. The interference response (Fig. 7.6.7a) for a pinhole of $v_d = 0.34$ ($r_d = 2.5$ μm), approximately corresponding to a point detector, shows a visibility of 0.96 as imaging in this case is almost purely coherent. When optical fibres are used (Fig. 7.6.7b), the visibility of the fringes is unity, although the envelope of the interference axial response becomes broad, as expected. This agrees with the theoretical prediction. For the confocal system with a large pinhole of $v_d = 5.4$ ($r_d = 40$ μm), the interference fringes are degraded with a visibility of 0.48 (Fig. 7.6.7c). When compared with Figs. 7.6.2a, 7.6.2c and 7.6.6, the measured interference axial responses in Fig. 7.6.7 qualitatively demonstrate the theoretical predictions. The discrepancy between the measured data and the theoretical predictions may be caused by spherical aberration in the system. This aberration also leads to the asymmetry of the observed interference axial response.[7.3, 7.12]

The interference axial images were also observed in a FOCSM including a fibre coupler,[7.39] as shown in Fig. 7.1.2. The interference axial response can be used to derive the aberration of the objective as well as of the system.[7.18, 7.39]

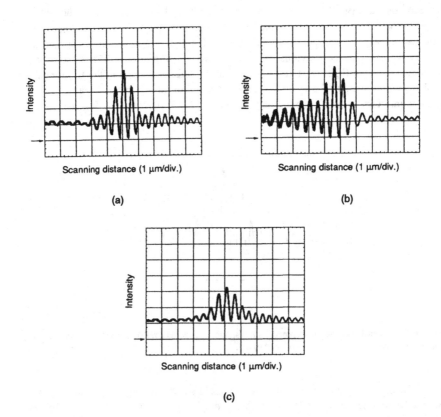

Fig. 7.6.7   Intensity of the measured interference axial response to a perfect reflector: (a) in a confocal microscope with a small detector for $v_d = 2.5$ ($p = 1$); (b) in a FOCSM for $A_1 = 0.43$ and $A_2 = 11.2$ ($p = 1$); (c) in a confocal microscope with a finite-sized detector for $v_d = 5.4$ ($p = 0.5$). The arrows indicate the position of the zero intensity.

7.7 Fluorescence Imaging

If the sample in a FOCSM is labelled with fluorescent materials, the purely-coherent nature is completely destroyed due to the interaction of the light with the sample. As a result, imaging performance in the fluorescence FOCSM is different from that discussed in Sections 7.2-7.6. We assume that the objective and collector lenses have the same focal length. Only single-photon fluorescence imaging is considered and the fluorescence wavelength is assumed to be identical to the incident wavelength. Readers can use the following method and the description presented in Chapter 6 to analyse the imaging properties in two-photon fluorescence FOCSM.

7.7.1 Incoherent Image Formation

Let $f_1(x_0, y_0)$ be the amplitude profile on the output end of the fibre F_1. The intensity distribution at a point $r_1(x_1, y_1, z_1)$ in the object space may be expressed, from Eq. (7.2.1), as

$$I_o(r_1) = \left| \int_{-\infty}^{\infty} f_1(x_0, y_0)\delta(z_0)\exp[ik(z_0 - z_1)]h_1(r_0 + M_1 r_1)dr_0 \right|^2. \tag{7.7.1}$$

Here $r_0(x_0, y_0, z_0)$ is shown in Fig. 7.2.1.

If a thick object can emit fluorescence light with a strength $o_f(r)$, the total fluorescence strength emitted from the point r_1 of the object under the illumination $I_o(r_1)$ can be expressed as

$$I_1(r_1) = \left| \int_{-\infty}^{\infty} f_1(x_0, y_0)\delta(z_0)\exp[ik(z_0 - z_1)]h_1(r_0 + M_1 r_1)dr_0 \right|^2 o_f(r_s - r_1), \tag{7.7.2}$$

where the scanning movement of the object has been introduced by including a scan position $r_s(x_s, y_s, z_s)$.

On the other hand, according to Eqs. (7.2.4)-(7.2.6), for a single point object at r_1 in the FOCSM, the detected intensity is given, after the signal passes through the collector and the fibre F_2, by

$$I_1(r_1) = \left| \int_{-\infty}^{\infty} f_2^*(x_2, y_2)\delta(z_2)\exp[ik(\pm z_1 - z_2)]h_2(r_1 + M_2 r_2)dr_2 \right|^2, \tag{7.7.3}$$

where $f_2(x_2, y_2)$ denotes the amplitude profile on the input end of the fibre F_2.

Therefore, the total intensity from the object with the fluorescence strength $I_1(r_1)$, detected by an incoherent detector D, is the integration of the product of $I_1(r_1)$ and $I_2(r_1)$ with respect to r_1, i. e., the detected intensity from the scan point r_s is given by

$$I(r_s) = \int_{-\infty}^{\infty} \left[\left| \int_{-\infty}^{\infty} f_1(x_0, y_0)\delta(z_0)\exp[ik(z_0 - z_1)]h_1(r_0 + M_1 r_1)dr_0 \right|^2 o_f(r_s - r_1) \right.$$

$$\left. \left| \int_{-\infty}^{\infty} f_2^*(x_2, y_2)\delta(z_2)\exp[ik(\pm z_1 - z_2)]h_2(r_1 + M_2 r_2)dr_2 \right|^2 \right] dr_1, \tag{7.7.4}$$

which can be rewritten as a form of the 3-D convolution operation:

$$I(r_s) = h_i(r_s) \otimes_3 o_f(r_s),$$ (7.7.5)

where $\otimes_3$ denotes the 3-D convolution operation and $h_i(r_s)$ is the 3-D effective intensity point spread function in the FOCSM given by

$$h_i(r) = \left| f_1(M_1 x, M_1 y) \otimes_2 h_1(M_1 r) \right|^2 \left| f_2^*(M_1 x, M_1 y) \otimes_2 h_2(r) \right|^2,$$ (7.7.6)

which is the modulus squared of the 3-D effective amplitude point spread function given in Eq. (7.2.10). Here $\otimes_2$ represents the 2-D convolution operation in the transverse plane. Eqs. (7.7.4) and (7.7.5) indicate the incoherent imaging nature in a fibre-optical confocal fluorescence microscope.

7. 7. 2   Optical Transfer Function

Eq. (7.7.5) has the same form as Eq. (5.1.6), so that the 3-D optical transfer function (OTF) is given by the 3-D inverse Fourier transform of the effective intensity point spread function, as described in Eq. (5.2.3). Consequently, the 3-D OTF, $C(m)$, in the FOCSM may be derived as

$$C(m) = F_3 \left\{ \left| f_1(M_1 x, M_1 y) \otimes_2 h_1(M_1 r) \right|^2 \left| f_2^*(M_1 x, M_1 y) \otimes_2 h_2(r) \right|^2 \right\},$$ (7.7.7)

which can be rewritten as

$$C(m) = C_1(m) \otimes_3 C_2(m),$$ (7.7.8)

where

$$C_1(m) = F_3 \left\{ \left| f_1(M_1 x, M_1 y) \otimes_2 h_1(M_1 r) \right|^2 \right\},$$ (7.7.9)

and

$$C_2(m) = F_3 \left\{ \left| f_2^*(M_1 x, M_1 y) \otimes_2 h_2(r) \right|^2 \right\}.$$ (7.7.10)

It can be seen that $C_1(m)$ and $C_2(m)$ are related to the objective and collector lenses, respectively, and termed the 3-D effective OTF for objective and collector lenses.

The significance of $C_1(m)$ (or $C_2(m)$) is that it can be considered to be the 3-D OTF for a single circular lens of a pupil function P_1 (or P_2) which has a shaded function $\tilde{f}_1$ (or $\tilde{f}_2$). Thus the effective pupil function for the lens falls off in the radial direction, which effectively reduces the size of the pupil function and accordingly reduces the values of the OTF at higher spatial frequencies propagating at larger angles with respect to the optical axis. The larger the fibre spot size (i. e., the value of A_1 (or A_2)), the stronger the reduction of the OTF. Eventually, when A_1 (or A_2) approaches infinity, corresponding to a point pupil function for the lens, the OTF becomes a δ-function at the origin in Fourier space. Physically, this means that only the light propagating along the optical axis can be imaged.

Let us consider the solution of Eq. (7.7.9) first. For circular fibres, Eq. (7.7.9) can be reduced, by considering the cylindrical symmetry, to

$$C_1(l,s) = \int_{-\infty}^{\infty} C_1(l,z)\exp(-2\pi izs)dz, \tag{7.7.11}$$

where $l = (m^2 + n^2)^{1/2}$ denotes the radial spatial frequency. Here $C_1(l, z)$ is given by

$$C_1(l,z) = \left[\tilde{f}_1\left(\frac{l}{M_1}\right)P_1(\lambda d_2 l, -M_1^2 z)\right] \otimes_2 \left[\tilde{f}_1^*\left(\frac{l}{M_1}\right)P_1^*(\lambda d_2 l, -M_1^2 z)\right], \tag{7.7.12}$$

which can be expressed as

$$C_1(l,u) = \left[\tilde{f}_1\left(\frac{la}{\lambda d_1}\right)P_1(l,u)\right] \otimes_2 \left[\tilde{f}_1^*\left(\frac{la}{\lambda d_1}\right)P_1^*(l,u)\right] \tag{7.7.13}$$

if the normalization factors in Eqs. (2.4.8) and (2.4.9) are used. It is clear that Eq. (7.7.13) is identical to the 3-D CTF presented in Eq. (7.2.28) for a transmission FOCSM with two equal fibres if the pupil function for the collector lens is the same as the objective, i. e., if $P_1^*(l,u) = P_2(l,-u)$. After understanding this relation, one can derive the solution to Eq. (7.7.13) for an annular lens as

$$C_1(l,s) = \frac{\sqrt{\pi}\exp\left\{-A_1\left[\frac{l^2}{4} + \left(\frac{s}{l}\right)^2\right]\right\}}{\sqrt{A_1}\,l}\left\{\text{erf}\left[\sqrt{A_1}\,\text{Re}(y_0)\right] - \text{erf}\left[\sqrt{A_1}\,\text{Re}(y_1)\right]\right\}, \tag{7.7.14}$$

where $|s| \le (1-\varepsilon_1^2)/2$, A_1 is defined in Eq. (7.2.27a) and

$$y_0 = \sqrt{1 - \left(\frac{|s|}{l} + \frac{l}{2}\right)^2} , \tag{7.7.15}$$

$$y_1 = \sqrt{\varepsilon_1^2 - \left(\frac{|s|}{l} - \frac{l}{2}\right)^2} . \tag{7.7.16}$$

In a similar way, we can derive the 3-D OTF $C_2(m)$ for an annular collector lens and a circular fibre. In particular, if the fluorescence wavelength is the same as the incident wavelength and the outer radius of the collector lens is the same as that of objective lens, one has

$$C_2(l,z) = \left[\tilde{f}_2^*\left(\frac{l}{M_1}\right) P_2(\lambda d_2 l, z)\right] \otimes_2 \left[\tilde{f}_2\left(\frac{l}{M_1}\right) P_2^*(\lambda d_2 l, z)\right] \tag{7.7.17}$$

or

$$C_2(l,u) = \left[\tilde{f}_2^*\left(\frac{la}{\lambda d_1}\right) P_2(l,u)\right] \otimes_2 \left[\tilde{f}_2\left(\frac{la}{\lambda d_1}\right) P_2^*(l,u)\right], \tag{7.7.18}$$

so that $C_2(l, s)$ can be analytically expressed as

$$C_2(l,s) = \frac{\sqrt{\pi}\exp\left\{-A_2\left[\frac{l^2}{4} + \left(\frac{s}{l}\right)^2\right]\right\}}{\sqrt{A_2}\, l}\left\{\mathrm{erf}\left[\sqrt{A_2}\,\mathrm{Re}(y_0)\right] - \mathrm{erf}\left[\sqrt{A_2}\,\mathrm{Re}(y_2)\right]\right\} \tag{7.7.19}$$

with $|s| \le (1 - \varepsilon_2^2)/2$. Here A_2 is given by Eq. (7.2.27b) and

$$y_2 = \sqrt{\varepsilon_2^2 - \left(\frac{|s|}{l} - \frac{l}{2}\right)^2} . \tag{7.7.20}$$

Finally, the 3-D optical transfer function $C(m)$ for the fluorescence FOCSM can be expressed, for different annular objective and collector lenses and different circular illumination and collection fibres, as

$$C(l,s) = \iiint_V \frac{\pi}{\sqrt{A_1}\sqrt{A_2}} \frac{1}{l_1 l_2}$$

$$\exp\left\{-A_1\left[\frac{l_1^2}{4} + \left(\frac{s'-s/2}{l_1}\right)^2\right]\right\}\left\{\text{erf}\left[\sqrt{A_1}\,\text{Re}(y_1')\right] - \text{erf}\left[\sqrt{A_1}\,\text{Re}(y_2')\right]\right\}$$

$$\exp\left\{-A_2\left[\frac{l_2^2}{4} + \left(\frac{s'+s/2}{l_2}\right)^2\right]\right\}\left\{\text{erf}\left[\sqrt{A_2}\,\text{Re}(y_3')\right] - \text{erf}\left[\sqrt{A_2}\,\text{Re}(y_4')\right]\right\} dm'\, dn'\, ds',$$

$$(7.7.21)$$

which is cut off at $|s| = 1 - (\varepsilon_1^2 + \varepsilon_2^2)/2$ like Eq. (5.2.16). Hence V, l_1 and l_2 have the same definition as those in Eq. (5.2.16). In Eq. (7.7.21), we have

$$y_1' = \sqrt{1 - \left(\frac{|s'-s/2|}{l_1} + \frac{l_1}{2}\right)^2},$$

$$(7.7.22a)$$

$$y_2' = \sqrt{\varepsilon_1^2 - \left(\frac{|s'-s/2|}{l_1} - \frac{l_1}{2}\right)^2},$$

$$(7.7.22b)$$

$$y_3' = \sqrt{1 - \left(\frac{|s'+s/2|}{l_2} + \frac{l_2}{2}\right)^2},$$

$$(7.7.22c)$$

$$y_4' = \sqrt{\varepsilon_2^2 - \left(\frac{|s'+s/2|}{l_2} - \frac{l_2}{2}\right)^2}.$$

$$(7.7.22d)$$

It is seen that the 3-D OTF is a function of the radii of the central obstruction and the fibre spot sizes. Eq. (7.7.21) is a general expression for the 3-D OTF in the FOCSM with different fibre spot sizes for fibres F_1 and F_2 and different central obstructions for annular lenses P_1 and P_2.

It is important to note that the 3-D OTF in Eq. (7.7.21) is always a positive function because the fibre mode has been assumed to be a Gaussian function which gives a positive Fourier spectrum. In this respect, it is different from that in the fluorescence confocal scanning microscope with a finite-sized detector (Chapter 5). The latter exhibits a negative tail in the 3-D OTF when the detector size is larger. Therefore, introducing optical fibres into a confocal system removes negative values in the 3-D OTF, thus avoiding imaging artefacts. In addition, the 3-D OTF in Eq. (7.7.21) does not change value on exchange of ε_1 and ε_2, if $A_1 = A_2$, implying that the fluorescence imaging property in the FOCSM remains unchanged when exchanging the objective and the collector. This is understandable because the system has the optical reciprocity when $A_1 = A_2$. However in the fluorescence confocal microscope with a point source and a finite-sized detector only the axial imaging property is unchanged on exchanging the objective and the collector (see Eq. (5.2.26)).

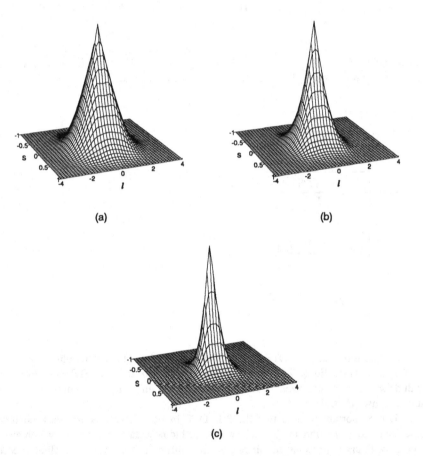

(a)   (b)

(c)

Fig. 7.7.1   3-D optical transfer function for a fluorescence FOCSM consisting of two identical fibres: (a) $A = 1$; (b) $A = 5$; (c) $A = 10$.

7. 7. 3   Effect of Fibre Spot Size

To understand the effect of the fibre spot size on the 3-D OTF, we depict the 3-D OTF, normalized to unity at the origin, for different values of A in Fig. 7.7.1, where two fibres for illumination and collection are assumed to be the same ($A_1 = A_2 = A$). As expected, the size of the 3-D OTF is gradually decreased as A increases. This is because the effective pupil functions of the objective and collector lenses are shaded by a Gaussian function. The larger the value of A the stronger the shading effect. As a result, the effective cut-off spatial frequencies in the transverse and axial directions are reduced, as shown in Fig. 7.7.2.

One of the important features in Fig. 7.7.1 is that there is no missing cone of spatial frequencies in the axial direction when A increases. This property contrasts with the confocal fluorescence microscope with a finite-sized detector (see Figs. 5.3.8a and 5.3.11b), which can be understandable from Eq. (7.7.21). The 3-D OTF for the fluorescence FOCSM is given by the convolution of two 3-D OTFs effectively contributed from two lenses shaded by $\tilde{f}_1$ and $\tilde{f}_2$, respectively. If $A_1 = A_2 = A$, these two OTFs are identical as shown in Fig. 7.2.5. The final OTF for the fluorescence FOCSM does not show the missing cone of spatial frequencies because both lenses play an equal role in confocal imaging. However, the function of the collector lens in a confocal fluorescence microscope with a finite-sized detector is degraded by the Fourier transform of the detector sensitivity as described in Eq. (5.2.4). Therefore, the objective and collector lenses play an unequal role.

On the other hand, if A_1 is not equal to A_2 and the value for one of them is quite large, the lens with a strong shaded function becomes less important for imaging. Consequently, the 3-D OTF for the FOCSM is mainly determined by one lens with the

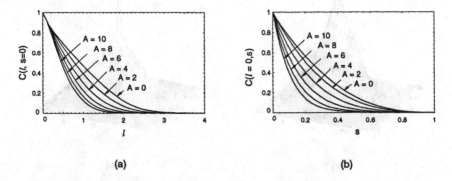

(a)   (b)

Fig. 7.7.2   Transverse (a) and axial (b) cross-sections of the 3-D OTF for a fluorescence FOCSM consisting of two identical fibres.

less shaded function. As an example, the 3-D OTF for $A_1 = 1$ and $A_2 = 20$ is shown in Fig. 7.7.3. As expected, a missing cone of spatial frequencies is apparent in the axial direction and the size of the 3-D OTF is reduced.

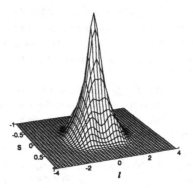

Fig. 7.7.3   3-D optical transfer function for a fluorescence FOCSM consisting of two unequal fibres $(A_1 = 1$ and $A_2 = 20)$.

7. 7. 4   Effect of Annular Pupils

To reveal the dependence of the 3-D OTF on the radii of the central obstruction,

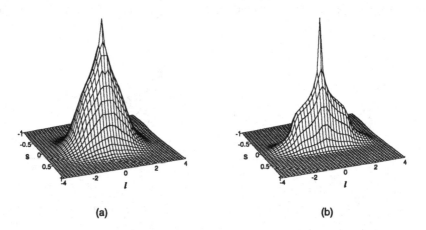

Fig. 7.7.4   3-D optical transfer function for a fluorescence FOCSM consisting of two identical fibres $(A = 1)$ and of two equal annular lenses: (a) $\varepsilon = 0.5$; (b) $\varepsilon = 0.75$.

we assume, for simplicity, that $A_1 = A_2 = A$. All 3-D OTFs in this sub-section are normalized by $C(l = 0, s = 0)$.

We first consider the FOCSM in which both the objective and the collector have the same radius of central obstruction, that is $\varepsilon_1 = \varepsilon_2 = \varepsilon$. For $A = 1$, Fig. 7.7.4a shows the 3-D OTF for $\varepsilon_1 = \varepsilon_2 = 0.5$, while Fig. 7.7.4b represents the 3-D OTFs for $\varepsilon_1 = \varepsilon_2 = 0.75$. Comparing Fig. 7.7.4 with Figs. 5.3.1c and 5.3.1d in Chapter 5, we find that they are quite similar for a given value of ε. This implies that when the fibre spot size is small the fluorescence imaging properties in the FOCSM are similar to those in the confocal fluorescence scanning system with a point source and a point detector. Therefore, the axial and transverse cut-off spatial frequencies are approximately 1 and 4 for $A = 1$ and for a small value of ε. As the radius of the central obstruction increases, the area where the 3-D OTF is appreciable becomes narrower and is cut off at $1 - (\varepsilon_1^2 + \varepsilon_2^2)/2$ in the s direction, while it produces a superior response at high spatial frequencies in the transverse direction. The transverse resolution in the annular system is therefore improved, as expected.

As the fibre spot size increases, the responses of the OTF at higher axial and transverse spatial frequencies become weaker, so that the cut-off spatial frequencies are effectively decreased in both axial and transverse directions. Nevertheless, by increasing the radius of the central obstruction the transverse response of the OTF can be improved (Figs. 7.7.5a and 7.7.5b).

It should be noted from Figs. 7.7.4 and 7.7.5 that there is no missing cone of spatial frequencies even for larger values of the fibre spot size, as expected. Hence, in the fluorescence FOCSM, more axial information can be imaged, compared with a confocal

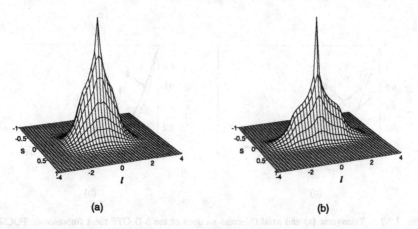

(a)   (b)

Fig. 7.7.5   3-D optical transfer function for a fluorescence FOCSM consisting of two identical fibres ($A = 10$) and of two equal annular lenses: (a) $\varepsilon = 0.5$; (b) $\varepsilon = 0.75$.

fluorescence microscope with a finite-sized detector.

The variations of the cross-sections of the 3-D OTF for $s = 0$ and $l = 0$ with the parameter A (fibre spot size) are displayed in Fig. 7.7.6. It is seen that for a given radius of the central obstruction, for example for $\varepsilon_1 = \varepsilon_2 = 0.5$, both transverse and axial cross-sections of the 3-D OTF become poor on increasing the value of A.

For a given value of A, the transverse cross-section of the OTF is improved by increasing the radius of the central obstruction (Fig. 7.7.7a). The larger the value of the radius of the central obstruction the better the response at higher transverse spatial frequencies. The transverse resolution can therefore be improved by using annular lenses. However, the axial cross-section of the OTF (Fig. 7.7.7b) behaves in a more

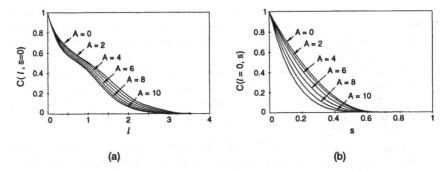

Fig. 7.7.6   Transverse (a) and axial (b) cross-sections of the 3-D OTF for a fluorescence FOCSM consisting of two identical fibres and of two equal annular lenses ($\varepsilon = 0.5$).

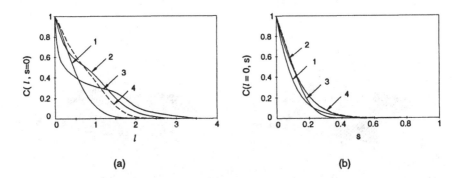

Fig. 7.7.7   Transverse (a) and axial (b) cross-sections of the 3-D OTF for a fluorescence FOCSM consisting of two identical fibres and of annular lenses for $A = 10$. Curves 1, 2 and 3 correspond to $\varepsilon_1 = \varepsilon_2 = 0$, 0.5, and 0.75, respectively, while curve 4 represents $\varepsilon_1 = 0$ and $\varepsilon_2 = 0.5$.

complicated way than that in the transverse direction. When $\varepsilon_1 = \varepsilon_2 = 0.5$, the axial response is better than that for $\varepsilon_1 = \varepsilon_2 = 0$, showing a higher effective cut-off spatial frequency. But the axial response for $\varepsilon_1 = \varepsilon_2 = 0.75$ is poorer than that for $\varepsilon_1 = \varepsilon_2 = 0$, showing a lower effective cut-off spatial frequency.

In order to discuss this phenomenon further, we can introduce the half width $\Delta s_{1/2}$ of the axial cross-section, $C(0, s)$, of the 3-D OTF. A comparison of $\Delta s_{1/2}$ as a function of A for $\varepsilon_1 = \varepsilon_2 = 0$ with that for $\varepsilon_1 = \varepsilon_2 = 0.5$ is shown in Fig. 7.7.8. Interestingly, for small values of A, $\Delta s_{1/2}$ for $\varepsilon_1 = \varepsilon_2 = 0$ is larger than that for $\varepsilon_1 = \varepsilon_2 = 0.5$, which is in agreement with Fig. 7.7.4, while the former is smaller than the latter when A is larger than about 2.9. Both Fig. 7.7.7b and Fig. 7.7.8 suggest that for a given value of A, the

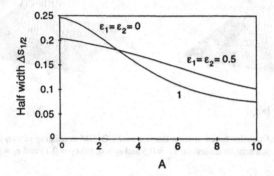

Fig. 7.7.8   Half width of the axial cross-section, $\Delta s_{1/2}$, of the 3-D OTF for a fluorescence FOCSM consisting of two identical fibres and of two equal annular lenses.

axial response of the 3-D OTF can be improved, compared with that for circular pupils, by altering the radius of the central obstruction to a value which is related to A. In other words, for a given fibre spot size axial resolution (or the strength of optical sectioning) can be made improved by altering the central obstruction to an optimum value. The optimum relationship of the axial resolution to the sizes of the central obstruction and the fibre spot will be discussed in Section 7.7.5.

Let us consider an alternative system where either the objective or the collector is an annular lens. Fig. 7.7.9 represents the 3-D OTFs when an annular objective and a circular collector are used, i. e., when $\varepsilon_1 = 0.5$ and $\varepsilon_2 = 0$ (Fig. 7.7.9a), and $\varepsilon_1 = 0.75$ and $\varepsilon_2 = 0$ (Fig. 7.7.9b), respectively. As has been pointed out before, Fig. 7.7.9 is also applicable for a system with a circular objective and an annular collector, i. e., for $\varepsilon_1 = 0$ and $\varepsilon_2 = 0.5$, and $\varepsilon_1 = 0$ and $\varepsilon_2 = 0.75$. Again, it is seen that by increasing the radius of one of the obstructions, the transverse cross-section (see the dashed curve in Fig. 7.7.7) is improved, compared with that for $\varepsilon_1 = \varepsilon_2 = 0$. The axial cut-off spatial frequency is now $1 - \varepsilon_1^2/2$ (if $\varepsilon_1 \neq 0$). It is also found that for a given value of A, by altering the radius of one of the obstructions to a value which is related to A, the axial cross-section is improved, in comparison with that for $\varepsilon_1 = \varepsilon_2 = 0$. In addition, no missing cone of spatial frequencies occurs in the 3-D OTF even for a larger value of A.

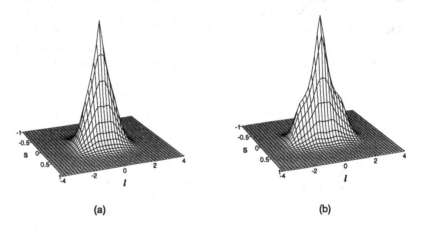

(a)   (b)

Fig. 7.7.9   3-D optical transfer function for a fluorescence FOCSM consisting of two identical fibres ($A = 10$) and of two unequal annular lenses: (a) $\varepsilon_1 = 0.5$ and $\varepsilon_2 = 0$; (b) $\varepsilon_1 = 0.75$ and $\varepsilon_2 = 0$.

Finally, it should be pointed out that the normalization parameter $C(l = 0, s = 0)$ is a function of the fibre spot size and the central obstruction size of the lenses. If the 3-D OTFs are not normalized by $C(l = 0, s = 0)$, they give the relative signal strength for different spatial frequency components for given values of the fibre spot size and the central obstruction size. Using $C(l = 0, s = 0)$ and the method presented in Section 5.4, we can calculate the signal level as a function of the sizes of the fibre spot and the central obstruction for various objects. For a given fibre spot size, the introduction of the annular pupils in both illumination and collection lenses may reduce the signal level significantly. However, one may expect that like the situation in confocal fluorescence microscopy with a finite-sized detector (Section 5.4), there may be an optimum value of the central obstruction if only the illumination lens is annular, at which the signal level and the axial behaviour of the 3-D OTF are both improved compared with those for a system with circular pupils.

7. 7. 5   Axial Resolution

We consider the object to be an infinitely-thin fluorescent sheet scanned along the axial direction. The axial response to the sheet, i. e., the strength of optical sectioning, can be expressed as

$$I(u) = K_f \int_{-\infty}^{\infty} C(l = 0, s) \exp(ius) ds, \tag{7.7.23}$$

where $C(l = 0, s)$ is the axial cross-section of the 3-D OTF for the FOCSM. According to Eq. (7.7.21), we may express $C(l = 0, s)$, when $A_1 = A_2 = A$, as

$$C(l = 0, s) = \iint_{\sigma} \frac{2\pi^2}{Al}$$

$$\exp\left\{-A\left[\frac{l^2}{4} + \left(\frac{s'-s/2}{l}\right)^2\right]\right\}\left\{\mathrm{erf}\left[\sqrt{A}\,\mathrm{Re}(y_1^{\cdot})\right] - \mathrm{erf}\left[\sqrt{A}\,\mathrm{Re}(y_2^{\cdot})\right]\right\} \tag{7.7.24}$$

$$\exp\left\{-A\left[\frac{l^2}{4} + \left(\frac{s'+s/2}{l}\right)^2\right]\right\}\left\{\mathrm{erf}\left[\sqrt{A}\,\mathrm{Re}(y_3^{\cdot})\right] - \mathrm{erf}\left[\sqrt{A}\,\mathrm{Re}(y_4^{\cdot})\right]\right\} dl ds',$$

where σ denotes the overlapped area for non-zero functions in the integrand, and

$$y_1^{\cdot} = \sqrt{1 - \left(\frac{|s'-s/2|}{l} + \frac{l}{2}\right)^2}, \tag{7.7.25a}$$

$$y_2^{\cdot} = \sqrt{\varepsilon_1^2 - \left(\frac{|s'-s/2|}{l} - \frac{l}{2}\right)^2}, \tag{7.7.25b}$$

$$y_3^{\cdot} = \sqrt{1 - \left(\frac{|s'+s/2|}{l} + \frac{l}{2}\right)^2}, \tag{7.7.25c}$$

$$y_4^{\cdot} = \sqrt{\varepsilon_2^2 - \left(\frac{|s'+s/2|}{l} + \frac{l}{2}\right)^2}. \tag{7.7.25d}$$

As expected, $C(0, s)$ in Eq.(7.7.24) is a symmetric function on exchange of ε_1 and ε_2. The axial response to the thin sheet in fluorescence imaging is determined by both ε_1 and ε_2, unlike the axial response to a perfect reflector in non-fluorescence imaging, in which case it is determined only by the larger value of ε_1 and ε_2 (see Eq. (7.3.2)).

The variations of the axial response in Eq. (7.7.23) with A are displayed in Fig. 7.7.10 for $\varepsilon_1 = \varepsilon_2 = 0$ and $\varepsilon_1 = \varepsilon_2 = 0.5$. In both cases, the axial resolution is degraded as A increases. Fig. 7.7.11 shows that when $A = 0$, the axial response for annular pupils ($\varepsilon_1 = \varepsilon_2 = 0.5$) is broader than that for circular pupils ($\varepsilon_1 = \varepsilon_2 = 0$). However, for non-zero values of A, the situation may be reversed. For example, for $A = 10$, the axial response

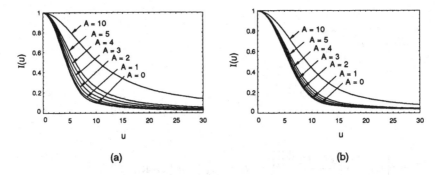

Fig. 7.7.10   Intensity of the axial response to a thin fluorescent sheet for different values of A: (a) $\varepsilon_1 = \varepsilon_2 = 0$; (b) $\varepsilon_1 = \varepsilon_2 = 0.5$.

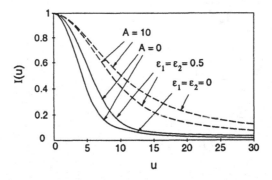

Fig. 7.7.11   Intensity of the axial response to a thin fluorescent sheet for a given value of A.

for annular pupils is narrower than that for circular pupils (dashed curves in Fig. 7.7.11). This property is confirmed in Fig. 7.7.12a, where the half width $\Delta u_{1/2}$ as a function of A is given for $\varepsilon_1 = \varepsilon_2 = 0$ and $\varepsilon_1 = \varepsilon_2 = 0.5$, respectively. This conclusion suggests that for a given fibre spot size, an improved axial resolution can be obtained by altering the central obstruction radius to a value which is determined by the fibre spot size. It is consistent with the property of $\Delta s_{1/2}$ from the 3-D OTF presented in Fig. 7.7.8.

The optimum relationship of the central obstruction radius to the parameter A is shown in Fig. 7.7.12b. It shows the variations of the half width $\Delta u_{1/2}$ with the radius of the central obstruction for $\varepsilon = \varepsilon_1 = \varepsilon_2$, corresponding to two equal annular lenses, for a given value of A. It is seen that when $A = 0$, the value of $\Delta u_{1/2}$ monotonically increases with ε. This implies that for a point detector and a point source, no improvement in axial resolution can be achieved by using annular lenses. For non-zero values of A, which occurs in practice because any fibres have finite cross-section, a minimum value of $\Delta u_{1/2}$ occurs when $\varepsilon = \varepsilon_o$. We call the point at $\varepsilon = \varepsilon_o$ the optimum point as the minimum value of $\Delta u_{1/2}$ is obtained for a given value of A. After the optimum point, the half width increases with ε, and eventually approaches infinity. The point at $\varepsilon = \varepsilon_c$ where the value of $\Delta u_{1/2}$ equals that at $\varepsilon = 0$ is called the critical point. Therefore, when $0 < \varepsilon < \varepsilon_c$, the axial resolution is improved compared with that for $\varepsilon = 0$. In our examples shown in Fig. 7.7.12b, for $A = 5$, ε_o and ε_c are about 0.2 and 0.38, respectively, while for $A = 10$, they are about 0.42 and 0.7, respectively.

At the optimum point, the relationships of ε_o with the parameter A and $Du_{1/2}$, which is defined as the difference of the half widths between the critical point and the optimum point, are displayed in Fig. 7.7.13. In practice, for a given value of A, an

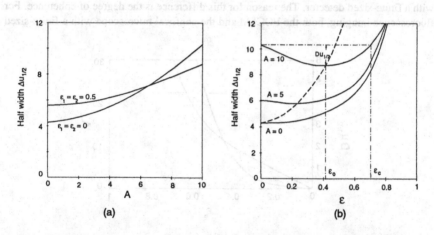

Fig. 7.7.12   Half width of the axial response, $\Delta u_{1/2}$, to a thin fluorescent sheet (a) as a function of A and (b) as a function of ε.

optimum axial resolution can be obtained by altering the radius of the central obstruction to an optimum value ε_o.

In general, the radius of the central obstruction of the objective may not be equal to that of the collector. In that case, for a given value of A, the improvement in axial resolution for fluorescence imaging can still be obtained by altering one of the central obstruction radii. Optimum relationships similar to those discussed in Figs. 7.7.13 can also be achieved, but the quantitative relation may be different. In particular, use of an obstruction only in the objective may result in both a higher signal collection efficiency and an improved axial resolution, as we have discussed in Section 5.4.

The result that the axial resolution can be improved by altering the central obstruction radii is of particular significance in the fibre-optical confocal scanning microscope. In practice, any fibre has finite cross-section, so that the parameter A always has a non-zero value. Accordingly, as we have shown, the axial resolution in fluorescence is degraded compared with that for $A = 0$ if circular lenses ($\varepsilon = 0$) are used (Fig. 7.7.12). However, the axial resolution can be improved by altering the central obstruction radius to an optimum value which is related to A (Figs. 7.7.11 and 7.7.12). This method results in a simultaneous improvement in transverse resolution.

Another significance of Fig. 7.7.12 is that a larger value of the half width $\Delta u_{1/2}$ can be achieved by altering either the radius of the central obstruction or the value of A. This is sometimes needed to obtain more signal from a thick specimen.

Finally, it should be mentioned that the improvement in axial resolution in the fluorescence FOCSM is similar to that in confocal fluorescence microscopy with a finite-sized detector discussed in Chapter 5. However, a bright-field FOCSM does not exhibit this property of improvement, which contrasts with confocal bright-field microscopy with a finite-sized detector. The reason for this difference is the degree of coherence. For fluorescence imaging, both the FOCSM and the confocal microscope with a finite-sized

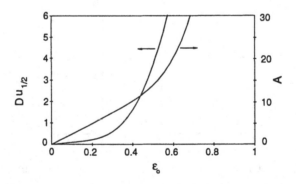

Fig. 7.7.13   Dependence of $Du_{1/2}$ and A as a function of ε_o at the optimum point.

detector are incoherent. However, imaging in a bright-field FOCSM is fully coherent even for a finite fibre spot size and a confocal bright-field microscope with a finite-sized detector is partially coherent. In the case of coherent imaging, the detected intensity is the modulus squared of the integrated amplitude and imaging is thus linear with respect to the amplitude. But imaging in incoherent or partially-coherent cases does not show this linearity. It is this difference that results in there being no improvement in axial resolution in the bright-field FOCSM using annular lenses.

References

7.1.   M. Gu, C. J. R. Sheppard, and X. Gan, *J. Opt. Soc. Am. A*, **8** (1991) 1755 .

7.2.   S. Kimura and T. Wilson, *Applied Optics*, **30** (1991) 2143.

7.3.   M. Gu and C. J. R. Sheppard, *Micron*, **24** (1993) 557.

7.4.   J. Benshchop and G. van Rosmalen, *Applied Optics*, **30** (1991) 1179.

7.5.   Tim Dabbs and Monty Glass, *Applied Optics*, **31** (1992) 705.

7.6.   Tim Dabbs and Monty Glass, *Applied Optics*, **31** (1992) 3030.

7.7.   K. Ghiggino, M. R. Harris, and P. G. Spizzirri, *Rev. Sci. Instrum.*, **63** (1992) 2999.

7.8.   P. M. Delaney, M. R. Harris and R. G. King, *Applied Optics*, **33** (1994) 573.

7.9.   A. Snyder and J. Love, *Optical Waveguide Theory* (Chapman and Hill, London, 1983).

7.10. C. J. R. Sheppard and T. Wilson, *Optica Acta*, **25** (1978) 315.

7.11. M. Gu and C. J. R. Sheppard, *J. Modern Optics*, **41** (1994) 1701.

7.12   M. Gu and C. J. R. Sheppard, *Opt. Commun.*, **100** (1993) 79.

7.13   M. Gu, X. Gan, and C. J. R. Sheppard, *J. Opt. Soc. Am. A*, **8** (1992) 1019.

7.14   M. Gu and C. J. R. Sheppard, *Opt. Commun.*, **94** (1993) 485.

7.15   M. Gu and C. J. R. Sheppard, *J. Modern Optics*, **38** (1991) 1621.

7.16   M. Gu and C. J. R. Sheppard, *Opt. Commun.*, **88** (1992) 27.

7.17. C. J. R. Sheppard, Confocal interference microscopy, in *Confocal Microscopy*, ed. T. Wilson (Academic, London, 1990), p. 389-411.

7.18. H. Zhou, M. Gu, and C. J. R. Sheppard, *J. Modern Optics*, **42** (1995) 627.

7.19. H. J. Matthews. D. K. Hamilton and C. J. R. Sheppard, *J. Modern Optics*, **36** (1989) 233.

7.20. C. J. R. Sheppard and Y. Gong, *Optik*, **87** (1991) 129.

7.21. R. Juskaitis and T. Wilson, *Opt. Commun.*, **92** (1992) 315.

7.22. R. Juskaitis and T. Wilson, *Opt. Commun.*, **99** (1993) 105.

7.23. T. Wilson, R. Juskaitis, N. P. Rea, and D. K. Hamilton, *Opt. Commun.*, **110** (1994) 1.

7.24. M. Gu and C. J. R. Sheppard, *J. Opt. Soc. Am. A*, **9** (1992) 1991.

7.25. X. Gan, M. Gu and C. J. R. Sheppard, *J. Modern Optics*, **39** (1992) 825.

7.26. R. Juskaitis and T. Wilson, *Applied Optics*, **31** (1992) 898.

7.27. R. Juskaitis and T. Wilson, *Applied Optics*, **31** (1992) 4569.

7.28.   T. Wilson, *Opt. Commun.*, **96** (1993) 133.

7.29.   T. Wilson, *J. Opt. Soc. Am. A*, **10** (1993) 1535.

7.30.   A. F. Gmitro and D. Aziz, *Opt. Lett.*, **18** (1993) 565.

7.31.   H. Ooki and J. Iwasaki, *Opt. Commun.*, **85** (1991) 177.

7.32.   B. R. Bracewell, *The Fourier Transform and Its Applications* (McGraw-Hill, New York, 1965).

7.33.   M. Gu and C. J. R. Sheppard, *J. Modern Optics*, **40** (1993) 1255.

7.34.   M. Gu and C. J. R. Sheppard, *J. Modern Optics*, **39** (1992) 783.

7.35.   I. S. Gradstein and I. Ryshik, *Tables of Series, Products, and Integrals* (Herri Deutsch, Frankfurt, 1981).

7.36.   A. Yariv, *Quantum Electronics* (John Wiley & Son, New York, 1975).

7.37.   T. Wilson and S. J. Hewlett, *Opt. Lett.*, **16** (1992) 1062.

7.38.   C. J. R. Sheppard and T. Wilson, *Applied Optics*, **18** (1979) 3764.

7.39.   M. Gu, H. Zhou, and C. J. R. Sheppard, in *19th ACOFT Proceedings* (IREE Society, Sydney, 1994), p. 318.

Chapter 8

CONFOCAL MICROSCOPY
UNDER ULTRASHORT PULSE ILLUMINATION

Recently, there has been an increasing interest in applying an ultrashort pulsed laser beam into confocal scanning optical microscopy.[8.1-8.15] The combination of the ultrashort pulses with confocal scanning microscopy promises more wide applications including, for example, nonlinear optical microscopy resulting from the high peak power of the pulsed beam,[8.4-8.6] time-resolved imaging of objects with depth structures,[8.4, 8.7] and time-of-flight imaging through highly-scattering media.[8.8-8.12] The third application is of particular importance in the early medical diagnosis of human tumours while they are still small to be cured, and is currently an active research field.[8.12]

Analyses on image formation in a confocal microscope under ultrashort laser-pulse illumination have been initiated very recently.[8.7, 8.13-15] There are many issues which have not been addressed so far. For example, the present theory[8.7, 8.13-8.15] is based on the assumption that scattering of the pulsed beam in the object under inspection is not strong, so that the theoretical results cannot be applicable for explaining experimental phenomena in highly-scattering media. In this chapter, the theory of three-dimensional (3-D) image formation in confocal scanning microscopy using a femtosecond pulsed beam, which has been understood so far, is summarized.

Because a femtosecond pulsed beam corresponds to a finite spectral distribution, material dispersion which has been neglected in Chapter 2 should be considered in the diffraction theory of light. The modification[8.16-8.18] of the 3-D amplitude point spread function for a thin lens is described in Section 8.1. Two imaging processes in confocal scanning bright-field microscopy with ultrashort pulse illumination, time-resolved and time-averaged imaging modes, are analyzed in Section 8.2. In Section 8.3, we discuss confocal single-photon (1-p) and two-photon (2-p) fluorescence microscopy under ultrashort pulse illumination. Throughout this chapter, the object function is assumed to be time-independent. In fact, this assumption is not necessarily true in time-resolved measurement. However, the derived results can be used when the response time of the object is much longer than the period of the pulsed beam.

8. 1 Fourier Optics for a Thin Lens under Ultrashort Pulse Illumination

As mentioned above, a pulsed laser beam is not monochromatic but has a finite spectral distribution. When the temporal width of the pulse is shorter than 100 fs, the spectral bandwidth becomes pronounced, compared with the central wavelength.[8.13] Because of the frequency dependence of the refractive index of the lens, material dispersion cannot be neglected when a pulsed beam of femtoseconds propagates through a lens. As a result, the transmittance of a thin lens and the 3-D point spread function for a thin lens, derived in Chapter 2, requires some modifications.

8. 1. 1 Transmittance of a Thin Lens

Suppose that a pulsed beam is of a temporal amplitude $U_0(t)$. Its Fourier transform, the amplitude spectrum of the pulsed beam, is denoted by $V_0(\omega)$. If $\Delta\omega = \omega - \omega_0$ is used, where ω_0, called the central frequency of the pulse, is the frequency at which the light field has the maximum value, $V_0(\omega)$ can be denoted by $V_0(\Delta\omega)$. When the pulsed beam passes through a thin lens, the transmittance of the lens is not given by Eq. (2.1.9) but can be expressed, from Eq. (2.1.6), as

$$t(x,y) = P(x,y)\exp(-ik\bar{n}\overline{D}_0)\exp\left[ik(\bar{n}-1)\frac{(x^2+y^2)}{2}\left(\frac{1}{R_1}-\frac{1}{R_2}\right)\right]. \tag{8.1.1}$$

Because of material dispersion, the refractive index of the lens is a function of the frequency ω, i. e., $\bar{n} = \bar{n}(\omega)$. The wave number k in vacuum is also frequency-dependent, i. e., $k = k(\omega) = \omega/c$, where c is the speed of light in vacuum. We use $\bar{n}_0$ and k_0 to denote the values of $\bar{n}$ and k at the central frequency ω_0.

The dependence of the refractive index $\bar{n}$ on the frequency ω is determined by material structures. However, a general dependence of the refractive index on the frequency, $\bar{n} = \bar{n}(\omega)$, can be expanded about the central frequency:

$$\bar{n}(\omega) = \bar{n}(\omega_0) + \Delta\omega\frac{d\bar{n}}{d\omega}\bigg|_{\omega=\omega_0} + \frac{1}{2}(\Delta\omega)^2\frac{d^2\bar{n}}{d\omega^2}\bigg|_{\omega=\omega_0} +..., \tag{8.1.2}$$

where the first and the second derivatives of $\bar{n}$ can be found from standard handbooks of lens design. With Eq. (8.1.2), an approximate expression for the factor $k\bar{n}$ in the first phase factor of Eq. (8.1.1) can be derived as

$$k\bar{n} = k_0\bar{n}_0[1 + \tilde{a}_1\Delta\omega + \tilde{a}_2(\Delta\omega)^2], \tag{8.1.3}$$

where only the terms up to the second order in $\Delta\omega$ have been kept. This assumption holds for most practical imaging lenses.[8.17] Here

$$\tilde{a}_1 = \frac{1}{\omega_0} + \frac{1}{\bar{n}_0}\frac{d\bar{n}}{d\omega}\bigg|_{\omega=\omega_0},$$

$$\tilde{a}_2 = \frac{1}{\bar{n}_0\omega_0}\frac{d\bar{n}}{d\omega}\bigg|_{\omega=\omega_0} + \frac{1}{2\bar{n}_0}\frac{d^2\bar{n}}{d\omega^2}\bigg|_{\omega=\omega_0}. \tag{8.1.4}$$

Similarly, we can derive an approximate expression for $k(\bar{n} - 1)$:

$$k(\bar{n}-1) = k_0(\bar{n}_0 - 1)[1 + \tilde{b}_1\Delta\omega + \tilde{b}_2(\Delta\omega)^2], \tag{8.1.5}$$

where

$$\tilde{b}_1 = \frac{1}{\omega_0} + \frac{1}{(\bar{n}_0 - 1)}\frac{d\bar{n}}{d\omega}\bigg|_{\omega=\omega_0},$$

$$\tilde{b}_2 = \frac{1}{(\bar{n}_0 - 1)\omega_0}\frac{d\bar{n}}{d\omega}\bigg|_{\omega=\omega_0} + \frac{1}{2(\bar{n}_0 - 1)}\frac{d^2\bar{n}}{d\omega^2}\bigg|_{\omega=\omega_0}. \tag{8.1.6}$$

Finally, making use of Eqs. (8.1.1), (8.1.3) and (8.1.5) leads to the transmittance of the thin lens under pulse illumination:

$$t(x,y) = P(x,y)\exp[-ik_0\bar{n}_0\overline{D}_0(\Delta\omega)(\tilde{a}_1 + \tilde{a}_2\Delta\omega)]$$

$$\exp\left\{i\frac{k_0}{2f_0}[1 + \tilde{b}_1\Delta\omega + \tilde{b}_2(\Delta\omega)^2](x^2 + y^2)\right\}, \tag{8.1.7}$$

where a constant phase term $\exp(ik_0\tilde{n}_0\overline{D}_0)$ has been neglected, and f_0 is the focal length for the central frequency:

$$\frac{1}{f_0} = (\tilde{n}_0 - 1)\left(\frac{1}{R_1} - \frac{1}{R_2}\right).$$

(8.1.8)

It is clear that when the spectral bandwidth is small, corresponding a long pulse in time, Eq. (8.1.7) reduces to Eq. (2.1.9).

8. 1. 2   Temporal and Spectral Distributions of a Pulsed Beam

The results in the last section apply for the illumination of a particular frequency. As we have mentioned, an ultrashort pulsed beam comprises a distribution of the frequency. As an example, let us consider that the pulsed illumination is a Gaussian-shaped pulse given by

$$U_0(t) = \exp(-i\omega_0 t)\exp\left[-\left(\frac{t}{T}\right)^2\right],$$

(8.1.9)

the corresponding Fourier spectrum of which is

$$V_0(\Delta\omega) = \sqrt{\pi}T\exp\left[-\left(\frac{T\Delta\omega}{2}\right)^2\right],$$

(8.1.10)

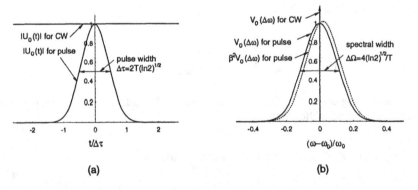

(a)   (b)

Fig. 8.1.1   Temporal (a) and spectral (b) distributions for CW illumination and ultrashort pulsed illumination ($\Delta\tau = 10$ fs). The dashed curve corresponds to the effective spectral distribution of a pulsed beam in the focus of a lens.

where t is the local time coordinate. The pulse width $\Delta\tau$ and the spectral width $\Delta\Omega$ of the pulse, which are defined as the total bandwidth between two positions at which the amplitude drops to one half of its peak value (i. e., FWHM), are equal to

$$\Delta\tau = 2T\sqrt{\ln 2} \qquad (8.1.11)$$

and

$$\Delta\Omega = 4\sqrt{\ln 2}\,/\,T, \qquad (8.1.12)$$

respectively. When $\Delta\tau = 10$ fs and $\lambda_0 = 0.8$ μm for an input pulse, the spectral width $\Delta\Omega$, normalized by the central frequency ω_0, is approximately 0.235. In this case, the temporal and spectral distributions are depicted in Fig. 8.1.1, which includes those for continuous wave (CW) illumination for comparison.

In practice, the pulse width is sometimes defined from its intensity shape, which is denoted by $\Delta\tau'$ (FWHM) here. $\Delta\tau'$ and $\Delta\tau$ satisfy the following relations:

$$\Delta\tau' = \frac{\sqrt{2}}{2}\Delta\tau \approx 0.707\,\Delta\tau \qquad (8.1.13)$$

for a Gaussian-shaped pulse. For a transform-limited pulsed beam, the temporal shape of a pulse is a sech-function defined as

$$U_0(t) = \exp(-i\omega_0 t)\mathrm{sech}\!\left(\frac{t}{T}\right), \qquad (8.1.14)$$

which accordingly gives

$$\Delta\tau' = \frac{\mathrm{sech}^{-1}(\sqrt{2}/2)}{\mathrm{sech}^{-1}(1/2)}\Delta\tau \approx 0.669\,\Delta\tau. \qquad (8.1.15)$$

8. 1. 3 Three-Dimensional Point Spread Function

Eq. (8.1.7) is applicable to a particular frequency ω within an ultrashort pulsed beam. Therefore, for a given frequency ω, the coherent image formation for a thin lens under ultrashort pulsed-beam illumination can be developed using the method presented in Sections 2.2 and 2.3 and using Eq. (8.1.7). For a system given in Fig. 2.2.1, the image field of a thick object is, for a given frequency ω,

$$U_3(x_3, y_3, z_3, \Delta\omega) = V_0(\Delta\omega)\exp[-ik(d_{10} + d_{20})]\exp\left[-\frac{ikM}{2d_{10}}(x_3^2 + y_3^2)(1 + M)\right]$$

$$\text{(8.1.16)}$$

$$\iiint_{-\infty}^{\infty} o(x_1, y_1, z_1)\exp[ik(z_1 - z_3)]h\left(x_1 + Mx_3, y_1 + My_3, z_1 - M^2 z_3\right)dx_1 dy_1 dz_1,$$

where d_{10} and d_{20} satisfy the lens law at the central frequency:

$$\frac{1}{f_0} = \frac{1}{d_{10}} + \frac{1}{d_{20}} \qquad\qquad (8.1.17)$$

and the 3-D space-invariant spectrum-dependent amplitude point spread function is

$$h(x, y, z, \Delta\omega) = \frac{M}{d_1^2 \lambda^2} \iint_{-\infty}^{\infty} P(x_2, y_2)\exp\left[-ik_0 \tilde{n}_0 \overline{D}_0(\Delta\omega)(\tilde{a}_1 + \tilde{a}_2 \Delta\omega)\right]$$

$$\exp\left[\frac{ik_0(x_2^2 + y_2^2)}{2f_0}\Delta\omega\left(\frac{1}{\tilde{n}_0 - 1}\frac{d\tilde{n}}{d\omega}\bigg|_{\omega=\omega_0} + \tilde{b}_2 \Delta\omega\right)\right]\exp\left[-\frac{ik}{2}\left(\frac{1}{d_{10}}\right)^2 z(x_2^2 + y_2^2)\right] \quad \text{(8.1.18)}$$

$$\exp\left[\frac{ik}{d_{10}}(x_2 x + y_2 y)\right]dx_2 dy_2,$$

where $k_0 = 2\pi/\lambda_0$, and λ_0, corresponding to the central frequency, is the wavelength in vacuum. Therefore the image of a single point in the case of a single circular lens of radius a is[8.16]

$$U_3(v, u, \Delta\omega) = \frac{M\omega_0^2\left(1 + \dfrac{\Delta\omega}{\omega_0}\right)^2 V_0(\Delta\omega)\exp\left[-ik_0\left(1 + \dfrac{\Delta\omega}{\omega_0}\right)\left(d_{10} + d_{20} + \dfrac{d_{20}^2 u}{k_0 a^2}\right)\right]}{2\pi d_1^2 c^2}$$

$$\exp\left[-\frac{iv^2}{4\pi N}\left(1 + \frac{\Delta\omega}{\omega_0}\right)(1 + M)\right]\int_0^1 P(\rho)J_0\left[\rho v\left(1 + \frac{\Delta\omega}{\omega_0}\right)\right]\exp\left[i\frac{\rho^2 u}{2}\left(1 + \frac{\Delta\omega}{\omega_0}\right)\right]$$

$$\exp[-i(\Delta\omega)^2(\delta - \delta\rho^2)]\exp\{-i\Delta\omega[\tau - \tau\rho^2]\}\rho d\rho. \qquad\qquad (8.1.19)$$

Here $P(\rho)$ is the pupil function of the lens and J_0 is a Bessel function of the first kind of order zero. The transverse and axial optical coordinates v and u are defined as

$$v = \frac{k_0 r_3 a}{d_{20}} \approx k_0 r_3 \sin \alpha_i,$$

(8.1.20)

$$u = \frac{k_0 z_3 a^2}{d_{20}^2} \approx k_0 z_3 \sin^2 \frac{\alpha_i}{2}.$$

(8.1.21)

N is the Fresnel number, defined as

$$N = \frac{a^2}{\lambda_0 d_{20}}.$$

(8.1.22)

The definitions of other parameters are expressed as follows:

$$\delta = \frac{a^2 k_0}{2 f_0} \tilde{b}_1 = \frac{a^2 k_0}{2 f_0 (\tilde{n}_0 - 1)} \left(\frac{1}{\omega_0} \frac{d\tilde{n}}{d\omega}\bigg|_{\omega=\omega_0} + \frac{1}{2} \frac{d^2 \tilde{n}}{d\omega^2}\bigg|_{\omega=\omega_0} \right),$$

$$\delta' = k_0 n_0 \overline{D}_0 \tilde{a}_2 = k_0 \overline{D}_0 \left(\frac{1}{\omega_0} \frac{d\tilde{n}}{d\omega}\bigg|_{\omega=\omega_0} + \frac{1}{2} \frac{d^2 \tilde{n}}{d\omega^2}\bigg|_{\omega=\omega_0} \right),$$

$$\tau = \frac{a^2 k_0}{2 f_0 (\tilde{n}_0 - 1)} \frac{d\tilde{n}}{d\omega}\bigg|_{\omega=\omega_0},$$

(8.1.23)

$$\tau' = k_0 \tilde{n}_0 \overline{D}_0 a_1 = k_0 \overline{D}_0 \left(\frac{\tilde{n}_0}{\omega_0} + \frac{d\tilde{n}}{d\omega}\bigg|_{\omega=\omega_0} \right).$$

The factors δ and δ' represent the group-velocity dispersion in the lens, resulting in frequency chirps. The former gives a chirp that is dependent on the radial coordinate over the lens aperture, whereas the latter is a constant chirp. τ and τ' are responsible for time delays in the propagation of the pulse. τ plays a more important role in imaging with a pulse than τ' does because the delay caused by τ is radius-dependent, which is termed the propagation time difference.[8.17, 8.18] The influence of τ' may be neglected as it affects only the time shift. Therefore δ and τ produce chromatic aberration for a single lens. However, the influence of δ and τ may be removable by using an achromatic lens doublet.[8.17] δ' can also be neglected if the incident pulse is pre-chirped by a certain amount.

The total time-dependent image field of a point object can be obtained by performing the inverse Fourier transform of Eq. (8.1.12):

$$U_3(v,u,t) = \int_{-\infty}^{\infty} U_3(v,u,\Delta\omega)\exp(-i\Delta\omega t)d\Delta\omega. \qquad (8.1.24)$$

It is useful to discuss the role of the pre-factors in Eq. (8.1.19). The pre-quadratic phase term

$$\exp\left[-\frac{iv^2}{4\pi N}\left(1+\frac{\Delta\omega}{\omega_0}\right)(1+M)\right]$$

in Eq. (8.1.19) can be neglected as the Fresnel number N is large in an imaging system. Eq. (8.1.19) also includes other two frequency-dependent terms which may affect the image quality when a pulsed beam is used. The first term is

$$\exp\left[-ik_0\left(1+\frac{\Delta\omega}{\omega_0}\right)\left(d_{10}+d_{20}+\frac{d_{20}^2 u}{k_0 a^2}\right)\right], \qquad (8.1.25)$$

in which the phase factor

$$\exp\left[-ik_0\left(1+\frac{\Delta\omega}{\omega_0}\right)(d_{10}+d_{20})\right] \qquad (8.1.26)$$

is not of interest as it causes only a constant time shift of $U_3(v, u, t)$, but the phase factor

$$\exp\left[-ik_0\left(1+\frac{\Delta\omega}{\omega_0}\right)\frac{d_{20}^2 u}{k_0 a^2}\right] \qquad (8.1.27)$$

does give a phase associated with the defocus distance u. It is the term in Eq. (8.1.27) that plays an important role in time-resolved confocal imaging using a pulsed beam, as will be discussed in Section 8.2. The second term is

$$\frac{M\omega_0^2\left(1+\frac{\Delta\omega}{\omega_0}\right)^2 V_0(\Delta\omega)}{2\pi d_1^2 c^2}, \qquad (8.1.28)$$

which is again determined by the spectral width of the pulsed laser beam. This factor determines the relative magnitude of the 3-D APSF for a given frequency component ω and therefore results in an asymmetric spectrum with respect to ω_0 in the focus of a lens, as is shown by the dashed curve in Fig. 8.1.1b. The asymmetric spectrum may result in a stronger effect at high frequencies than that at low frequencies.

As an example, let us consider an objective to be achromatic (i. e., without material dispersion). In practice, this assumption may hold for a well-corrected achromatic objective. We further assume $\Delta \tau = 10$ fs and $\lambda_0 = 0.8$ μm for an input pulse, which is currently obtainable in an ultrashort pulsed Ti: sapphire laser.[8.19-8.21] For a Gaussian pulsed beam, the temporal shape of the amplitude at the focus ($u = v = 0$) can be calculated by the Fourier transform (see Eq. (8.1.24)). The effect of Eq. (8.1.26) has been neglected since it contributes only a constant time shift. The temporal shapes before the lens and at the focus are depicted in Fig. 8.1.2, showing that a real input pulse is converted into a complex pulse due to the effect of Eq. (8.1.28).

It should be mentioned that the effects associated with the spectral width $\Delta \Omega$ of the pulsed beam may not be appreciable for a pulse of a duration of picoseconds, i. e., for $\Delta \Omega / \omega_0 << 1$.[8.13, 8.17] However, if the pulse duration is only a few femtoseconds, the effects can be significant. Further, Eqs. (8.1.27) and (8.1.28) originate from the diffraction of light through a lens. In other words, even for an achromatic lens of doublets, in which case δ and τ are equal to zero, the effects caused by Eqs. (8.1.27) and (8.1.28) still exist.

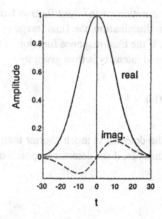

Fig. 8.1.2   Temporal amplitude of a laser pulse ($\Delta \tau = 10$ fs) before the lens (solid curve) and in the focal plane (dashed curve). The time is normalized by $1/\omega_0$.

8. 2 Bright-Field Imaging

Schematic diagrams of the confocal systems can be refereed to Fig. 3.1.1. In the present case, the point source should be considered to give a pulsed beam of a temporal amplitude $U_0(t)$. The spectrum of the source is denoted by $V_0(\Delta\omega)$. For a given frequency component ω, the image amplitude can be expressed, if we consider bright-field microscopy and a point detector, as

$$U(r_s,\Delta\omega) = V_0(\Delta\omega)\int_{-\infty}^{\infty} o(r_s - r_1)\exp[ik(\pm z_1 - z_1)]h_1(M_1 r_1,\Delta\omega)h_2(r_1,\Delta\omega)dr_1, \quad (8.2.1)$$

where Eqs. (3.1.7) and (8.1.16) have been used. In Eq. (8.2.1), we have assumed that the Fresnel number is large and that the constant phase term associated with $d_{10} + d_{20}$ can be neglected. Here M_1 is defined in Eq. (3.1.5), r_s is the scan position vector, and $o(r)$ denotes the reflectivity in reflection microscopy or the transmittance in transmission microscopy. h_q (q =1, 2) represents the 3-D amplitude point spread function at frequency ω for objective ($q = 1$) and collector ($q = 2$) lenses, respectively, and is given by Eq. (8.1.18).

The total time-dependent image amplitude is given by performing the inverse Fourier transform of Eq. (8.2.1) with respect to $\Delta\omega$, that is

$$U(r_s,t) = \int_{-\infty}^{\infty} U(r_s,\Delta\omega)\exp(-i\Delta\omega t)d\Delta\omega, \quad (8.2.2)$$

which implies a coherent summation of the contribution from the different frequency components. For pulsed beam illumination, the final image intensity is dependent on the response time of the detector. If the time-response function of the detector for intensity is assumed to be $f_t(t)$, the measured intensity is then given by

$$I(r_s) = \int_{-\infty}^{\infty} |U(r_s,t)|^2 f_t(t)dt. \quad (8.2.3)$$

When the response time of the detector is much shorter than the temporal pulse width, the corresponding image intensity is, if the detector responds only at $t = t_0$,

$$I(r_s,t_0) = |U(r_s,t_0)|^2. \quad (8.2.4)$$

It is clear that Eq. (8.2.4) gives a time-resolved image of the 3-D object of interest. Alternatively, when a detector with a response time longer than the temporal pulse width, the measured intensity is virtually averaged over many pulses. In this case, the image intensity may be expressed as[8.22]

$$I(r_s) = \int_{-\infty}^{\infty} |U(r_s,t)|^2 \, dt. \tag{8.2.5}$$

To understand the imaging properties of the system, one can investigate the point spread function of the system.[8.13, 8.16] An alternative description of the image formation is based on the concept of the transfer function as introduced in previous chapters. Introducing a 3-D inverse Fourier transform of the object function, $O(m)$, called the 3-D spatial spectrum of the object, we have

$$o(r) = \int_{-\infty}^{\infty} O(m)\exp(2\pi i r \bullet m) \, dr. \tag{8.2.6}$$

Here m represents the spatial frequency vector with two transverse components m and n, and one axial component s. Substituting Eq. (8.2.6) into (8.2.1) yields

$$U(r_s,\Delta\omega) = V_0(\Delta\omega)\int_{-\infty}^{\infty} c(m,\Delta\omega)O(m)\exp(2\pi i r_s \bullet m) \, dm. \tag{8.2.7}$$

Here

$$c(m,\Delta\omega) = V_0(\Delta\omega)\int_{-\infty}^{\infty} h_a(r,\Delta\omega)\exp(-2\pi i r \bullet m) \, dr, \tag{8.2.8}$$

where

$$h_a(r,\Delta\omega) = \exp[ik(\pm z - z)]h_1(M_1 r,\Delta\omega)h_2(r,\Delta\omega), \tag{8.2.9}$$

which is the 3-D effective amplitude point spread function at ω. Eq. (8.2.8) is the 3-D inverse Fourier transform of the product of the two point spread functions and is called the 3-D spectrum-dependent coherent transfer function (CTF). It gives the strength with which the spatial frequency m in objects of interest is imaged under the illumination of frequency ω. Substituting Eq. (8.2.7) into Eq. (8.2.2) and using Eq. (8.2.8), one has

$$U(r_s,t) = \int_{-\infty}^{\infty} O(m)c(m,t)\exp(2\pi i r_s \bullet m) \, dm, \tag{8.2.10}$$

where

$$c(m,t) = \int_{-\infty}^{\infty} c(m,\Delta\omega)\exp(-i\Delta\omega t) \, d\Delta\omega \tag{8.2.11}$$

is the inverse Fourier transform of $c(m, \Delta\omega)$ with respect to $\Delta\omega$ and termed the 3-D time-dependent CTF. Using Eq. (8.2.10) in Eq. (8.2.3) we have

$$I(r_s) = \iint_{-\infty}^{\infty} O(m)O^*(m')c(m;m')\exp[2\pi i r_s \bullet (m - m')]dmdm'. \qquad (8.2.12)$$

Here

$$c(m;m') = \int_{-\infty}^{\infty} c(m,t)c^*(m',t)f_t(t)dt, \qquad (8.2.13)$$

which becomes

$$c(m;m') = c(m,t_0)c^*(m',t_0) \qquad (8.2.14)$$

for time-resolved imaging, and

$$c(m;m') \equiv C(m;m') = \int_{-\infty}^{\infty} c(m,\Delta\omega)c^*(m',\Delta\omega)d\Delta\omega \qquad (8.2.15)$$

for time-averaged imaging. Finally, the image intensity corresponding to these two cases is, respectively,

$$I(r_s,t_0) = \left| \int_{-\infty}^{\infty} O(m)c(m,t_0)\exp(2\pi i r_s \bullet m)dm \right|^2 \qquad (8.2.16)$$

and

$$I(r_s) = \iint_{-\infty}^{\infty} O(m)O^*(m')C(m;m')\exp[2\pi i r_s \bullet (m - m')]dmdm'. \qquad (8.2.17)$$

Note that the time-resolved imaging process is purely coherent because it obeys the superposition principle of amplitude, so that it is fully determined by the 3-D time-dependent CTF. However, the time-averaged imaging method is controlled by the transfer function $C(m; m')$. In general, $C(m; m')$ cannot be separable and is therefore called the transmission cross-coefficient (TCC) in terms of the terminology used in Chapter 4. Thus Eq. (8.2.17) corresponds to a partially-coherent imaging process. According to Eq. (8.2.15), $C(m; m')$ is an arithmetic summation of the product of two 3-D spectrum-dependent CTFs. In general, the 3-D time-dependent CTF is a complex function even for a real spectrum-dependent CTF, whereas the TCC $C(m; m')$ can be real if the 3-D spectrum-dependent CTF is real.

8. 2. 1   3-D Point Spread Function

For a confocal system containing two identical circular lenses of radius a, the 3-D effective amplitude point spread function at ω, h_a, can be expressed, by taking the cylindrical symmetry into account in Eq. (8.2.9), as

$$h_a(v,u,\Delta\omega) = h_1(v,u,\Delta\omega)h_2(v,u,\Delta\omega). \qquad (8.2.18)$$

Here v and u are radial and axial optical coordinates in the object space of the confocal microscope, defined at the central frequency:

$$v = \frac{2\pi}{\lambda_0}r\frac{a}{d_{20}} \approx \frac{2\pi}{\lambda_0}r\sin\alpha_o, \qquad (8.2.19a)$$

and

$$u = \frac{2\pi}{\lambda_0}z\frac{a^2}{d_{20}^2} \approx \frac{8\pi}{\lambda_0}z\sin^2(\alpha_o/2). \qquad (8.2.19b)$$

In terms of Eq. (8.1.18), h_1 can be given by

$$h_1(v,u,\Delta\omega) = K\beta^2 \exp(-is_0u)\exp\{i[\Delta\omega\tau' +(\Delta\omega)^2\delta']\}$$
$$\int_0^1 P(\rho)J_0(\rho v\beta)\exp\left[i\left(\beta\frac{\rho^2u}{2}-w\rho^2\right)\right]\rho d\rho, \qquad (8.2.20)$$

where

$$w = \Delta\omega\tau + (\Delta\omega)^2\delta, \qquad (8.2.21)$$

$$\beta = (1+\frac{\Delta\omega}{\omega_0}), \qquad (8.2.22)$$

$$s_0 = \frac{\beta d_{20}^2}{a^2} \approx \frac{\beta}{4\sin^2(\alpha_o/2)}. \qquad (8.2.23)$$

K is the constant independent of ω. The definitions of the parameters τ', τ, δ' and δ resulting from material dispersion are given in Eq. (8.1.23). Note that the phase factor

$w\rho^2$ in Eq. (8.2.20) acts as an effective defocus aberration and w is therefore called the effective defocusing coefficient, or the chromatic aberration coefficient. It is now clear that the effect of the lens dispersion leads to a frequency-dependent defocus aberration. In fact, the focal plane is axially shifted to $u = 2w/\beta$. β, the normalized frequency, denotes the effect of the spectrum of the pulsed beam. Because of the existence of β, the 3-D amplitude point spread function, h_1, may be distorted compared with that under monochromatic (or CW) illumination in Eq. (3.1.15). The pre-phase term

$$\exp(-is_0u) \tag{8.2.24}$$

and the pre-amplitude factor β^2 are two frequency-dependent factors and can result in significant effects on 3-D imaging. We have shown, in Fig. 8.1.2 , the effect of β^2 on the temporal distribution of a pulsed beam in the focal region of a lens.

To understand the effect of β^2 on the 3-D effective point spread function h_a, we can calculate the time-averaged intensity of a point object and it can be expressed, in terms of Eq. (8.2.5), as

$$I(v,u) = K\int_{-\infty}^{\infty} \left|V_0(\Delta\omega)h_a(v,u,\Delta\omega)\right|^2 d\Delta\omega. \tag{8.2.25}$$

The transverse and axial cross-sections of $I(v, u)$ are displayed in Fig. 8.2.1 for a 10 fs pulse. The responses for the pulsed illumination are slightly narrower than those for CW illumination, which is purely caused by β^2. This feature implies a 3-4% improvement in resolution in the transverse and axial directions.

The half width of $I(v, u)$, $v_{1/2}$ and $u_{1/2}$, in the transverse and axial directions as a function of the pulse width $\Delta\tau$ is given in Fig. 8.2.2. It is noted that there is no

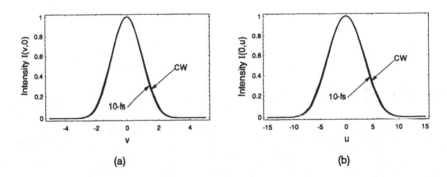

(a)   (b)

Fig. 8.2.1   Transverse (a) and axial (b) cross-sections of the image intensity of a point object in a confocal microscope under ultrashort pulsed-laser illumination ($\Delta\tau = 10$ fs).

pronounced difference in the half width between 10 fs and 40 fs illumination. In fact, when a pulse is shorter than 10 fs, the improvement increases appreciably. An improvement in resolution of more than 10% is obtainable when the illumination pulse width becomes 5 fs or shorter. As an example, an illumination with 3 fs pulses has approximately 25% better resolution than with CW illumination.

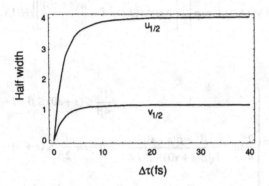

Fig. 8.2.2   Half width of the image intensity of a point object in a confocal microscope as a function of the pulse width $\Delta\tau$.

8. 2. 2   3-D Coherent Transfer Function in Reflection

We now consider 3-D imaging in terms of the 3-D CTF. For a reflection-mode confocal microscope using two identical circular lenses, $h_2(v, u, \Delta\omega) = h_1(v, u, \Delta\omega)$ as the light reflected backward from the sample is collected, so that the 3-D spectrum-dependent CTF can be expressed, according to Eq. (8.2.8), as

$$c_r(l, s, \Delta\omega) = V_0(\Delta\omega) \int_{-\infty}^{\infty} \left\{ \int_{-\infty}^{\infty} [h_1(v, u, \Delta\omega)]^2 J_0(vl) v \, dv \right\} \exp(-ius) du. \qquad (8.2.26)$$

Here $l = (m^2 + n^2)^{1/2}$ and s, called the radial and axial spatial frequencies, have been normalized by

$$\frac{\sin\alpha_o}{\lambda_0} \qquad (8.2.27)$$

and

$$\frac{4\sin^2(\alpha_o/2)}{\lambda_0},\tag{8.2.28}$$

respectively. An analytical solution of Eq. (8.2.26) can be derived in terms of the method presented in Section 3.2 and is given by

$$c_r(l,s,\Delta\omega) = V_0(\Delta\omega)\beta^3\exp\left\{2i\left[\Delta\omega\tau' + (\Delta\omega)^2\delta' - \frac{w(s+s0)}{\beta}\right]\right\}H(l,s,\Delta\omega),\tag{8.2.29}$$

where

$$H(l,s,\Delta\omega) = \begin{cases} 1, & \dfrac{l^2}{4\beta} \le (s+s0) \le \beta - l + \dfrac{l^2}{2\beta}, \\[2ex] \dfrac{2}{\pi}\sin^{-1}\dfrac{\beta^2 - \beta(s+s0)}{l\sqrt{\beta(s+s0) - l^2/4}}, & \beta - l + \dfrac{l^2}{2\beta} \le (s+s0) \le \beta, \\[2ex] 0, & otherwise, \end{cases}\tag{8.2.30}$$

and

$$s0 = 2s_0\tag{8.2.31}$$

is an axial shift of the CTF in spatial frequency space and is dependent on the frequency of the illumination and on the numerical aperture of the objective. For the central frequency ω_0, the shift becomes $s0' = 1/[2\sin^2(\alpha_o/2)]$. The function H has been normalized to unity at $l = 0$ and $s + s0 = 0$ for $\omega = \omega_0$. Eqs. (8.2.29) and (8.2.30) imply an important property that the 3-D spectrum-dependent CTF at different frequencies exhibits different axial shifts. Thus according to Eqs. (8.2.11) and (8.2.15), the time-dependent CTF and the TCC may have larger areas in which the values of the transfer functions are appreciable, compared with those under CW illumination.

Most objectives used in optical microscopes are usually well-corrected achromatically. In this case, we can assume $\tau = \delta = 0$, i. e., $w = 0$. Parameters τ' and δ' can be justified to be zero for both illumination and collection if a pre-frequency-chirped pulse is used and a post frequency-chirping process is employed. Thus $c_r(l, s, \Delta\omega)$ reduces to

$$c_r(l,s,\Delta\omega) = V_0(\Delta\omega)\beta^3 H(l,s,\Delta\omega).\tag{8.2.32}$$

To calculate the 3-D time-dependent CTF and the TCC we use the same example as adopted in Fig. (8.1.2). Consider an objective of numerical aperture $1/\sqrt{2}$. Under these conditions, the passbands of Eq. (8.2.32) for three particular frequencies ω_0, $\omega_0 + \Delta\Omega/2$ and $\omega_0 - \Delta\Omega/2$ are depicted in Fig. 8.2.3a. The 3-D spectrum-dependent CTF cuts off at 2β and $s0$ in the transverse and axial directions, respectively. It is difficult to present the 3-D spectrum-dependent CTF in an illustrative way as it is a three-variable function. However, the axial cross-section of Eq. (8.2.32) is depicted in Fig 8.2.4, showing a wavelength-dependent axial offset. The modulus of the complete

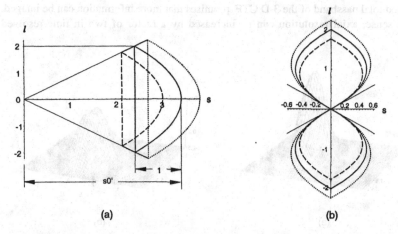

(a)   (b)

Fig. 8.2.3 Passband of the 3-D spectrum-dependent CTF in reflection (a) and transmission (b). The dashed, solid, and dotted curves correspond to the passbands at $\omega_0 - \Delta\Omega/2$, ω_0, and $\omega_0 + \Delta\Omega/2$ ($\Delta\tau = 10$ fs).

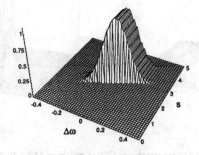

Fig. 8.2.4 Axial cross-section of the 3-D spectrum-dependent CTF at $l = 0$ in reflection ($\Delta\tau = 10$ fs).

3-D time dependent CTF is shown in Fig. 8.2.5, in which the 3-D CTF is symmetric with respect to the time t and has been normalized to unity for the peak value at $t = 0$. We have included the 3-D CTF for CW illumination for comparison.

As is expected, the passband of the 3-D time-dependent CTF at $t = 0$ (Fig. 8.2.5a) is larger than that for the CW case (Fig. 8.2.5d). As time proceeds, the effective passband is reduced in the transverse direction but does not change appreciably in the axial direction. This property is further illustrated in Fig. 8.2.6a, in which the axial cross-section of the time-dependent 3-D CTF is given. Compared with CW illumination (Fig. 8.2.6b), the axial passband is twice as large as that for CW illumination,[8.7, 8.23] although the half width of the axial cross-sections are almost the same in both cases. The increased total passband of the 3-D CTF promises that more information can be imaged. In that sense, axial resolution can be increased by a factor of two in time-resolved

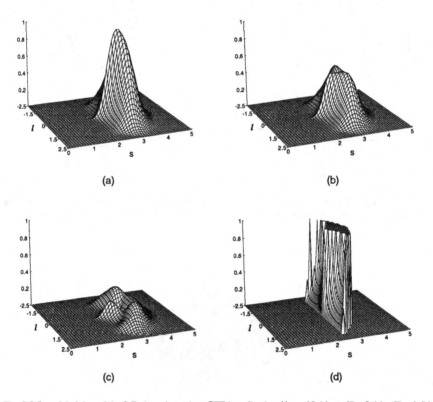

Fig. 8.2.5   Modulus of the 3-D time-dependent CTF in reflection ($\Delta \tau = 10$ fs) at $t/T = 0$ (a), $t/T = 1$ (b) and $t/T = 2$ (c). The plot in (d) is the 3-D CTF for CW illumination.

confocal reflection microscopy. It should be pointed out that the improvement in axial resolution for imaging of an axial line object is stronger than that for a 3-D object, because the 1-D CTF in the former case is a projection of the 3-D CTF on the axis.[8.15]

With respect to time-averaged imaging, the corresponding TCC at $l = l' = 0$ is a four-variable function, so that only the axial cross-section of the TCC, normalized to unity at the peak value, is shown in Fig. 8.2.7a, while the axial cross-section of the TCC for CW illumination is depicted in Fig. 8.2.7b. Again, the non-zero area of the TCC in the former case is larger compared with that in the latter, which is an advantage over the CW illumination in the sense that more information can be transferred.

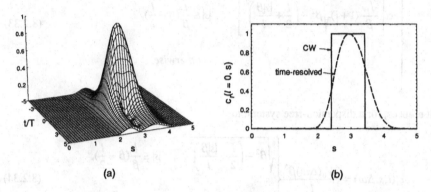

(a)   (b)

Fig. 8.2.6   Axial cross-section of the modulus of the 3-D time-dependent CTF ($\Delta\tau = 10$ fs) at $l = 0$: (a) time dependence; (b) comparison of pulsed and CW illumination.

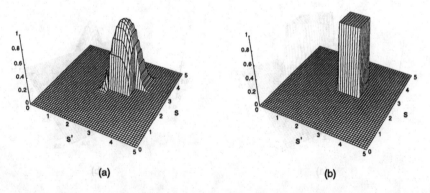

(a)   (b)

Fig. 8.2.7   Axial cross-section of the TCC in reflection for pulse illumination ($\Delta\tau = 10$ fs) (a) and for CW illumination (b).

8. 2. 3　3-D Coherent Transfer Function in Transmission

Consider a transmission-mode confocal microscope using two identical circular lenses. The 3-D amplitude point spread function for the collector is $h_2(v,\ u,\ \Delta\omega) = h_1(v,\ -u,\ \Delta\omega)$. In a similar way, the 3-D spectrum-dependent CTF can be derived as

$$c_t(l,s,\Delta\omega) = \frac{\sqrt{\pi}V_0(\Delta\omega)\beta^4}{(1+i)l\sqrt{w}}\exp\left\{2i\left[\Delta\omega\tau' + (\Delta\omega)^2\delta - \frac{w}{\beta^2}\left(\left(\frac{s\beta}{l}\right)^2 + \frac{l^2}{4}\right)\right]\right\}$$

$$\begin{cases} \mathrm{erf}\left[\dfrac{\sqrt{w}}{\beta}(1+i)\sqrt{\beta^2 - \left(\dfrac{l}{2} + \dfrac{|s|\beta}{l}\right)^2}\right] & ,\quad |s| \le \dfrac{l}{\beta}(\beta - \dfrac{l}{2}), \\[3ex] 0 & ,\qquad otherwise. \end{cases}$$
(8.2.33)

It reduces, for a dispersion-free system, to

$$c_t(l,s,\Delta\omega) = \frac{2V_0(\Delta\omega)\beta^3}{l}\begin{cases} \sqrt{\beta^2 - \left(\dfrac{l}{2} + \dfrac{|s|\beta}{l}\right)^2} & ,\quad |s| \le \dfrac{l}{\beta}(\beta - \dfrac{l}{2}), \\[3ex] 0 & ,\qquad otherwise. \end{cases}$$
(8.2.34)

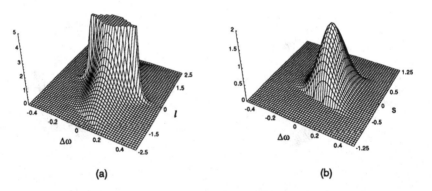

(a)　　　　　　　　　　　　　(b)

Fig. 8.2.8　Cross-section of the modulus of the 3-D spectrum-dependent CTF in transmission ($\Delta\tau = 10$ fs): (a) $s = 0$;　(b) $l = 1$.

The cut-off transverse and axial spatial frequencies of Eq. (8.2.34) are, respectively, 2β and $\beta/2$, as can be seen from the passband shown in Fig. 8.2.3b, which is calculated for the same conditions as in reflection microscopy. Two cross-sections of the 3-D spectrum-dependent CTF at $s = 0$ and $l = 1$ are depicted in Fig. 8.2.8, the former describing imaging of a thick object which has no axial structures. Time-resolved confocal imaging is described by the time-dependent CTF shown in Fig. 8.2.9, normalized to unity at $l = 1$ and $s = 0$ at $t = 0$. It is noted that the CTF at $t = 0$ (Fig. 8.2.9a) exhibits the largest passband of the spatial frequency content, compared with that under CW illumination (Fig. 8.2.9d). The transverse cross-section at $s = 0$ and the axial cross-section at $l = 1$ are shown in Fig. 8.2.10. Note that the passband in the transverse direction decreases as time proceeds while that in the axial direction does not.

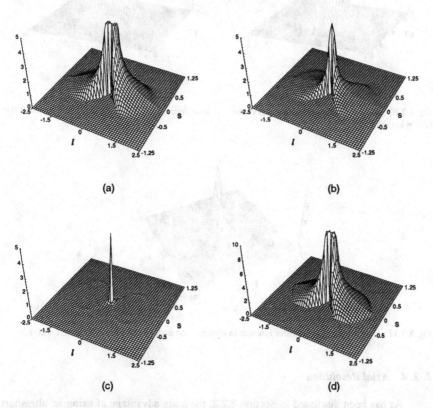

(a)　　　　　　　　　　　　　　　(b)

(c)　　　　　　　　　　　　　　　(d)

Fig. 8.2.9   Modulus of the 3-D time-dependent CTF in transmission ($\Delta\tau = 10$ fs) at $t/T = 0$ (a), $t/T = 1$ (b), and $t/T = 2$ (c). The plot in (d) is the 3-D CTF for CW illumination.

For time-averaged imaging, the complete TCC is a four-variable function.[8.7] However, a special form of the TCC for a weakly-scattering object, termed the weak-object transfer function (WOTF), can be derived using the method described in Section 4.6. The 3-D WOTF shown in Fig. 8.2.11 has a similar behaviour to the 3-D CTF for CW illumination except in the region of the origin.

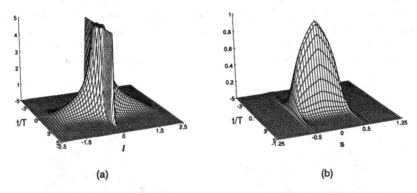

(a)	(b)

Fig. 8.2.10   Cross-section of the modulus of the 3-D time-dependent CTF in transmission ($\Delta\tau = 10$ fs): (a) $s = 1$; (b) $l = 1$.

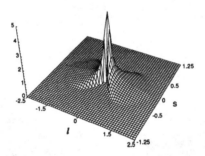

Fig. 8.2.11   3-D weak-object transfer function in transmission under pulse illumination ($\Delta\tau = 10$ fs).

8. 2. 4   Axial Resolution

As has been discussed in Section 8.2.2, the main advantage of using an ultrashort pulsed beam is the increased passband of the transfer function in the axial direction. To understand the axial imaging property further in a reflection confocal system, we consider the image of a perfect reflector scanned in the axial direction. The intensity of the image can be calculated by the modulus squared of the Fourier transform of $c_r(0, s, t)$, in terms of Eq. (8.2.16):

$$I(u,t) = \left| \int_{-\infty}^{\infty} V_0(\Delta\omega)\beta^4 \exp\left[-\frac{iu\cot^2(\alpha/2)\beta}{2} \right] \frac{\sin(\beta u/2)}{(\beta u/2)} \exp(-i\Delta\omega t)d\Delta\omega \right|, \quad (8.2.35)$$

As a comparison, we give the image intensity for the same object in the time-averaged case from Eq. (8.2.17):

$$I(u) = \int_{-\infty}^{\infty} [V_0(\Delta\omega)]^2 \beta^8 \left\{ \frac{\sin(\beta u/2)}{(\beta u/2)} \right\}^2 d\Delta\omega . \quad (8.2.36)$$

For time-resolved imaging, the axial response is time-dependent, as is shown in Fig. 8.2.12a. A direct comparison of the strength of optical sectioning is displayed in Fig. 8.2.12b, where, for time-resolved imaging, only the strength of optical sectioning at $t = 0$ is plotted. It is shown that the half width of the responses is approximately 2 and 2.65 for time-resolved and time-averaged imaging methods, respectively, exhibiting that the axial resolution is improved by approximately 28% and 4-5%, respectively, when compared with the value of 2.78 for CW illumination of frequency ω_0. This conclusion is of importance for time-resolved confocal microscopy because this technique provides not only information on the dynamic behaviour of the sample but also its 3-D structure with

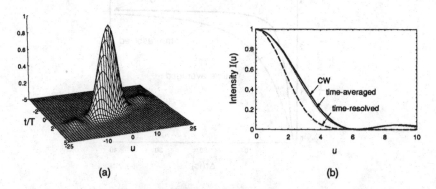

(a)   (b)

Fig. 8.2.12   Image intensity of a perfect reflector ($\Delta\tau = 10$ fs): (a) time-resolved imaging; (b) comparison of time-resolved imaging, time-averaged imaging, and CW illumination.

an axial resolution higher than the current limit in a confocal microscope under CW illumination.

In Fig. 8.2.13, the relationship of the half width of the axial response, $\Delta u_{1/2}$, to the pulse width $\Delta\tau$ is revealed.[8.23] The axial resolution can be improved dramatically when the pulse width is less than 10 fs. There should be, however, a physical limitation on the shortest pulse width. This shortest pulse will result in the highest axial resolution in practical time-resolved confocal imaging. Although the results presented in this section is for an objective with an assumed value for numerical aperture, the conclusion concerning the improvement in axial resolution is generally true for any objective. In fact, the smaller the numerical aperture the larger the amount of improvement. However, the absolute axial resolution of the image is degraded as the numerical aperture becomes smaller.

It is necessary to discuss the physical reason underlying the improvement in axial resolution in time-resolved confocal imaging. As has been mentioned above, the image of a point source has a defocus-related phase shown in Eq. (8.2.24). It is frequency-dependent and is purely a result of light diffraction. Since time-resolved reflection-mode confocal microscopy uses a point-like detector, it is a purely-coherent imaging system. Therefore, under ultrashort pulse illumination, the contributions from the defocused phase resulting from different frequency components can be superposed together in a coherent manner in time-resolved imaging. Eventually, the resulting interference image gives a narrower axial response from a perfect reflector, i. e., higher resolution.

The purely-coherent nature of the confocal system may be destroyed when a time-averaged process is employed and then the contribution from the defocus-related phase is lost in recording the image. Consequently, the measured signal is simply the summation

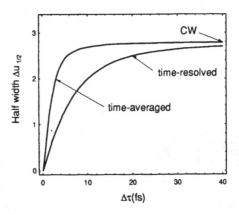

Fig. 8.2.13   Half width $\Delta u_{1/2}$ of the image intensity of a perfect reflector as a function of the pulse width $\Delta\tau$.

of the intensity from different frequency components, as shown in Eq. (8.2.35). The slight improvement in axial resolution in this case mainly originates from the term β^2 in the point spread function shown in Eq. (8.2.20). This term actually results from the factor $1/\lambda$ in the diffraction integral (see Eq. (2.1.14))[8.22] used for the derivation of Eq. (8.2.9). Because of the existence of β^2, higher frequency components within the spectrum $V_0(\Delta\omega)$ contribute more to the axial response than lower frequency components. Therefore, the resolution improvement described here is purely a result of diffraction of a pulsed beam without using other methods such as interference. In that sense, it is a fundamental effect.

Let us briefly discuss transverse resolution in time-dependent confocal imaging. Fig. 8.2.14 gives the projection of the 3-D CTF and TCC in the focal plane and is the 2-D in-focus CTF and TCC, thus providing the information on transverse resolution. Because the cut-off transverse spatial frequency of the spectrum-dependent CTF is 2β, the cut-off spatial frequencies of the 2-D time-dependent CTF and the TCC are, for the same example used above, approximately 10% larger than those for CW illumination. From the information-content point of view, this result implies a slight improvement in transverse resolution.

According to the above discussion, it is understood that the imaging properties in confocal microscopy under ultrashort pulse illumination can be considerably affected by the spectral width and spectral distribution, if a 10 fs pulsed laser beam is used. The spectral shape of the pulse can be altered using many optical methods including, for example, optical fibres, in order to obtain a preferred distribution.

In order to perform the time-dependent imaging described above, very fast gating is necessary. An alternative way of achieving similar results is to use interferometric methods with ultrashort pulses. If the interferometric term is extracted and integrated by a broadband detector the result may be similar to that for the time-dependent imaging for $t = 0$.

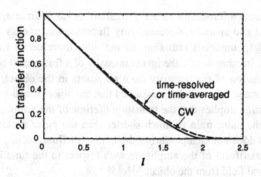

Fig. 8.2.14   2-D in-focus CTF for pulsed ($\Delta\tau = 10$ fs) and CW illumination.

8. 2. 5   Effect of Chromatic Aberration

We have assumed that the dispersion coefficients τ' and δ' are zero in the above numerical calculations. The term τ' represents just a time-shift of the response and hence results only in a change in the zero of the time coordinate. The term δ' results in pulse-broadening, which we have claimed can be compensated for by pre-chirping of the pulses. Strictly this need be performed both for the illumination and also for the detection (post-chirping), although if the compensation is performed completely on the incoming beam the result is only to leave some pulse-broadening in the specimen region, which affects the time-resolution of the imaging but not the spatial resolution. This argument relies on the assumption that the interaction with object is linear. In a reflection system, the pre- and post-chirping of pulses can be performed by inserting, for example, a pair of prisms in the optical path before the light beam enters the microscope objective.

The remaining dispersion terms can be easily included in the calculation of the CTF and the TCC, as they have the effect of a defocus, or higher order, aberrations.[8.24, 8.25] In general the wavefront aberration can be expressed as

$$\Phi(\rho, \Delta\omega) = a_{00} + a_{10}\, \Delta\omega + a_{20}\, (\Delta\omega)^2 + \dots + a_{j0}(\Delta\omega)^j \dots$$

$$+ a_{01}\, \rho^2 + a_{11}\, \Delta\omega\, \rho^2 + a_{21}(\Delta\omega)^2\rho^2 + a_{02}\, \rho^4 + a_{12}\,\Delta\omega\, \rho^4 + \dots \qquad (8.2.37)$$

where $a_{10} = \tau'$ is the first order dispersion, $a_{20} = \delta'$ is the second order dispersion, $a_{01} = \beta u/2$ is the defocus, $a_{11} = -\tau$ is the primary chromatic aberration, $a_{21} = -\delta$ is the secondary chromatic aberration, a_{02} is the spherical aberration, and a_{12} is the spherochromatism. For a well-corrected objective the spherical aberration is balanced by defocus for the central frequency, while the zonal dispersion is also balanced.

8. 3   Fluorescence Imaging

For fluorescence microscopy under ultrashort pulse illumination, we should distinguish temporal and spatial coherence. Any fluorescence process is related to an emitting process of spontaneous radiation, so that the fluorescence light is spatially incoherent radiation. In other words, the image intensity of a fluorescent object at a given point is the superposition of the intensity from all points in the object, if the imaging system is space-invariant. Since we have assumed that the object function is independent of time, this assumption implies that the transition lifetime of the sample is much longer than the period of the pulse train or much shorter than the pulse duration, so that the temporal coherence does not change during the course of fluorescence radiation. As a result, the Fourier transform of the amplitude with respect to the time is applicable to obtaining the temporal field from the object.[8.13, 8.14]

8. 3. 1   Single-Photon Fluorescence

For a given frequency component ω, the illumination field from a point source in the object space is, from Eq. (8.1.16),

$$V_0(\Delta\omega)\exp(-ikz_1)h_1(M_1r_1,\Delta\omega). \tag{8.3.1}$$

In the case of a single-photon fluorescent object, let us consider that the object is a single fluorescent point placed at r_1 in the object space. Because the single-photon fluorescence radiation is produced by the first-order polarization in the sample under the external field, in which case the emitted field is proportional to the incident field, and because the temporal coherence does not change during the interaction of light with the sample, the fluorescence field is reasonably proportional to the incident amplitude for a given frequency of ω, i. e., proportional to Eq. (8.3.1). Therefore, the image field of the point fluorescent object is, if the incident and fluorescence wavelengths are identical,

$$U_{1-p}(r_s,\Delta\omega) = V_0(\Delta\omega)\int_{-\infty}^{\infty} \delta(r_s - r_1)\exp[ik(\pm z_1 - z_1)]h_1(M_1r_1,\Delta\omega)h_2(r_1,\Delta\omega)dr_1. \tag{8.3.2}$$

The total image field of the point object under ultrashort pulse illumination is given by the inverse Fourier transform of Eq. (8.3.2) with respect to $\Delta\omega$:

$$U_{1-p}(r_s,t) = \int_{-\infty}^{\infty} V_0(\Delta\omega)\exp[ik(\pm z_s - z_s)]h_1(M_1r_s,\Delta\omega)h_2(r_s,\Delta\omega)\exp(-i\Delta\omega t)d\Delta\omega. \tag{8.3.3}$$

The intensity of the fluorescent point object is the modulus squared of Eq. (8.3.3). Because of the spatially-incoherent nature, the image intensity of any 3-D fluorescent object is the superposition of the intensity of all points in the object:

$$I(r_s,t) = \int_{-\infty}^{\infty} o_f(r_s - r_1)|U_{1-p}(r_1,t)|^2 dr_1, \tag{8.3.4}$$

where $o_f(r)$ is the fluorescence strength of the 3-D object.

Eq. (8.3.4) is also the measured intensity obtained in the time-resolved imaging mode when the response time of the detector is much shorter than the pulse duration. $|U_{1-p}(r_1,t)|^2$ is of course the 3-D effective intensity point spread function in the present case. Therefore the 3-D time-dependent optical transfer function OTF can be introduced and is given by

$$C(m,t) = \int_{-\infty}^{\infty} |U_{1-p}(r,t)|^2 \exp(-2\pi ir \cdot m)dr, \tag{8.3.5}$$

so that

$$I(r_s,t) = \int_{-\infty}^{\infty} C(m,t)O_f(m)\exp(2\pi ir_s \bullet m)dm, \tag{8.3.6}$$

where $O_f(m)$ is the inverse Fourier transform of $o_f(r)$.

Using Eq. (8.3.3) in Eq. (8.3.5) yields

$$C(m,t) = c(m,t) \otimes_3 c^*(-m,t), \tag{8.3.7}$$

where Eqs. (8.2.8) and (8.2.9) have been used. Clearly, $c(m, t)$ is the 3-D CTF in time-resolved imaging for non-fluorescent objects and is given by Eq. (8.2.11). As a result, time-resolved fluorescence confocal microscopy in reflection may be different from that in transmission. In this respect, this feature contrasts with confocal fluorescence microscopy with CW illumination, discussed in Chapter 5.

Alternatively, in the case of time-averaged imaging, the measured intensity is the average of Eq. (8.3.4) over a long time period and can be expressed as

$$I(r_s) = \int_{-\infty}^{\infty} o_f(r_s - r_1)\left[\int_{-\infty}^{\infty}|U_{1-p}(r_1,t)|^2 dt\right]dr_1, \tag{8.3.8}$$

which can be described by

$$I(r_s) = \int_{-\infty}^{\infty} C(m)O_f(m)\exp(2\pi ir_s \bullet m)dm. \tag{8.3.9}$$

Here

$$C(m) = \int_{-\infty}^{\infty} C(m,t)dt \tag{8.3.10}$$

is called the 3-D time-averaged OTF and can be derived, in terms of Eq. (8.3.5), as

$$C(m) = \int_{-\infty}^{\infty} |V_0(\Delta\omega)|^2 C_1(m,\Delta\omega) \otimes_3 C_2(m,\Delta\omega)d\Delta\omega, \tag{8.3.11}$$

which implies the superposition of the OTF contributed from different frequency components. This property is ensured by the fact that the temporal coherence is destroyed by the process of time-averaging. Here $C_1(m, \Delta\omega)$ and $C_2(m, \Delta\omega)$ are the 3-D spectrum-dependent OTF for the objective and the collector lenses given by

$$C_1(m,\Delta\omega) = \int_{-\infty}^{\infty} \left| h_1(M_1 r, \Delta\omega) \right|^2 \exp(-2\pi i r \bullet m) dr \tag{8.3.12}$$

and

$$C_2(m,\Delta\omega) = \int_{-\infty}^{\infty} \left| h_2(r, \Delta\omega) \right|^2 \exp(-2\pi i r \bullet m) dr, \tag{8.3.13}$$

respectively. Eqs. (8.3.11)-(8.3.13) indicate that $C(m)$ is the same in reflection and transmission confocal microscopy, whereas they are different in the time-resolved imaging method. The difference results from that fact that the temporal coherence is retained in time-resolved imaging, whereas it is completely destroyed in the process of time-averaged imaging.

For two identical circular lenses and for equal incident and fluorescence wavelengths, we have the 3-D time-dependent OTF:

$$C(l,s,t) = c(l,s,t) \otimes_3 c*(l,-s,t), \tag{8.3.14}$$

where $c(l, s, t)$ is given by the inverse Fourier transform, with respect to $\Delta\omega$, of Eq. (8.2.27) for reflection or of Eq. (8.2.31) for transmission, and the 3-D time-averaged OTF:

$$C(l,s) = \int_{-\infty}^{\infty} \left| V_0(\Delta\omega) \right|^2 [C(l,s,\Delta\omega) \otimes_3 C(l,s,\Delta\omega)] d\Delta\omega, \tag{8.3.15}$$

where $C(l, s, \Delta\omega)$ is given by

$$C(l,s,\Delta\omega) = \frac{2\beta^3 \exp\left(-i\dfrac{2ws}{\beta}\right)}{l} \begin{cases} \sqrt{\beta^2 - \left(\dfrac{l}{2} + \dfrac{|s|\beta}{l}\right)^2} & , \quad |s| \le \dfrac{l}{\beta}(\beta - \dfrac{l}{2}), \\ 0 & , \quad \text{otherwise.} \end{cases} \tag{8.3.16}$$

Here l and s have been normalized by Eqs. (8.2.27) and (8.2.28). Notice that the 3-D spectrum-dependent OTF for a lens differs from the 3-D spectrum-dependent CTF in transmission if the lens has chromatic aberration.

It is noticed that Eq. (8.3.16) does not include the effect of τ' and δ, which have cancelled due to the spatially-incoherent imaging nature and the time-averaging process. As a result, it is only affected by the radius-dependent chromatic aberration τ and δ. As mentioned before, any commercial microscope objective consisting of lens doublets may not exhibit strong chromatic aberration.[8.17] Therefore, τ and δ are zero, i. e., $w = 0$ corresponding to a zero-shift of the focal plane. We conclude that the imaging performance in a practical confocal fluorescence microscope under ultrashort pulse illumination may not be affected by material dispersion and that the finite spectral bandwidth of the pulsed beam may slightly improve the image quality because of the pre-factor β^3 in Eq. (8.3.16) and the spectral distribution of the incident pulse.

To understand the effects of the spectral bandwidth and the factor β^3, we assume that the pulse temporal profile is of a Gaussian temporal shape and that the temporal pulse width $\Delta\tau$ is 10 fs.

Fig. 8.3.1 is the 3-D time-averaged OTF normalized to unity at the origin. Clearly, it has a similar behaviour to the 3-D OTF for CW illumination shown in Fig. 5.3.1a. In fact, it is slightly broadened, which can be clearly observed from the dashed curves shown in Fig. 8.3.2 for the transverse and axial cross-sections of the 3-D OTF. The slight broadening of the 3-D OTF corresponds to an improvement in resolution. In particular, the strength of optical sectioning, i. e., the axial response to a thin fluorescent layer, in a confocal fluorescence microscope is a measure of the axial resolution of 3-D imaging and can be calculated by the Fourier transform of the axial cross-section of the 3-D OTF. Fig. 8.3.3a exhibits the comparison of the axial response for pulse and CW illumination (dashed curves). The magnitude of the improvement in the former case is approximately 7%. However, if the pulse is shorter than 10 fs, the improvement becomes large (see Fig. 8.3.3b). For example, when $\Delta\tau = 5$ fs, one can improve the axial resolution by 15% because the effect of the pre-factor β^3 and the spectral width becomes pronounced.

While the improvement in resolution for confocal fluorescence under pulse illumination is not large for a 10 fs pulse, the corresponding 3-D OTF has the cut-off spatial frequencies larger than those for CW illumination despite the fact that the values of the 3-D OTF for spatial frequencies higher than the cut-offs for CW illumination are small. The enlarged cut-off spatial frequencies may prove advantageous for post-image processing when an inverse filter is used.

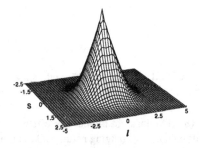

Fig. 8.3.1   3-D time-averaged OTF for confocal single-photon fluorescence microscopy ($\Delta\tau = 10$ fs).

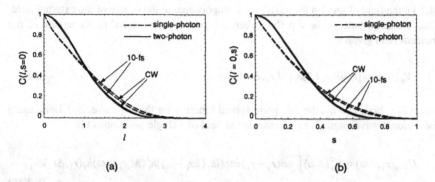

(a)   (b)

Fig. 8.3.2   Transverse (a) and axial (b) cross-sections of the 3-D time-averaged OTF for confocal single-photon and two-photon fluorescence imaging for pulsed ($\Delta \tau = 10$ fs) and CW illumination.

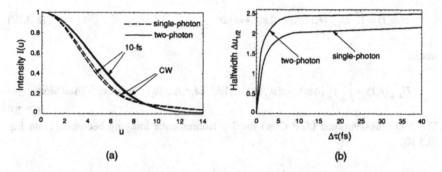

(a)   (b)

Fig. 8.3.3   Axial response to a thin fluorescent layer in confocal single-photon and two-photon fluorescence imaging for pulsed and CW illumination: (a) intensity of the axial response ($\Delta \tau = 10$ fs); (b) half width $\Delta u_{1/2}$ of the axial response as a function of the pulse width $\Delta \tau$.

8. 3. 2   Two-Photon Fluorescence

Another spatially-incoherent confocal imaging method is based on the nonlinear two-photon (2-p) fluorescence process.[8.5] This is particularly relevant for investigation of ultrashort pulse effects, as pulses are used in particular to increase the strength of the nonlinear signal. In this case, we assume that the wavelength of the fluorescence light, λ_f, is one half of the incident wavelength λ. Because the 2-p fluorescence process is caused by the second-order polarization in the sample under the external field of Eq. (8.3.1), the field of the 2-p fluorescence is proportional to the square of the external field. Thus the amplitude of the 2-p fluorescence signal is proportional to the square of the incident field given by

$$V_0(\Delta\omega)\exp(-ik_f z_1 / 2)h_1(M_1 r_1, \Delta\omega),$$

(8.3.17)

where $k_f = 2\pi/\lambda_f$ and h_1, the 3-D point spread function for the objective, is defined using the incident wavelength λ, so that the image field of a single point object is

$$U_{2-p}(r_s, \Delta\omega) = V_0^2(\Delta\omega)\int_{-\infty}^{\infty}\delta(r_s - r_1)\exp[ik_f(\pm z_1 - z_1)]h_1^2(M_1 r_1, \Delta\omega)h_2(r_1, \Delta\omega)dr_1,$$

(8.3.18)

where h_2, the 3-D point spread function for the collector, is defined using the 2-p fluorescence wavelength λ_f. In a similar way to the 1-p case, the 3-D time-dependent OTF $C(m, t)$ is given by Eq. (8.3.7). But $c(m, t)$ is now given by

$$c(m,t) = \int_{-\infty}^{\infty} U_{2-p}(r,t)\exp(-2\pi i r \bullet m)dr,$$

(8.3.19)

where

$$U_{2-p}(r,t) = \int_{-\infty}^{\infty} V_0^2(\Delta\omega)\exp[ik_f(\pm z - z)]h_1^2(M_1 r, \Delta\omega)h_2(r, \Delta\omega)\exp(-i\Delta\omega t)d\Delta\omega.$$

(8.3.20)

The 3-D time-averaged OTF $C(m)$ for 2-p fluorescence imaging becomes, from Eq. (8.3.10)

$$C(m) = \int\int_{-\infty}^{\infty}\left|U_{2-p}(r, \Delta\omega)\right|^2\exp(-2\pi i r \bullet m)dr d\Delta\omega,$$

(8.3.21)

which can be expressed as

$$C(m) = \int_{-\infty}^{\infty}\left|V_0(\Delta\omega)\right|^4[C_1(m, \Delta\omega) \otimes_3 C_2(m, \Delta\omega)]d\Delta\omega.$$

(8.3.22)

Here $C_2(m, \Delta\omega)$ is the 3-D spectrum-dependent OTF for the collector lens given by Eq. (8.3.13) but $C_1(m, \Delta\omega)$ is defined as

$$C_1(m, \Delta\omega) = \int_{-\infty}^{\infty}\left|h_1(r, \Delta\omega)\right|^4\exp(-2\pi i r \bullet m)dr,$$

(8.3.23)

which is the 3-D spectrum-dependent OTF for confocal 1-p fluorescence microscopy with two identical lenses and with the wavelength $2\lambda_f$.

For two identical circular lenses, Eq. (8.3.22) becomes

$$C(l,s) = \int_{-\infty}^{\infty} |V_0(\Delta\omega)|^4 [C_1(l,s,\Delta\omega) \otimes_3 C_1(l,s,\Delta\omega) \otimes_3 C_2(l,s,\Delta\omega)] d\Delta\omega, \qquad (8.3.24)$$

where $C_2(l, s, \Delta\omega)$ is the same as Eq. (8.3.16) but $C_1(l, s, \Delta\omega)$ becomes

$$C_1(l,s,\Delta\omega) = \frac{\beta^3 \exp\left(-i\dfrac{4ws}{\beta}\right)}{l} \begin{cases} \sqrt{\beta^2 - \left(l + \dfrac{|s|\beta}{l}\right)^2} & , \quad |s| \le \dfrac{l}{\beta}(\beta - l), \\ \\ 0 & , \quad otherwise. \end{cases} \qquad (8.3.25)$$

Therefore we conclude that there is no difference of imaging performance between reflection and transmission and that the chromatic aberration coefficients τ' and δ do not affect the time-averaged imaging process. It should be mentioned that the normalization factors for l and s are given by Eqs. (8.2.27) and (8.2.28), but defined at λ_f.

For the same condition as used in 1-p fluorescence, the 3-D time-averaged OTF for confocal 2-p fluorescence microscopy under pulse illumination exhibits little change in comparison with that for CW illumination. The transverse and axial cross-sections of the 3-D OTF for a 10 fs pulse are depicted in Fig. 8.3.2 (solid curves). The small change is caused by the nonlinear dependence of the fluorescence radiation, which makes the effective spectral bandwidth narrow despite the fact that the pre-factor now becomes β^9 in Eq. (8.3.24). Fig. 8.3.3a also gives the axial response to a thin 2-p fluorescent layer, $I(u)$, which is the Fourier transform of $C(0, s)$ in Fig. 8.3.2b. As expected, the improvement is only 2% in comparison with CW illumination. Even for a pulse of 5 fs, the change is still not pronounced (Fig. 8.3.3b).

As mentioned above, Eqs. (8.3.9) and (8.3.24) imply that the temporal coherence is destroyed due to the time-averaging process. Therefore the results presented in Figs. 8.3.1-8.3.3 are also applicable to the white-light illumination which has no temporal coherence. In both cases, the resolution improvement results from the finite spectral shape and the pre-factor β^j, where $j = 6$ in Eq. (8.3.15) and $j = 9$ in Eq. (8.3.24). The pre-factor β^j originates from the pre-factor β in the diffraction integral. β^j causes an asymmetric spectral distribution: the peak of the spectrum shifts towards high frequencies. As a result, frequency components higher than the central frequency contribute more strongly than those lower, so that resolution under pulse illumination of the central frequency ω_0 is slightly better than that for the CW illumination of frequency

ω_0. For a pulse shorter than 10 fs, the spectral width becomes comparable to the central frequency ω_0. In fact, $\Delta\Omega/\omega_0$ can be 0.45 for a 5 fs pulse. Consequently, the contributions from the frequency components higher than ω_0 become significantly strong, leading to a considerable improvement in axial resolution.

As the spectral width becomes broader for a pulse shorter than 10 fs, an objective corrected for the high-order terms of dispersion may be needed in order to achieve the aberration-free condition. Otherwise, the effect of δ and τ must be incorporated in the 3-D OTF. However, in any case, the results presented in this section give the limiting values one can achieve in practice.

Before we leave this section, it is worthwhile to mention that the assumption of the time-independent fluorescent object function requires that the temporal coherence of the incident light does not change during the course of fluorescence radiation. Of course, in practice, any fluorescent object function must be time-dependent and accordingly the temporal coherence changes somewhat. A time-dependent object function is needed for the description of the imaging performance. For example, the fluorescent object function may be expressed as

$$o_f(r,t) = o_f(r)\exp\left(-\frac{t}{t_f}\right), \tag{8.3.26}$$

where t_f can be considered to be the lifetime of the fluorescent object. Imaging of the object expressed in Eq. (8.3.26), in particular, in time-resolved imaging, is of importance, as it gives the dynamic behaviour of the sample, and should be explored in future. However, the results presented in this section are applicable if the fluorescence lifetime is either much longer than the pulse period or much shorter than the pulse period. The former condition may not be satisfied for some lasers. But one can reduce the repetition rate of the pulsed laser so that the latter condition is satisfied.

References

8.1.   K. P. Ghiggino, M. R. Harris, and P. G. Spizzirri, *Rev. Sci. Instrum.*, **63** (1992) 2999.

8.2.   C. G. Morgan, A. C. Mitchell, and J. G. Murray, *J. Microscopy*, **165** (1992) 49.

8.3.   E. P. Buurman, R. Sanders, A. Draaijer, H. C. Gerritsen, J. J. F. van Veen, P. M. Houpt, and Y. K. Levine, *Scanning*, **14** (1992) 155.

8.4.   C. J. R. Sheppard, Scanning optical microscopy, in *Advances in Optical and Electron Microscopy*, Vol. 10, eds. R. Barer and V. E. Cosslett (Academic, London, 1987), p. 58; p. 84-87.

8.5.   E. H. K. Stelzer, S. Hell, S. Lindek, R. Stricker, R. Pick C. Storz, G. Ritter, and N. Salmon, *Opt. Commun.*, **104** (1994) 223.

8.6.   K. Sasaki, M. Koshioka, and H. Masuhara, *J. Opt. Soc. Am. A*, **9** (1992) 932.

8.7.   M. Gu and C. J. R. Sheppard, *J. Modern Optics*, **42** (1995) 747.

8.8.   G. E. Anderson, F. Lui, and R. R. Alfano, *Opt. Lett.*, **19** (1994) 981.

8.9.   M. Kempe, A. Thon, and W. Rudolph, *Opt. Commun.*, **110** (1994) 492.

8.10.   M. Kempe and W. Rudolph, *Opt. Lett.*, **19** (1994) 1919.

8.11.   M. R. Hee, J. A. Izatt, E. A. Swanson, and J. G. Fujimoto. *Opt. Lett.*, **18** (1994) 1107.

8.12.   B. Chance and R. R. Alfano, *Photon Migration and Imaging in Random Media and Tissues*, *SPIE Proc.*, **1888** (1993).

8.13.   M. Kempe and W. Rudolph, *J. Opt. Soc. Am. A*, **10** (1993) 240.

8.14.   M. Gu, T. Tannous, and C. J. R. Sheppard, *Opt. Commun.*, **117** (1995) 406.

8.15.   M. Gu, *Optik*, (1995), in press.

8.16.   M. Gu and C. J. R Sheppard, *J. Opt. Soc. Am. A*, **11** (1994) 2742.

8.17.   M. Kempe, U. Stamm, B. Wilhelmi, and W. Rudolph, *J. Opt. Soc. Am. B*, **9** (1992) 1158.

8.18.   M. Kempe, U. Stamm, and B. Wilhelmi, *Opt. Commun.*, **89** (1992) 119.

8.19.   M. T. Asaki, C-P Huang, D. Garvey, J. Zhou, H. C. Kapteyn, and M. M. Murnane, *Opt. Lett.*, **18** (1993) 977.

8.20.   J. Zhou, G. Taft, C. P. Huang, M. M. Murnane, H. C. Kapteyn, and I. P. Christov, *Opt. Lett.*, **19** (1994) 1149.

8.21   A. Stingl, C. Spielmann, F. Krausz, and R. Szipocs, *Opt. Lett.*, **19** (1994) 204.

8.22.   M. Born and E. Wolf, *Principles of Optics* (Pergamon, New York, 1980).

8.23.   M. Gu, T. Tannous, and C. J. R. Sheppard, *Zoological Studies*, **34**, suppl. (1995) 76.

8.24.   M. Kempe and W. Rudolph, *Opt. Lett.*, **18** (1993) 137.

8.25.   M. Kempe and W. Rudolph, *Phys. Rev. A*, **48** (1993) 4721.

Chapter 9

CONFOCAL MICROSCOPY

WITH HIGH-APERTURE OBJECTIVES

Three-dimensional (3-D) transfer functions developed in the previous chapters are based on the paraxial approximation. This assumption may not hold for an objective of a numerical aperture higher than $1/\sqrt{2}$. In this chapter, transfer functions for confocal microscopy of high-aperture objectives are studied. We only consider a confocal system consisting of two identical circular lenses and of a point source and a point detector.

In Section 9.1, the 3-D coherent transfer function for confocal bright-field microscopy is developed and axial resolution is discussed. Section 9.2 deals with the 3-D optical transfer function in confocal single-photon and two-photon fluorescence microscopy. Imaging properties in recently proposed 4Pi confocal microscopy which uses two high-aperture objectives are studied in terms of the 3-D transfer functions in Section 9.3.

In this chapter, a scalar theory is used to describe the imaging system. This description means that the effect of apodization for a high-aperture optical system is included but that the vectorial property of the light near the focal plane is neglected. Two kinds of the apodization functions are taken into account: one is for a lens obeying the sine condition[9.1, 9.2] which is approximately the case for commercial objectives, and the other is for a lens satisfying the Herschel condition, i. e., the uniform angular illumination condition.[9.1, 9.2]

9. 1 Confocal Bright-Field Imaging

According to the scalar theory for an optical system of a high-aperture circular lens,[9.1-9.3] the image field of a single point placed on the axis, $h(v, u)$, i. e., the 3-D amplitude point spread function (APSF) in image space, is given by

$$h(v,u) = \exp\left[-\frac{iu}{4\sin^2(\alpha_o/2)}\right]\int_0^{\alpha_o} P(\theta)J_0\left(\frac{v\sin\theta}{\sin\alpha_o}\right)\exp\left[\frac{iu\sin^2(\theta/2)}{2\sin^2(\alpha_o/2)}\right]\sin\theta d\theta,$$

(9.1.1)

where J_0 is the Bessel function of the first kind of order zero, θ is the angle of convergence of a ray, and $P(\theta)$ is an apodization function or the pupil function of the lens. For uniform angular illumination, i. e., for the Herschel condition,[9.2] $P(\theta)$ is a constant, whereas for a lens satisfying the sine condition,[9.2] it is

$$P(\theta) = \cos^{1/2}\theta.$$

(9.1.2)

The sine condition also means that the imaging system has the two-dimensional (2-D) transverse space-invariance, while the Herschel system represents the one-dimensional (1-D) axial space-invariance. The expressions for other apodization conditions including, for example, the uniform, tangent and parabolic conditions, can be found from Refs. 9.3 and 9.4.

Variables v and u in Eq. (9.1.1) are the transverse and axial optical coordinates, respectively, defined as

$$v = \frac{2\pi}{\lambda}r\sin\alpha_o,$$

(9.1.3a)

$$u = \frac{8\pi}{\lambda}z\sin^2(\alpha_o/2),$$

(9.1.3b)

where r and z are the real radial and axial coordinates in the image space, and $\sin\alpha_o$ is the numerical aperture of the objectives in the image space. λ is the wavelength of the incident light.

It should be emphasized that Eq. (9.1.1), in general, does not have 3-D space-invariance, unlike the 3-D APSF under the paraxial approximation shown in Eq. (2.3.10). However, the 2-D transverse space-invariance can be obtained if the lens obeys the sine condition, which is usually the case for a commercial objective.

The 3-D coherent transfer function (CTF) for a thin lens in the object space can be formally derived by performing the 3-D inverse Fourier transform of Eq. (9.1.1), which can be reduced to

$$c(l,s) = \int_{-\infty}^{\infty}\int_{0}^{\infty}\left[\int_{0}^{\alpha_o}\exp\left[-\frac{iu}{4\sin^2(\alpha_o/2)}\right]J_0\left(\frac{v\sin\theta}{\sin\alpha_o}\right)\exp\left[\frac{iu\sin^2(\theta/2)}{2\sin^2(\alpha_o/2)}\right]P(\theta)\sin\theta d\theta\right]$$

$$J_0(2\pi lr)\exp(-2\pi izs)2\pi rdrdz.$$

(9.1.4)

Normalizing l and s by

$$1/\lambda,$$

(9.1.5)

we have the 3-D CTF:

$$c(l,s) = \frac{P(l)}{\sqrt{1-l^2}}\delta\left(s+\sqrt{1-l^2}\right),$$

(9.1.6)

where $l = \sin\theta$ and $s = \cos\theta$. Therefore, the 3-D CTF for a high-aperture lens is, as expected, represented by a cap of a sphere,[9.5] given by the following function:

$$s = -\sqrt{1-l^2}.$$

(9.1.7)

The cap has an effective weighting function:

$$\frac{P(l)}{\sqrt{1-l^2}}.$$

(9.1.8)

Clearly, it cuts off at $l = \sin\alpha_0$ in the transverse direction, corresponding to the maximum convergence angle of the lens, and at $s = \cos\alpha_0$ and 1 in the axial direction. Thus the 3-D CTF is axially shifted. When the numerical aperture of the lens is small, the cap of a sphere can be approximately represented by a cap of a paraboloid, which is called the paraxial approximation. The 3-D CTF for a lens under the paraxial approximation is shown in Fig. 2.4.1.

In comparison with the 3-D CTF in Eq. (2.4.11) derived under the paraxial approximation, the definitions of l and s are different between the high-aperture and paraxial-approximation cases. The relationship of the two definitions is

$$(s \mp s_0)[4\sin^2(\alpha_o/2)] \Rightarrow s \mp 1,$$

(9.1.9)

$$l\sin\alpha_o \Rightarrow l.$$

The negative and positive signs correspond to the axial spatial frequency in the positive and negative regions, respectively. The variables s and l on the left-hand side of Eq. (9.1.8) are the axial and radial (transverse) spatial frequencies used under the paraxial approximation, whereas those on the right-hand side correspond to the case of a high-aperture objective. Using the definition on the right-hand side, one can express the cap of a sphere in a simple mathematical form.

9.1.1 Coherent Transfer Function

For a confocal bright-field microscope of a point source and a point detector, we have expressed the 3-D effective APSF $h_a(v, u)$ in Eq. (3.1.19), which is a product of two 3-D APSFs, $h_1(v, u)$ and $h_2(v, u)$, for the objective and collector lenses in the object space (see Fig. 3.1.1). Therefore we have

$$h_a(v, u) = h_1(v, u)\, h_2(v, u). \tag{9.1.10}$$

If an object scanning method is used, the field of a confocal image can be represented by the 3-D convolution of the 3-D object function with the 3-D effective point spread function $h_a(v, u)$ because the optical properties of the confocal system remains unchanged when the object is scanned. Thus, the confocal system exhibits 3-D space-invariance in this case. Even for a beam scanning instrument, object scanning is usually used for axial imaging so that imaging is space-invariant in the axial direction, Then for an objective obeying the sine condition imaging is again fully 3-D space-invariant. As a result, the image intensity in the present case is

$$I(v_x, v_y, u) = \left| h_a(v, u) \otimes_3 o(v_x, v_y, u) \right|^2, \tag{9.1.11}$$

where $o(v_x, v_y, u)$ is a 3-D object function expressed by transverse optical coordinates v_x and v_y, and one axial optical coordinate u.

Accordingly, the 3-D CTF for a confocal system can be introduced and is given by the 3-D inverse Fourier transform of Eq. (9.1.10):

$$c(l, s) = F_3[h_1(v, u)] \otimes_3 F_3[h_2(v, u)]. \tag{9.1.12}$$

Here F_3 denotes the 3-D inverse Fourier transform. For a reflection confocal system consisting of two identical lenses, we have

$$h_1(v, u) = h(v, u), \tag{9.1.13a}$$

$$h_2(v, u) = h(v, u), \tag{9.1.13b}$$

whereas we have

$$h_1(v, u) = h(v, u) , \tag{9.1.14a}$$

$$h_2(v, u) = h(v, -u) , \tag{9.1.14b}$$

for a transmission confocal system. Substituting Eqs. (9.1.13) and (9.1.14) into Eq. (9.1.12) yields

$$c_r(l,s) = \frac{P(l)}{\sqrt{1-l^2}} \delta\left(s + \sqrt{1-l^2}\right) \otimes_3 \frac{P(l)}{\sqrt{1-l^2}} \delta\left(s + \sqrt{1-l^2}\right) \tag{9.1.15}$$

for reflection, and

$$c_t(l,s) = \frac{P(l)}{\sqrt{1-l^2}} \delta\left(s + \sqrt{1-l^2}\right) \otimes_3 \frac{P(l)}{\sqrt{1-l^2}} \delta\left(s - \sqrt{1-l^2}\right) \tag{9.1.16}$$

for transmission. They correspond, respectively, to the auto-convolution of a cap of a sphere and to the convolution of a cap of a sphere with its axially inverted cap. The analytical solutions[9.6] to Eqs. (9.1.15) and (9.1.16) can be expressed, for a system satisfying the sine condition, as

$$c_r(l,s) = \begin{cases} \dfrac{2|s|}{\pi\sqrt{l^2+s^2}} E(\tilde{p}) & , \quad 2(l\sin\alpha_o + |s|\cos\alpha_o) \le (l^2+s^2) \le 4, |s| \ge 2\cos\alpha_o, \\[3mm] \dfrac{2|s|}{\pi\sqrt{l^2+s^2}} E(\beta_1, \tilde{p}) & , \quad (l^2+s^2) \le 2(l\sin\alpha_o + |s|\cos\alpha_o), |s| \ge 2\cos\alpha_o, \\[3mm] 0 & , \quad\quad otherwise, \end{cases} \tag{9.1.17}$$

and

$$c_t(l,s) = \begin{cases} \dfrac{2|s|}{\pi\sqrt{l^2+s^2}} (\tilde{p}^2-1)^{1/2} E\left[\beta_2, \dfrac{\tilde{p}}{(\tilde{p}^2-1)^{1/2}}\right] & , \quad (l^2+s^2) \le 2(l\sin\alpha_o - |s|\cos\alpha_o), \\[3mm] 0 & , \quad\quad otherwise. \end{cases} \tag{9.1.18}$$

Here $E(x, y)$ is an incomplete elliptic integral of the second kind[9.7] and $E(x)$ is a complete elliptic integral.[9.7] Variables β_1 and β_2 are defined as

$$\beta_1 = \sin^{-1}\left[\frac{1}{\tilde{p}}\left(1 - \frac{2\cos\alpha_o}{|s|}\right)\right], \tag{9.1.19}$$

$$\beta_2 = \cos^{-1}\left[\frac{1}{\tilde{p}}\left(\frac{2\cos\alpha_o}{|s|} + 1\right)\right], \tag{9.1.20}$$

where

$$\tilde{p} = \frac{2l}{|s|\sqrt{l^2 + s^2}}\left(1 - \frac{l^2 + s^2}{4}\right)^{1/2}. \tag{9.1.21}$$

In the case of the uniform angular illumination condition, the solutions in both cases are[9.6]

$$c_r(l,s) = \begin{cases} \dfrac{2}{\sqrt{l^2 + s^2}} & , \quad 2(l\sin\alpha_o + |s|\cos\alpha_o) \le (l^2 + s^2) \le 4, |s| \ge 2\cos\alpha_o, \\[2ex] \dfrac{4}{\pi\sqrt{l^2 + s^2}}\beta_1 & , \quad (l^2 + s^2) \le 2(l\sin\alpha_o + |s|\cos\alpha_o), |s| \ge 2\cos\alpha_o, \\[2ex] 0 & , \quad \qquad\qquad otherwise, \end{cases} \tag{9.1.22}$$

and

$$c_t(l,s) = \begin{cases} \dfrac{4}{\pi\sqrt{l^2 + s^2}}\beta_2 & , \quad (l^2 + s^2) \le 2(l\sin\alpha_o - |s|\cos\alpha_o), \\[2ex] 0 & , \quad \qquad otherwise. \end{cases} \tag{9.1.23}$$

Clearly, the non-zero region of the 3-D CTF in the reflection case is a spherical cap given by $l^2 + s^2 = 4$ and by $s = \pm 2\cos\alpha_o$.[9.8] It cuts off at $2\sin\alpha_o$ in the transverse direction, and has low and high cut-off axial spatial frequencies of $\pm 2\cos\alpha_o$ and ± 2. Hence, the 3-D CTF in this case is axially shifted like that derived under the paraxial approximation (Fig. 3.2.3). The 3-D CTF comprises two components shifted in the opposite axial directions. Either of the two axially shifted components denotes the 3-D

CTF in a reflection system, depending on the direction of the incident wave. The cut-off transverse spatial frequency in the transmission is the same as that in the reflection, while the axial cut-off spatial frequency in transmission is given by $\pm(1 - \cos\alpha_o)$.

Fig. 9.1.1 gives the plots of the 3-D CTF in reflection for the sine condition, whereas Fig. 9.1.2 gives the 3-D CTFs in reflection for the uniform angular illumination condition. It is noted that when the numerical aperture of the objective is larger than $\pi/3$, the 3-D CTFs behave differently between the two apodization conditions. This is a feature of the 3-D CTF for a high-aperture system. Accordingly, the 3-D imaging performance can be affected by the apodization of the objective. However, for a low-aperture objective, the 3-D CTFs shown in Figs. 9.1.1 and 9.1.2 approach the same

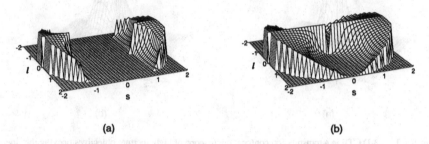

(a)   (b)

Fig. 9.1.1   3-D CTF in a reflection confocal microscope of high-aperture objectives obeying the sine condition: (a) $\alpha_o = 60°$; (b) $\alpha_o = 90°$.

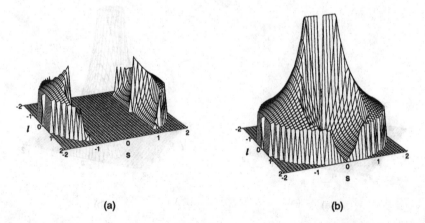

(a)   (b)

Fig. 9.1.2   3-D CTF in a reflection confocal microscope of high-aperture objectives obeying the Herschel condition: (a) $\alpha_o = 60°$; (b) $\alpha_o = 90°$.

limiting form derived under the paraxial approximation (see Fig. (3.2.3)) and are less affected by the apodization function.

It should be particularly mentioned that the 3-D CTF for a reflection microscope for the uniform angular illumination condition has a singularity at the origin when the numerical aperture is $\pi/2$, whereas that for the sine condition does not. This difference is

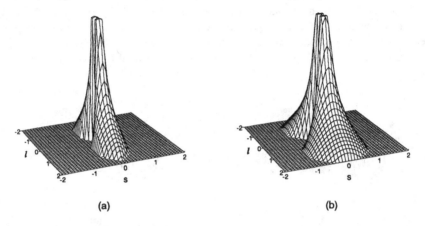

Fig. 9.1.3   3-D CTF in a transmission confocal microscope of high-aperture objectives obeying the sine condition: (a) $\alpha_o = 60°$; (b) $\alpha_o = 90°$.

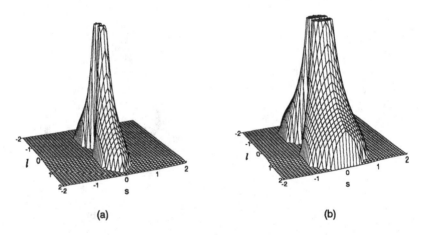

Fig. 9.1.4   3-D CTF in a transmission confocal microscope of high-aperture objectives obeying the Herschel condition: (a) $\alpha_o = 60°$; (b) $\alpha_o = 90°$.

purely caused by the apodization effect and results in the difference of axial resolution between the two cases.

The 3-D CTFs in transmission are shown in Figs. 9.1.3 and 9.1.4 for the sine and uniform angular illumination conditions, respectively. They exhibit a missing cone of spatial frequencies at the origin, as expected. The difference between the two apodization functions becomes pronounced only when the numerical aperture of the objective approaches $\pi/2$. These figures are also the 3-D optical transfer function (OTF) for a single lens of high aperture.[9.6]

9.1.2 Axial resolution

In order to characterize axial resolution of a confocal microscope of high aperture, we consider the object to be a thin uniform layer scanned in the axial direction. The axial response to the layer is then given by the Fourier transform of the axial cross-section of the 3-D CTF in reflection. Alternatively, from a general point of view, i. e., from Eq. (9.1.11), the intensity of the axial response can be expressed as the convolution of the 3-D effective APSF with the object function $\delta(u)$:

$$I(u) = \left| h_a(v,u) \otimes_3 \delta(u) \right|^2, \tag{9.1.24}$$

which can be evaluated as

$$I(u) = \left| K \exp\left[-\frac{iu}{2\sin^2(\alpha_o/2)} \right] \int_0^{\alpha_o} \exp\left[\frac{iu\sin^2(\theta/2)}{\sin^2(\alpha_o/2)} \right] P^2(\theta)\tan\theta d\theta \right|^2, \tag{9.1.25}$$

where K is a constant of normalization. For a system obeying the sine condition, it becomes

$$I(u) = \frac{\sin^2(u/2)}{(u/2)^2}, \tag{9.1.26}$$

and for the uniform angular illumination condition, one has

$$I(u) = \left| K \exp\left[\frac{-iu}{2\sin^2(\alpha_o/2)} \right] \int_0^{\alpha_o} \exp\left[\frac{iu\sin^2(\theta/2)}{\sin^2(\alpha/2)} \right] \tan\theta d\theta \right|^2 \tag{9.1.27}$$

which may be evaluated numerically.

However, it is difficult to prepare a thin uniform layer in practice. Instead, a perfect reflector is often used for characterising axial resolution.[9.9] It has been found that a perfect reflector is a dipole layer when it is illuminated by a high-aperture objective and has an object spectrum:[9.10]

$$O(s) = s,$$ (9.1.28)

which in turn gives an object function by its Fourier transform:

$$o(u) = \int_{-\infty}^{\infty} O(s) \exp\left[-\frac{isu}{4\sin^2(\alpha_o/2)}\right] ds,$$ (9.1.29)

where the angle-related factor in the exponent results from the definition of s (see Eq. (9.1.9)). Eq. (9.1.29) can be evaluated as

$$o(u) = \frac{id}{du}\delta(u),$$ (9.1.30)

where a constant factor has been omitted. Replacing $\delta(u)$ by Eq. (9.1.30) in Eq. (9.1.24) and using the following the property of the delta function:

$$\int_{-\infty}^{\infty} f(x)\delta'(x-x_0)dx = f'(x_0),$$ (9.1.31)

where a prime represents the first derivative, one can express the intensity of the axial response to the perfect reflector as

$$I(u) = \left| K \exp\left[-\frac{iu}{2\sin^2(\alpha_o/2)}\right] \int_0^{\alpha_o} \exp\left[\frac{iu\sin^2(\theta/2)}{\sin^2(\alpha_o/2)}\right] P^2(\theta)\sin\theta d\theta \right|^2.$$ (9.1.32)

In a confocal system satisfying the sine condition, Eq. (9.1.32) can be analytically expressed as[9.9]

$$I(u) = \left\{\tan^2(\alpha_o/2)\left(\frac{\sin(u/2)}{u/2}\right)^2 + [1-\tan^2(\alpha_o/2)]\left(\frac{\sin(u/2)}{u/2}\right)\cos(u/2)\right\}^2 +$$

$$\left\{\frac{1}{u/2}\left\{\tan^2(\alpha_o/2)\left[1-\left(\frac{\sin(u/2)}{u/2}\right)\cos(u/2)\right] - [1-\tan^2(\alpha_o/2)]\sin^2(u/2)\right\}\right\}^2$$

(9.1.33)

while for the uniform angular illumination condition, we have the same expression as Eq. (9.1.26), which is also the identical to the axial response to a perfect reflector under paraxial approximation. In fact, Eqs. (9.1.27) and (9.1.33) reduce to Eq. (9.1.24) when the numerical aperture is small, showing that a perfect reflector and a thin uniform layer are the same object for characterizing the axial resolution for a low-aperture objective. Eqs. (9.1.26), (9.1.27) and (9.1.33) are shown in Fig. 9.1.5.

In practice, a planar reflector may not be uniform but have an angle-dependent reflectivity. In this case, the axial response to the reflector is given by[9.9]

$$I(z) = \left| K \int_0^{\alpha_o} R(\theta) P^2(\theta) \exp(2ikz\cos\theta)\sin\theta d\theta \right|, \qquad (9.1.34)$$

where $R(\theta)$ is the reflectivity of the reflector. It reduces to Eq. (9.1.32) for a perfect reflector. Comparing Eq. (9.1.34) with Eq. (9.1.25) and recalling the relation between u and z (see Eq. (9.1.3b)) we can find that the reflectivity for a thin uniform layer is

$$R(\theta) = 1/\cos\theta, \qquad (9.1.35)$$

which is purely a result for the objective of high aperture. In fact, when the angle α_o is small, Eq. (9.1.35) approaches a constant, as expected. For multi-layer structures, $R(\theta)$ can be a complex function.[9.11, 9.12]

Another practical effect is spherical aberration which can be produced when the refractive index of the immersion material of the objective does not match that of the specimen.[9.13, 9.14] The spherical aberration function can be derived, if the refractive-index mismatch is small, as

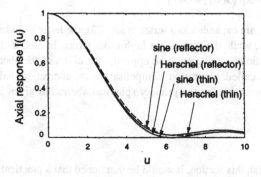

Fig. 9.1.5   Intensity of the axial response to a perfect reflector and to a thin layer. The numerical aperture of the objective is 1.4.

$$\Phi_{A_r} = A_r \sec\theta, \qquad\qquad\qquad\qquad\qquad\qquad (9.1.36)$$

where A_r is a dimensionless parameter defined as[9.13]

$$A_r = kL_d \Delta\bar{n} . \qquad\qquad\qquad\qquad\qquad\qquad (9.1.37)$$

Here L_d is the thickness of the immersion layer and $\Delta\bar{n}$ is the difference of the refractive indices between the immersion material and the specimen. k is the modulus of the wavelength vector in vacuum. In this case, the effect of the spherical aberration can be described by a complex pupil function (or a complex apodization function) of the lens, given by

$$P(\theta) = \cos^{1/2}\theta \exp(i\Phi_{A_r}) \qquad\qquad\qquad\qquad\qquad (9.1.38)$$

for the sine condition. It has been shown that the axial response in the presence of the spherical aberration Φ_{A_r} becomes asymmetric with pronounced sidelobes. The intensity of the central peak is reduced, while its width is accordingly broadened.[9.13, 9.14]

   Another spherical aberration source in confocal systems is caused by the alteration of the effective tube length at which the objective lens is operated, which can be expressed, in the case of the sine condition, as[9.3]

$$\Phi_{B_t} = B_t \sin^4\theta/2 , \qquad\qquad\qquad\qquad\qquad\qquad (9.1.39)$$

where B_t is a normalized parameter including optical parameters of the lens.[9.3] The complex pupil function for the lens is now given by

$$P(\theta) = \cos^{1/2}\theta \exp [i(\Phi_{A_r} + \Phi_{B_t})]. \qquad\qquad\qquad\qquad (9.1.40)$$

   If Φ_{A_r} and Φ_{B_t} are expanded as a series in $\sin(\theta/2)$, the latter includes only primary spherical aberration, while the former has high-order terms. In fact, if the signs of the two aberration functions are chosen to be opposite, the effect of the aberrations can be reduced, which is called aberration compensation or aberration balance.[9.13] The condition for compensating for the primary spherical aberration given in Eq. (9.1.39) is[9.13]

$$A_r = -3B_r. \qquad\qquad\qquad\qquad\qquad\qquad\qquad (9.1.41)$$

   Before we finish this section, it should be mentioned that a practical objective may not precisely obey the sine condition as the measured aberration function deviates from Eq. (9.1.39) appreciably[9.15] and includes the high-order terms of the spherical aberration caused by a change of tube length. The deviation from the sine condition may be caused by the optimization of the whole performance of the objective when it is designed or by the Fresnel loss on each surface of the glass elements within the objective.

9. 2 Confocal Fluorescence Imaging

In order to reveal the effect of the numerical aperture of the objective on 3-D confocal fluorescence imaging, we consider, in this section, the 3-D optical transfer function for confocal single-photon (1-p) and two-photon (2-p) fluorescence microscopy of high-aperture objectives.

9. 2. 1 Single-Photon Fluorescence

Let us recall that the 3-D effective intensity point spread function (IPSF) for a confocal 1-p fluorescence microscope of a point source and a point source is given by Eq. (4.2.7) which is a product of the 3-D intensity point spread functions for the objective and collector lenses in the object space. This conclusion still holds for an object scanning system of high-aperture lenses. If the two lenses are identical circular lenses and the incident and fluorescence wavelengths are the same, the 3-D effective IPSF for a confocal 1-p fluorescence microscope is given by

$$h_i(v,u) = |h(v,u)|^4. \tag{9.2.1}$$

Therefore, the 3-D optical transfer function (OTF) is the 3-D inverse Fourier transform of Eq. (9.2.1):

$$C(l,s) = F_3\left[|h(v,u)|^4\right]. \tag{9.2.2}$$

Here F_3 denotes the 3-D inverse Fourier transform.

It is easy to find out that the region where the value of the 3-D OTF is not zero[9.16] is given by

$$s = \begin{cases} \pm[\sqrt{4-(l-2\sin\alpha_o)^2} - 2\cos\alpha_o] & , \quad 2\sin\alpha_o \le l \le 4\sin\alpha_o, \\ \\ \pm(2-2\cos\alpha_o) & , \quad 0 \le l < 2\sin\alpha_o. \end{cases} \tag{9.2.3}$$

Therefore the cut-off spatial frequencies are $4\sin\alpha_o$ and $\pm2(1-\cos\alpha_o)$ in the transverse and axial directions, respectively. The difference of the cut-offs between the transverse and axial directions is caused by the limited illumination aperture, i. e., by the fact that the numerical aperture is less than $\pi/2$. When the illumination aperture becomes larger than $\pi/2$, the axial cut-off spatial frequency can increase significantly (see Section 9.3).

The 3-D OTFs are shown in Fig. 9.2.1 for the sine condition and in 9.2.2 for the uniform angular illumination condition. The 3-D OTFs are confined to the region

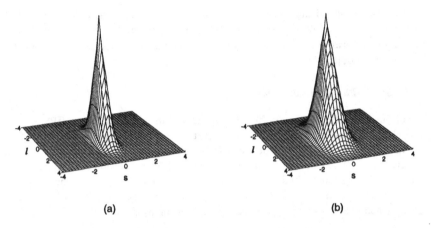

(a)   (b)

Fig. 9.2.1   3-D OTF for a confocal 1-p fluorescence microscope of high-aperture objectives obeying the sine condition: (a) $\alpha_o = 60°$; (b) $\alpha_o = 90°$.

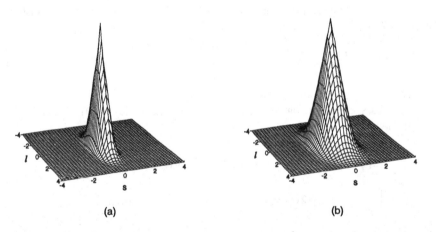

(a)   (b)

Fig. 9.2.2   3-D OTF for a confocal 1-p fluorescence microscope of high-aperture objectives obeying the Herschel condition: (a) $\alpha_o = 60°$; (b) $\alpha_o = 90°$.

described by Eq. (9.2.3). It is noticed that the 3-D OTFs show less difference between the sine and uniform angular illumination conditions even for the maximum value of α_o. In this respect, it contrasts with the 3-D CTF discussed in Section 9.1, which displays a significant difference between the two apodization functions when the numerical aperture of the lens is large. In comparison with the previous results of the 3-D OTF for the confocal 1-p fluorescence microscope under the paraxial approximation, discussed in Chapter 5, the 3-D OTF for $\alpha_o = \pi/3$ in Figs. 9.2.1a and 9.2.2a is very similar to that shown in Fig. 5.3.1a if the conversion of Eq. (9.1.9) is used. When the angle approaches $\alpha_o = \pi/2$, the 3-D OTF in Figs. 9.2.1 and 9.2.2 has cut-off spatial frequencies of 4 and 2 in the transverse and axial directions, respectively.

9. 2. 2 Two-Photon Fluorescence

For confocal two-photon (2-p) fluorescence microscopy, we can express the 3-D IPSF for the case of a point source and a point detector, using the expression of Eq. (6.2.6), as

$$h_i(v,u) = |h(v/2,u/2)|^4 |h(v,u)|^2, \tag{9.2.4}$$

where we have assumed that the confocal system has two identical circular lenses, and that the fluorescence and incident wavelengths are λ and 2λ. In Eq. (9.2.4), the first term is the modulus squared of the 3-D IPSF for the objective lens as a result of the fact that the strength of 2-p fluorescence is proportional to the square of the incident intensity,[9.17]

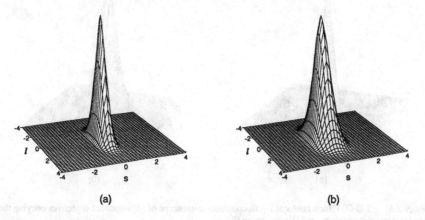

(a)   (b)

Fig. 9.2.3   3-D OTF for a confocal 1-p fluorescence microscope of high-aperture objectives obeying the sine condition: (a) $\alpha_o = 60°$; (b) $\alpha_o = 90°$.

and the second is the 3-D IPSF for the collector lens.

Therefore the 3-D OTF is given by

$$C(l,s) = F_3\left[|h(v/2,u/2)|^4\right] \otimes_3 F_3\left[|h(v,u)|^2\right]. \tag{9.2.5}$$

The first term represents the 3-D OTF for confocal fluorescence 1-p microscopy at wavelength 2λ, as shown in Figs. 9.2.1 and 9.2.2 for the sine and Herschel conditions, but its scale should be shrunk by a factor of two. This feature implies the existence of optical sectioning in conventional 2-p fluorescence imaging. The second term in Eq. (9.2.5) denotes the 3-D OTF for conventional 1-p fluorescence microscopy of a high-aperture objective and is given by Eq. (9.1.18) for the sine condition and by Eq. (9.1.23) for the uniform angular illumination condition.

The passband of Eq. (9.2.5) is as same as that of Eq. (9.2.3) for confocal 1-p fluorescence microscopy. Fig. 9.2.3 gives the 3-D OTF for confocal 2-p fluorescence microscopy obeying the sine condition. The 3-D OTF for the confocal 2-p fluorescence microscope obeying the uniform angular illumination condition is shown in Fig. 9.2.4. As expected, the slope of the 3-D OTF is zero at the origin, resulting in the enhanced response at low spatial frequencies like the situation observed in the paraxial approximation (Chapter 6). The results for the two apodization functions are almost identical even for $\alpha_o = 90°$. This property shows that the effect of apodization on 2-p fluorescence imaging is not so significant.

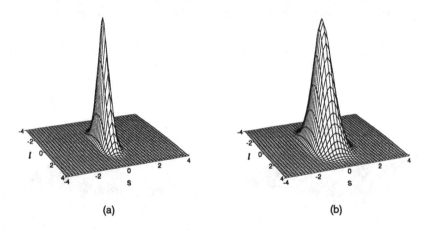

(a)   (b)

Fig. 9.2.4   3-D OTF for a confocal 2-p fluorescence microscope of high-aperture objectives obeying the Herschel condition: (a) $\alpha_o = 60°$; (b) $\alpha_o = 90°$.

9. 3 Confocal 4Pi Imaging

According to the transfer functions (CTF and OTF) presented in the last two sections, the area where the value of the transfer function is non-zero increases with the numerical aperture of the imaging lens. In general, the resolution, in both transverse and axial directions, of an optical imaging system, is improved as the aperture of the imaging lens is increased. This is because the larger aperture can accept light which has been scattered through larger angles, corresponding to fine structure in the object.[9.2] There is, however, a physical limit to the aperture which can be obtained in a microscope objective corresponding to a complete hemisphere of light collection, giving a numerical aperture of unity for a dry objective. For a conventional imaging system with a numerical aperture of unity the passband of the 3-D CTF for axial spatial frequencies is just one half of that for transverse spatial frequencies, as can be seen from Eq. (9.1.6). As the aperture is increased further so that more than a hemisphere is collected, the spatial frequency bandwidth for transverse imaging remains constant, but that for axial imaging continues to increase until, in the limiting case of a complete sphere of collection, symmetry indicates that the imaging in the axial direction is identical to that in the transverse direction.

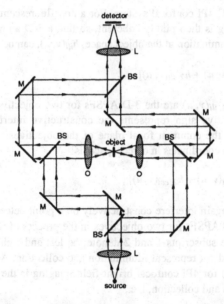

Fig. 9.3.1   Schematic geometry of a type-C 4Pi confocal microscope. M: mirrors; L: lenses; BS: beam splitters.

In practice an increase in the total aperture can be obtained by using two objectives, one on either side of the object under inspection. This arrangement then approximates to a complete sphere and has been termed 4Pi microscopy by Hell and Stelzer.[9.18-9.22] They proposed three types of 4Pi confocal microscope systems, type-A, type-B and type-C 4Pi confocal systems. In a type-A microscope, an object is illuminated constructively from two opposite sides but the signal is detected only on one side, whereas the illuminating light is focused into a specimen only from one side and the signal is collected on both sides in a type-B system. For a type-C system, illumination and collection are operated on both sides of the sample.

The aim of this section is to provide a transfer-function analysis for 4Pi confocal bright-field and fluorescence systems in terms of the results of the transfer functions for a high-aperture system presented in Sections 9.1 and 9.2. A type-C 4Pi confocal scanning optical microscope is schematically depicted in Fig. 9.3.1. The light from a point source is divided into two beams that are constructively focused into the object from the opposite sides. The scattered light from the object is collected from the both sides and focused to a point detector. We assume that the two objectives are identical and circularly symmetric.

9. 3. 1   Bright-Field Imaging

Consider a type-C 4Pi confocal system. For a non-fluorescent object, the system for bright-field imaging is then purely coherent, so that a 3-D amplitude point spread function (APSF) for illumination in the object space, $h_{ill}(v, u)$, can be written as

$$h_{ill}(v, u) = [h_{1,\,ill}(v, u) + h_{2,\,ill}(v, u)] . \tag{9.3.1}$$

Here $h_{1,\,ill}(v, u)$ and $h_{2,\,ill}(v, u)$ are the 3-D APSFs for two objectives in the process of illumination. The positive sign represents the constructive interference of the two illuminating beams in the common focal plane of the objectives. The 3-D APSF for collection, expressed in coordinates in object space, is

$$h_{col}(v, u) = [h_{1,\,col}(v, u) + h_{2,\,col}(v, u)] , \tag{9.3.2}$$

where the two beams again interfere constructively on a point detector. $h_{1,\,col}(v, u)$ and $h_{2,\,col}(v, u)$ are the 3-D APSFs for two objectives in the process of illumination. In Eqs. (9.3.1) and (9.3.2), the subscripts 1 and 2 denote the left and right objectives in Fig. 9.3.1. Subscripts ill and col represent illumination and collection. As mentioned above, the 3-D APSF $h_a(v, u)$ for 4Pi confocal bright-field imaging is the product of the 3-D APSFs for illumination and collection, i. e.,

$$h_a(v, u) = [h_{1,\,ill}(v, u) + h_{2,\,ill}(v, u)][h_{1,\,col}(v, u) + h_{2,\,col}(v, u)] . \tag{9.3.3}$$

In terms of our scalar assumption, the 3-D APSF for a single lens is given by Eq. (9.1.1) and therefore we have

$$h_{1,\,ill}(v, u) = h_{1,\,col}(v, u) = h(v, u), \tag{9.3.4a}$$

$$h_{2,\,ill}(v, u) = h_{2,\,col}(v, u) = h(v, -u) = h^*(v, u), \tag{9.3.4b}$$

where * denotes the complex conjugate operation. Thus

$$h_a(v,u) = h^2(v,u) + h^2(v,-u) + 2|h(v,u)|^2. \tag{9.3.5}$$

The first and the second terms represent the 3-D APSFs for reflection-mode confocal systems from the two opposite sides and therefore are formally identical, while the last term is that for a transmission-mode confocal system. It is concluded that a type-C 4Pi confocal system actually consists of two reflection-mode and two transmission-mode microscopes that are used coherently.

The 3-D CTF, $c(l, s)$, for the 4Pi confocal system with non-fluorescent objects can be derived by performing the 3-D inverse Fourier transform of Eq. (9.3.5) with respect to v and u, given by

$$c(l,s) = F_3\left[h^2(v,u)\right] + F_3\left[h^2(v,-u)\right] + 2F_3\left[|h(v,u)|^2\right]. \tag{9.3.6}$$

Here $F_3\left[h^2(v,u)\right]$ and $F_3\left[h^2(v,-u)\right]$ are represented by $c_r(l, s)$ in Eq. (9.1.17) for the sine condition or in Eq. (9.1.22) for the Herschel condition, whereas $F_3\left[|h(v,u)|^2\right]$ is $c_t(l, s)$, given by Eqs. (9.1.18) and (9.1.23) for the two apodization functions, respectively.

Therefore, the 3-D CTF for a type-C 4Pi system is the sum of $c_r(l, s)$ and $c_t(l, s)$, and is plotted in Figs. 9.3.2 and 9.3.3 for different values of the numerical aperture, where the 3-D CTF has been normalized to unity at $l = 0$ and $s = 2$. As expected, the 3-D CTF for the 4Pi system is composed of three parts. When the numerical aperture is small, the 3-D CTF for the sine condition is similar to that for the uniform angular illumination condition. Further, there exist gaps between the three parts. The gaps, however, gradually vanish as the numerical aperture of the lens increases. The condition for the three parts to overlap is [9.16]

$$\alpha_{o1} \geq \cos^{-1}(1/3) \approx 70.5°. \tag{9.3.7}$$

The corresponding 3-D CTFs behave like those shown in Figs. 9.3.2b and 9.3.3b. Eventually, for $\alpha_o = \pi/2$, the CTFs behave very differently between the two apodization conditions. In this case, the 3-D CTF for the uniform angular illumination condition (Fig. 9.3.3c) is spherically symmetrical and can be expressed, analytically, as

$$c(l,s) = \frac{2}{\sqrt{l^2 + s^2}}, \quad 0 \le l \le 2\sin\alpha_o, \tag{9.3.8}$$

whereas that for the sine condition is not (see Fig. 9.3.2c) because of the effect of the apodization factor $\cos^{1/2}\theta$. Eq. (9.3.8) is simply the result of the convolution of two complete spheres given by Eq. (9.1.7).

In addition, it is seen that the cut-off spatial frequencies are in general different in the transverse and axial directions: the cut-off in the axial direction is 2, independent of

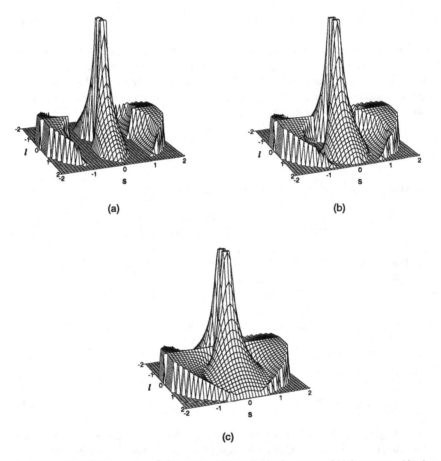

(a)   (b)

(c)

Fig. 9.3.2   3-D CTF for a type-C 4Pi confocal bright-field microscope of high-aperture objectives obeying the sine condition: (a) $\alpha_o = 60°$; (b) $\alpha_o = 71°$; (c) $\alpha_o = 90°$.

the lens aperture size, and the cut-off in the transverse direction is given by $2\sin\alpha_o$. It should be mentioned that these cut-off spatial frequencies are not directly comparable with those for the paraxial approximation, as the definitions of the spatial frequencies are different in the two cases (see Eq. (9.1.9)). The higher axial cut-off spatial frequency promises higher axial resolution, but the image of a single point can have strong side peaks as the 3-D CTF is not a smooth function along the axial direction.

In practice, the value of α_0 is determined by the numerical aperture of the objective, i. e., $\tilde{n}\sin\alpha_o = N. A.$, where $\tilde{n}$ denotes the refractive index of the immersion

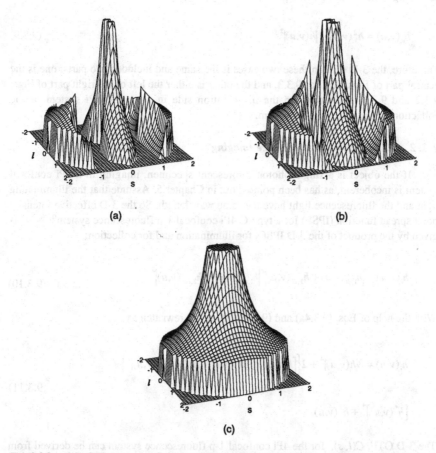

(a)   (b)

(c)

Fig. 9.3.3   3-D CTF for a type-C 4Pi confocal bright-field microscope of high-aperture objectives obeying the Herschel condition: (a) $\alpha_o = 60°$; (b) $\alpha_o = 71°$; (c) $\alpha_o = 90°$.

material. It is seen that for a given value of the numerical aperture, an objective with an immersion material of a small refractive index should be used in order to obtain a large value of α_o. For example, for an objective of numerical aperture 1.25, a water-immersion objective gives a larger angle α_o of about $71°$, which is close to the value shown in Eq. (9.3.7). This feature implies that for $N. A. = 1.25$ the water-immersion objectives may result in better images than oil-immersion objectives. The 3-D CTF for the sine and uniform angular illumination conditions at $\alpha_o = 71°$ are placed in Figs. 9.3.2b and 9.3.3b, respectively.

For type-A and type-B 4Pi confocal microscopes, the 3-D APSF is given, according to the working principle of the 4Pi system, by

$$h_a(v,u) = h^2(v,u) + |h(v,u)|^2. \tag{9.3.9}$$

Therefore, the 3-D CTF in these two cases is the same and includes two parts: one is the central part of Figs. 9.3.2 or 9.3.3, and the other is either the left or the right part of Figs. 9.3.2 and 9.3.3, depending on the illumination side in a type-A 4Pi system or the collection side in a type-B 4Pi system.

9. 3. 2 Single-Photon Fluorescence Imaging

If the object is a single-photon fluorescent specimen, imaging in a 4Pi confocal system is incoherent, as has been pointed out in Chapter 5. Assume that the illuminating light and the fluorescence light have the same wavelength. So the 3-D effective intensity point spread function (IPSF) for a type-C 4Pi confocal 1-p fluorescence system[9.16, 9.18] is given by the product of the 3-D IPSFs for illumination and for collection:

$$h_i(v,u) = \left|h_{1,ill}(v,u) + h_{2,ill}(v,u)\right|^2 \left|h_{1,col}(v,u) + h_{2,col}(v,u)\right|^2. \tag{9.3.10}$$

With the help of Eqs. (9.3.4a) and (9.3.4b), it can be rewritten as

$$h_i(v,u) = 3|h(v,u)|^4 + 2\left\{\left[h^*(v,u)\right]^2 |h(v,u)|^2 + h^2(v,u)|h(v,u)|^2\right\} + \tag{9.3.11}$$

$$\left[h^*(v,u)\right]^4 + h^4(v,u).$$

The 3-D OTF, $C(l, s)$, for the 4Pi confocal 1-p fluorescence system can be derived from the 3-D inverse Fourier transform of the 3-D IPSF:

$$C(l,s) = 3F_3\left[|h(v,u)|^4\right] + 2\left\{F_3\left\{[h^*(v,u)]^2|h(v,u)|^2\right\} + F_3\left[h^2(v,u)|h(v,u)|^2\right]\right\} +$$

$$(9.3.12)$$

$$F_3\left\{[h^*(v,u)]^4\right\} + F_3\left[h^4(v,u)\right].$$

It is seen that the 3-D OTF for the 4Pi confocal 1-p fluorescence system includes five terms. Each of them has a finite region where the value of the 3-D OTF is not zero. The size of these regions depends on the numerical aperture of the objectives. The boundary for the first term $F_3[|h(v,u)|^4]$ is given by Eq. (9.2.3), which is therefore the passband of the 3-D OTF for non-4Pi confocal 1-p fluorescence microscopy. The boundaries for the rest of the regions can be analytically expressed as[9.16]

$$s = \begin{cases} \pm[\sqrt{9-(l-\sin\alpha_o)^2} - \cos\alpha_o] & , & |s| \geq 2\cos\alpha_o, \\[2mm] \pm(-1+3\cos\alpha_o) & , & 0 \leq l < 3\sin\alpha_o, |s| \leq 2\cos\alpha_o, \\[2mm] \pm[\sqrt{1-(l-3\sin\alpha_o)^2} + 3\cos\alpha_o] & , & 3\sin\alpha_o \leq l \leq 4\sin\alpha_o, |s| \leq 2\cos\alpha_o, \end{cases}$$

$$(9.3.13)$$

for $F_3\left[h^2(v,u)|h(v,u)|^2\right]$ and $F_3\left\{[h^*(v,u)]^2|h(v,u)|^2\right\}$, and

$$s = \begin{cases} \pm\sqrt{16+l^2} & , & |s| > 4\cos\alpha_o, \\[2mm] \pm 4\cos\alpha_o & , & |s| = 4\cos\alpha_o, \end{cases}$$

$$(9.3.14)$$

for $F_3[h^4(v, u)]$ and $F_3\{[h^*(v, u)]^4\}$.

These boundaries are depicted in Fig. 9.3.4 for $\alpha_o = \pi/6$, $\pi/3$ and $\pi/2$. Note that for a small value of α_o, the non-zero regions do not overlap (Fig. 9.3.4a), so that the 3-D OTF for the 4Pi confocal 1-p fluorescence system has five peaks separating from each other. When the numerical aperture of the lens becomes large, they can overlap and are eventually combined into a single region. The condition for the solid and dashed regions to overlap is the same as that given by Eq. (9.3.7). The solid and dotted regions overlap when

$$\alpha_{o2} \geq \cos^{-1}(3/5) \approx 53.1°,$$

$$(9.3.15)$$

which is also the condition for the dotted and dashed regions to overlap. It can also be seen that the cut-off transverse and axial spatial frequencies are $4\sin\alpha_o$ and 4, respectively. Again the latter is independent of the numerical aperture and larger than the former for $\alpha_o < \pi/2$. This explains why the axial extension of the 3-D IPSF is smaller than that in the transverse direction[9.18].

The 3-D OTF for a type-C 4Pi confocal 1-p fluorescence microscope can be numerically calculated by the auto-convolution of the 3-D CTF for the 4Pi confocal

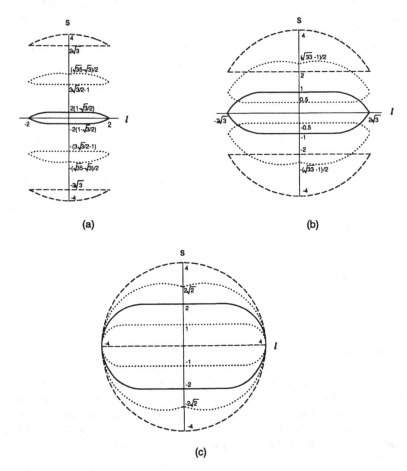

(a)   (b)

(c)

Fig. 9.3.4   Passband of the 3-D OTF for a type-C 4Pi confocal 1-p fluorescence microscope of high-aperture objectives. The solid, dotted, and dashed curves correspond to Eqs. (9.2.3), (9.3.13), and (9.3.14), respectively: (a) $\alpha_o = \pi/6$; (b) $\alpha_o = \pi/3$; (c) $\alpha_o = \pi/2$.

system shown in Figs. 9.3.2 and 9.3.3. Figs. 9.3.5 and 9.3.6 give the 3-D OTFs, normalized to unity at the origin, for different values of the numerical aperture. When the numerical aperture of the lens is small, the 3-D OTF for the 4Pi confocal 1-p fluorescence microscope shows five peaks, as is expected. The central peak represents the 3-D OTF for non-4Pi confocal 1-p fluorescence microscopy shown in Figs. 9.2.1 and 9.2.2. For $\alpha_o = \pi/3$, the 3-D OTF for the sine condition (Fig. 9.3.5a) is almost identical to that for the uniform angular illumination condition (Fig. 9.3.6a). The difference between the two apodization conditions becomes pronounced as the value of the numerical

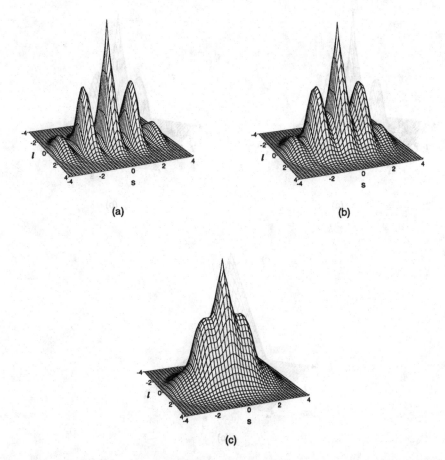

(a)

(b)

(c)

Fig. 9.3.5   3-D OTF for a type-C 4Pi confocal 1-p fluorescence microscope of high-aperture objectives obeying the sine condition: (a) $\alpha_o = 60°$; (b) $\alpha_o = 71°$; (c) $\alpha_o = 90°$.

aperture increases. For the limiting case when $\alpha_o = \pi/2$, the five-peak structure in the 3-D OTF disappears and the cut-off spatial frequencies are equal to 4 in both axial and transverse directions. It is noticed that the 3-D OTF for the sine condition is still not spherically symmetrical (Fig. 9.3.5c), but that for the uniform angular illumination condition is (Fig. 9.3.6c), as symmetry tells us it must be. The analytical expression for the 3-D OTF in the latter case can be written, by performing the auto-convolution of Eq. (9.3.8), as

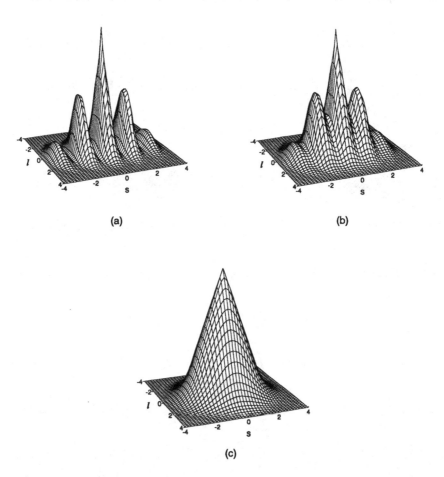

(a)   (b)

(c)

Fig. 9.3.6   3-D OTF for a type-C 4Pi confocal 1-p fluorescence microscope of high-aperture objectives obeying the Herschel condition: (a) $\alpha_o = 60°$; (b) $\alpha_o = 71°$; (c) $\alpha_o = 90°$.

$$C(l,s) = \begin{cases} 4 - \dfrac{3}{2}\sqrt{l^2 + s^2} \;, & 0 \le \sqrt{l^2 + s^2} \le 2, \\[4mm] \dfrac{(\sqrt{l^2 + s^2} - 4)^2}{2\sqrt{l^2 + s^2}} \;, & 2 < \sqrt{l^2 + s^2} \le 4, \end{cases} \qquad (9.3.16)$$

where $C(l, s)$ has been normalized to unity at the origin.

The transverse and axial cross-sections of the 3-D OTF through $s = 0$ and $l = 0$ are

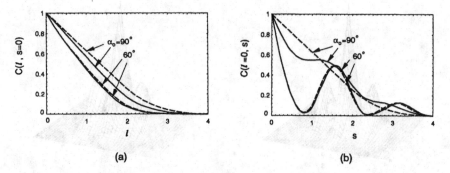

(a)　　　　　　　　　　　　(b)

Fig. 9.3.7　　Transverse (a) and axial (b) cross-sections of the 3-D OTF for a type-C 4Pi confocal 1-p fluorescence microscope of high-aperture objectives. The solid and dashed curves correspond to the sine and the Herschel conditions, respectively.

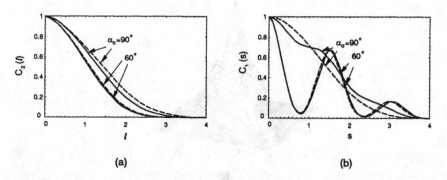

(a)　　　　　　　　　　　　(b)

Fig. 9.3.8　　2-D in-focus (a) and 1-D on-axis (b) OTFs for a type-C 4Pi confocal 1-p fluorescence microscope of high-aperture objectives. The solid and dashed curves correspond to the sine and the Herschel conditions, respectively.

shown in Fig. 9.3.7, which can be used to describe imaging of a thick object with only transverse variations or with only axial variations. Again, we can see that the cross-sections for the sine and uniform angular illumination conditions show only a small difference if the numerical aperture is small. The difference becomes pronounced when the angle α_0 approaches $\pi/2$. The five-peak structure can also be observed in the axial cross-section for $\alpha_0 = \pi/3$ (Fig. 9.3.7b).

The description of the 3-D OTF is a general method for 3-D imaging of any object. It can be also applied to two-dimensional (2-D) and one-dimensional (1-D) imaging

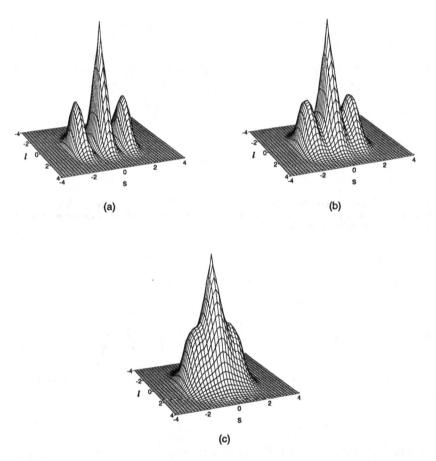

(a)   (b)

(c)

Fig. 9.3.9   3-D OTF for a type-A 4Pi confocal 1-p fluorescence microscope of high-aperture objectives obeying the sine condition: (a) $\alpha_o = 60°$; (b) $\alpha_o = 71°$; (c) $\alpha_o = 90°$.

(Chapter 10), for example if the object is a thin 2-D specimen in the focal plane or a thin line along the axial axis. The images of these objects may be modelled by the use of the 2-D in-focus OTF or the 1-D on-axis OTF. These two OTFs are shown in Fig. 9.3.8.

For a water-immersion objective of numerical aperture 1.25, the 3-D OTF is shown in Figs. 9.3.5b and 9.3.6b for the objective satisfying the sine and the uniform angular illumination conditions, respectively. Even at this largest practically achievable value for the angle α_o, the difference of the 3-D OTFs between the sine and uniform apodization functions is very small.

One can also obtain a type-A (or type-B) 4Pi confocal 1-p fluorescence system[9.18, 9.19] when detection (or illumination) is not performed in a coherent manner. The type-A and type-B confocal fluorescence microscopes have the same 3-D IPSF, if the wavelengths of the incident and fluorescence radiation are equal, which is given by

$$h_i(v,u) = \left| h(v,u) + h^*(v,u) \right|^2 \left| h(v,u) \right|^2$$

(9.3.17)

$$= 2|h(v,u)|^4 + \left[h^*(v,u) \right]^2 |h(v,u)|^2 + h^2(v,u)|h(v,u)|^2.$$

The 3-D inverse Fourier transform of Eq. (9.3.17) provides the 3-D OTF for the type-A (or type-B) system:

$$C(l,s) = F_3\left[\left| h(v,u) + h^*(v,u) \right|^2 \right] \otimes F_3\left[\left| h(v,u) \right|^2 \right],$$

(9.3.18)

in which the first factor is the 3-D CTF for a type-C 4Pi confocal bright-field system and

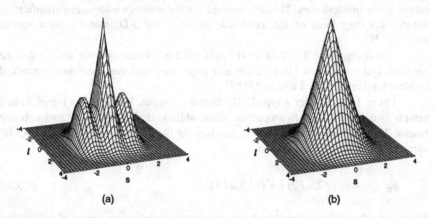

(a)   (b)

Fig. 9.3.10   3-D OTF for a type-A 4Pi confocal 1-p fluorescence microscope of high-aperture objectives obeying the Herschel condition: (a) $\alpha_o = 71°$; (b) $\alpha_o = 90°$.

the second is the 3-D OTF for a single lens in conventional 1-p fluorescence imaging. The 3-D OTF in Eq. (9.3.18) has only three peaks contributed from the regions given by Eqs. (9.2.3) and (9.3.13) and the second side peaks resulting from the region given by Eq. (9.3.14) do not appear. Figs. 9.3.9 and 9.3.10 show the 3-D OTFs for the two apodization functions at different angles. The 3-D OTF in both cases cuts off at $3 - \cos\alpha_o$ and $4\sin\alpha_o$ in the axial and transverse directions, respectively. Unlike the type-C confocal 1-p fluorescence microscope, the axial cut-off spatial frequency in Figs. 9.3.9 and 9.3.10 depends on the numerical aperture and is smaller than the transverse cut-off spatial frequency when

$$\alpha_o > 32.65°. \tag{9.3.19}$$

In addition, the difference of the 3-D OTF between the two apodization functions is noticeable when $\alpha_o = 90°$ (see Fig. 9.3.9c and 9.3.10b).

9. 3. 3   Two-Photon Fluorescence Imaging

A problem existing in a 4Pi confocal 1-p fluorescence system is the appearance of the strong side peaks of the intensity point spread function along the axial direction,[9.18, 9.19] which degrades the image quality. This issue is reflected by a pronounced five-peak structure in the corresponding 3-D OTF as shown in Figs. 9.3.5 and 9.3.6 for a type-C system or by a three-peak structure in Figs. 9.3.9 and 9.3.10 for a type-A (or type-B) system. In practice, it can be overcome by using two-photon (2-p) fluorescence microscopy,[9.20, 9.21] namely, the sample is excited using two incident photons and the image is recorded using fluorescence radiation, the photons of which have the twice energy of the incident ones. This 4Pi confocal 2-p fluorescence microscopy significantly reduces the magnitude of the axial side peaks in the 3-D intensity point spread function.[9.18]

To investigate the 3-D OTF in 4Pi confocal 2-p fluorescence, we assume that the confocal system has two identical circular objectives, and that the fluorescence and incident wavelengths are λ and 2λ.[9.23-9.25]

Let us first consider a type-C 4Pi confocal system, in which the signal from a sample is collected from two opposite sides while it is illuminated by two opposite beams. The 3-D intensity point spread function for illumination (IPSF) in a type-C 4Pi confocal 2-p fluorescence system is[9.20, 9.21]

$$h_{ill}(v,u) = \left| h(v/2, u/2) + h^*(v/2, u/2) \right|^4, \tag{9.3.20}$$

because the 2-p fluorescence light intensity is proportional to the square of the incident intensity.[9.17] Here the two incident beams illuminate the sample constructively. Since the fluorescence radiation from the sample is imaged, the 3-D IPSF for the collection process in the type-C system, is simply given by

$$h_{col}(v,u) = \left| h(v,u) + h^*(v,u) \right|^2, \tag{9.3.21}$$

which implies the requirements that two wave fronts constructively interfere in the detector and that the interference is synchronized with that of the excitation wave fronts. The 3-D IPSF for a confocal system is given by the product of the IPSFs for illumination and collection,[9.23-9.25] so that the 3-D IPSF for a type-C 4Pi confocal 2-p fluorescence system becomes

$$h_i(v,u) = \left| h(v/2,u/2) + h^*(v/2,u/2) \right|^4 \left| h(v,u) + h^*(v,u) \right|^2. \tag{9.3.22}$$

It is clear that because the incident wavelength is twice as large as the fluorescence wavelength, the position of the first axial side peak in Eq. (9.3.21) approximately corresponds to the position of the first axial minimum in Eq. (9.3.20). As a result, the 3-D effective IPSF for a type-C 4Pi 2-p fluorescence system in Eq. (9.3.22) can exhibit the reduced axial side peaks.[9.20]

The 3-D OTF, $C(l, s)$, for the type-C 4Pi confocal 2-p fluorescence system can be derived as

$$C(l,s) = F_3 \left[\left| h(v/2,u/2) + h^*(v/2,u/2) \right|^4 \right] \otimes_3 F_3 \left[\left| h(v,u) + h^*(v,u) \right|^2 \right]. \tag{9.3.23}$$

Compared with Eq. (9.3.12), the first factor in Eq. (9.3.23) is the 3-D OTF for a type-C 4Pi confocal 1-p fluorescence microscope at wavelength 2λ. For the sine and Herschel conditions, the shapes of the first factor are the same as those in Figs. 9.3.5 and 9.3.6 but are shrunk by a factor of two. The second factor of Eq. (9.3.23) is identical to Eq. (9.3.6) and is therefore the 3-D CTF for a type-C 4Pi confocal non-fluorescence microscope at wavelength λ, as depicted in Figs. 9.3.2 and 9.3.3 for the two apodization functions. Consequently, the cut-off axial and radial spatial frequencies of the 3-D OTF in Eq. (9.3.23) are 4 and $4\sin\alpha_o$, respectively, which are the same as those for type-C 4Pi confocal 1-p fluorescence microscopy described before.

Fig. 9.3.11 shows the calculated 3-D OTF for a type-C 4Pi confocal 2-p fluorescence microscope for different values of the numerical aperture. The objective is assumed to obey the sine condition. When the numerical aperture is small, the 3-D OTF exhibits nine pronounced peaks along the axial direction which result from the fact that the 3-D CTF described by the second factor in Eq. (9.3.23) has a gap in the axial direction (see Fig. 9.3.2) and that the 3-D OTF in the first factor has a discrete five-peak structure in the axial direction (see Fig. 9.3.5). However, compared with Fig. 9.3.5a, Fig.

9.3.11a gives a 3-D OTF with an enhanced response along the axial spatial frequency axis. This phenomenon implies that the axial extension of the 3-D intensity point spread function can be confined to a smaller region than that in a type-C 4Pi confocal 1-p fluorescence microscope, which is consistent with the experimental result of Hell et al.[9.21]

It is noted that the multi-peak structure in Fig. 9.3.11 quickly becomes less pronounced as the numerical aperture of the lens increases. In particular, when $\alpha_0 = 71°$, which corresponds to the maximum numerical aperture achievable in practice for a water

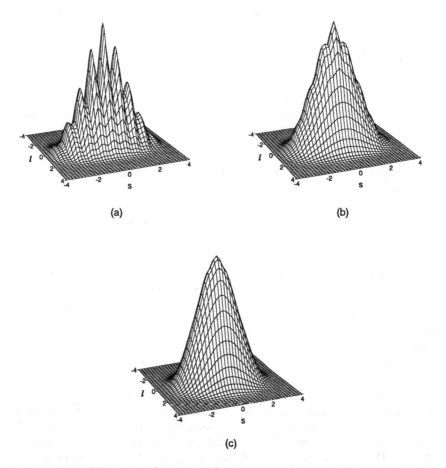

(a)   (b)

(c)

Fig. 9.3.11   3-D OTF for a type-C 4Pi confocal 2-p fluorescence microscope of high-aperture objectives obeying the sine condition: (a) $\alpha_o = 60°$; (b) $\alpha_o = 71°$; (c) $\alpha_o = 90°$.

immersion objective, the multi-peak structure in the 3-D OTF almost disappears (see Fig. 9.3.11b). Further increasing the numerical aperture results in little change in the OTF (Fig. 9.3.11c). This behaviour contrasts with that in a type-C 4Pi confocal 1-p fluorescence microscope in which the 3-D OTF has a pronounced multi-peak structure even for $\alpha_O = 71°$ (see Fig. 9.3.5b). It suggests that one can design a 4Pi confocal 2-p fluorescence imaging system in which the axial resolution can be approximately close to the limiting value for a true 4Pi system ($\alpha_O = 90°$) while this is impossible under 1-p excitation as the 3-D OTFs in the latter case for $\alpha_O = 71°$ and $\alpha_O = 90°$ exhibit a

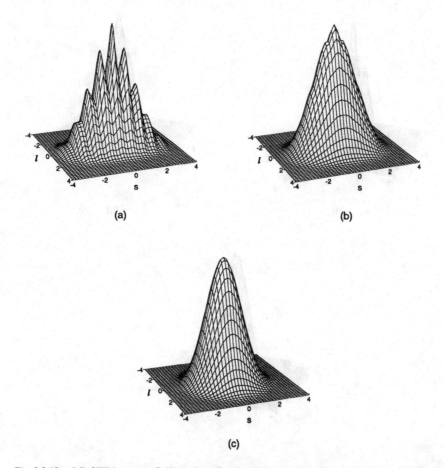

(a)   (b)

(c)

Fig. 9.3.12   3-D OTF for a type-C 4Pi confocal 2-p fluorescence microscope of high-aperture objectives obeying the Herschel condition: (a) $\alpha_o = 60°$; (b) $\alpha_o = 71°$; (c) $\alpha_o = 90°$.

considerable different behaviour (see Figs. 9.3.5b and 9.3.5c). In addition, the minima of the multi-peak structure in Fig. 9.3.11a do not go to zero, which is important in a sense that it is possible to use an electronic filter in image processing to enhance the strength of the OTF at some spatial frequencies at which the original OTF has a weak response. As expected, the slope of the 3-D OTF at the origin is zero, which is caused by the cooperative excitation using two incident photons.

The effect of apodization on 4Pi confocal 2-p fluorescence imaging can be derived from the 3-D OTF shown in Fig. 9.3.12 for a lens obeying the uniform angular

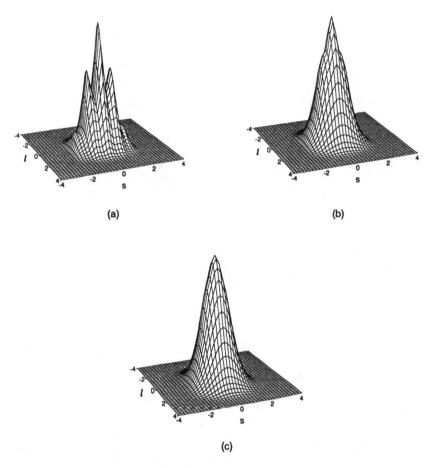

Fig. 9.3.13   3-D OTF for a type-A 4Pi confocal 2-p fluorescence microscope of high-aperture objectives obeying the sine condition: (a) $\alpha_o = 60°$; (b) $\alpha_o = 71°$; (c) $\alpha_o = 90°$.

illumination condition. A comparison of Fig. 9.3.11 and Fig. 9.3.12 shows that unlike 4Pi confocal 1-p fluorescence microscopy the effect of apodization is not crucial in 4Pi confocal 2-p fluorescence microscopy. The reason for this result is the nonlinear excitation using two incident photons described in Eq. (9.3.20). Therefore, the 3-D OTF for $\alpha_o = 90°$ in Fig. 9.3.12c is similar to that in Fig. 9.3.11c, although the former is spherically symmetric while the latter is slightly not.

It is interesting to compare the 3-D OTFs in Figs. 9.3.11 and 9.3.12 with those in Figs. 9.2.3 and 9.2.4 for non-4Pi confocal 2-p fluorescence microscopy. As expected, the OTFs in Figs. 9.2.3 and 9.2.4 are only the central peak of the nine-peak structure in Figs. 9.3.11 and 9.3.12. The difference of the axial cut-off spatial frequencies between 4Pi and non-4Pi confocal 2-p fluorescence microscopy is therefore $2 + 2\cos\alpha_o$. This result indicates that the use of a 4Pi geometry excitation increases significantly the information content in the imaging process.

Let us turn to the type-A 4Pi confocal 2-p fluorescence, which has been experimentally demonstrated.[9.21] The 3-D effective IPSF in this case becomes[9.23]

$$h_i(v,u) = \left| h(v/2,u/2) + h^*(v/2,u/2) \right|^4 |h(v,u)|^2. \tag{9.3.24}$$

It leads to the 3-D OTF for type-A 4Pi confocal 2-p fluorescence microscopy:

$$C(l,s) = F_3\left[\left| h(v/2,u/2) + h^*(v/2,u/2) \right|^4 \right] \otimes_3 F_3\left[|h(v,u)|^2 \right], \tag{9.3.25}$$

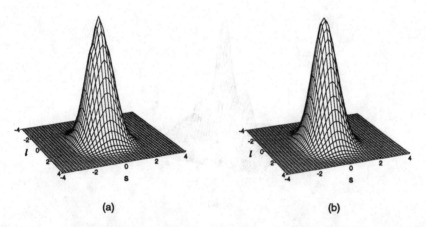

(a)   (b)

Fig. 9.3.14 3-D OTF for a type-A 4Pi confocal 2-p fluorescence microscope of high-aperture objectives obeying the Herschel condition: (a) $\alpha_o = 71°$; (b) $\alpha_o = 90°$.

which is shown in Fig. 9.3.13 for the sine conditions. The difference between Eqs. (9.3.23) and (9.3.25) is that the second factor in Eq. (9.3.25) is now the 3-D OTF for conventional 1-p fluorescence microscopy of high aperture and is of course the 3-D CTF for confocal transmission, presented in Eq. (9.1.17). As a result, Eq. (9.3.25) exhibits a five-peak structure, has a poorer response in the axial direction compared with Fig. 9.3.11b, and cuts off at $4\sin\alpha_o$ and $3-\cos\alpha_o$ in the transverse and axial spatial frequency directions, respectively. As the illumination of the 2-p excitation comes from two opposite sides, the effect of the apodization is not pronounced like the situation in type-C

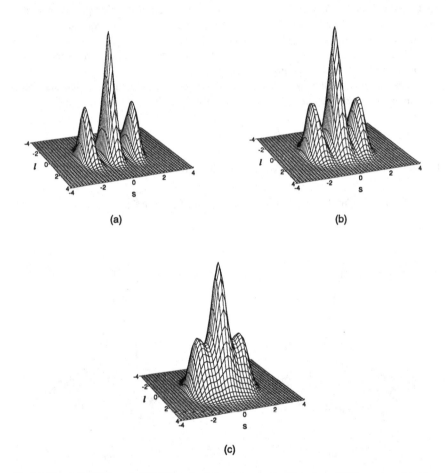

Fig. 9.3.15   3-D OTF for a type-B 4Pi confocal 2-p fluorescence microscope of high-aperture objectives obeying the sine condition: (a) $\alpha_o = 60°$; (b) $\alpha_o = 71°$; (c) $\alpha_o = 90°$.

confocal 2-p fluorescence microscopy. This is confirmed in Fig. 9.3.14 where the 3-D OTF for type-A confocal microscopy obeying the Herschel condition is depicted.

Unlike 4Pi confocal 1-p fluorescence microscopy, type-B 4Pi confocal 2-p fluorescence imaging is different from type-A 4Pi confocal 2-p fluorescence because the 2-p excitation is proportional to the square of the incident intensity. Its 3-D effective IPSF is given by

$$h_i(v,u) = |h(v/2, u/2)|^4 |h(v,u) + h^*(v,u)|^2, \tag{9.3.26}$$

which gives rise to the 3-D OTF for type-B 4P confocal 2-p fluorescence microscopy:

$$C(l,s) = F_3\left[|h(v/2, u/2)|^4\right] \otimes_3 F_3\left[|h(v,u) + h^*(v,u)|^2\right], \tag{9.3.27}$$

where the first term is the 3-D OTF for non-4Pi confocal 1-p fluorescence microscopy, and for the sine condition, gives the same shape as that in Fig. (9.2.1) with a shrinking factor of two. The second term is the 3-D CTF for 4Pi confocal non-fluorescence microscopy. Therefore, the 3-D OTF for type-B 4Pi confocal 2-p fluorescence microscopy has the same cut-off spatial frequencies as type-A 4Pi confocal 2-p fluorescence microscopy and shows only three peaks, as depicted in Figs. 9.3.15 and 9.3.16 for the sine and Herschel conditions, respectively. As can be seen, the difference between the two apodization functions becomes appreciable when $\alpha_o = \pi/2$, because the 2-p fluorescent sample is only excited from one side.

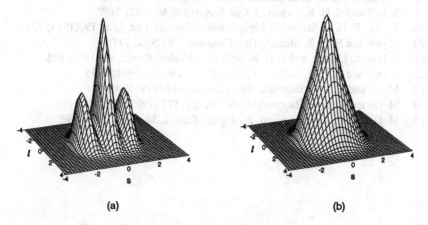

(a)   (b)

Fig. 9.3.16   3-D OTF for a type-B 4Pi confocal 2-p fluorescence microscope of high-aperture objectives obeying the Herschel condition: (a) $\alpha_o = 71°$; (b) $\alpha_o = 90°$.

References

9.1.   C. J. R. Sheppard and H. J. Matthews, *J. Opt. Soc. Am. A*, **4** (1987) 1354.

9.2.   M. Born and E. Wolf, *Principles of Optics* (Pergamon, New York, 1980).

9.3.   C. J. R. Sheppard and M. Gu, *J. Modern of Optics*, **40** (1993) 79.

9.4.   J. Stamnes, *Waves in Focal Regions* (Adam Hilgar, Bristal, 1986).

9.5.   C. W. McCutchen, *J. Opt. Soc. Am.*, **54** (1964) 240.

9.6.   C. J. R. Sheppard, Min Gu, Y. Kawata, and S. Kawata, *J. Opt. Soc. Am. A*, **11**, (1994) 593.

9.7.   I. S. Gradstein and I. Ryshik, *Tables of Series, Products, and Integrals* (Herri Deutsch, Frankfurt, 1981).

9.8.   C. J. R. Sheppard, *Optik*, **74** (1986) 128.

9.9.   C. J. R. Sheppard and T. Wilson, *Applied Phys. Lett*, **38** (1981) 858.

9.10. C. J. R. Sheppard and C. Cogswell, Three-dimensional imaging in confocal microscopy, in *Confocal Microscopy*, ed. T. Wilson (Academic, London, 1990), pp. 143-169.

9.11. C. J. R. Sheppard and M. Gu, *Opt. Commun.*, **88** (1992) 180.

9.12. C. J. R. Sheppard, T. J. Connolly, J. Lee, and C. Cogswell, *Applied Optics*, **33**, (1994) 631.

9.13. C. J. R. Sheppard and M. Gu, *Applied Optics*, **30** (1991) 3563.

9.14. C. J. R. Sheppard, M. Gu, K. Brain, and H. Zhou, *Applied Optics*, **33** (1994) 616.

9.15. M. Gu, Effect of apodization on axial resolution in confocal microscopy of high aperture, to be published.

9.16   M. Gu and C. J. R. Sheppard, *J. Opt. Soc. Am. A.*, **11** (1994) 1619.

9.17. W. Kaiser and C. G. B. Garrett, *Phys. Rev. Lett.*, **7** (1961) 229.

9.18. S. Hell and E. H. K. Stelzer, *J. Opt. Soc. Am. A*, **9** (1992) 2159.

9.19. S. Hell, E. H. K. Stelzer, S. Lindek, and C. Cremer, *Opt. Lett.*, **19**, (1994) 222 .

9.20. S. Hell and E. H. K. Stelzer, *Opt. Commun.*, **93** (1992) 277.

9.21. S. Hell, S. Lindek, and E. H. K. Stelzer, *J. Modern Optics*, **41** (1994) 675.

9.22. S. Hell and E. H. K. Stelzer, *Applied Phys. Lett.*, **64** (1994) 1335.

9.23. M. Gu and C. J. R. Sheppard, *Opt. Commun.*, **114** (1995) 45.

9.24. M. Gu and C. J. R. Sheppard, *J. Microscopy*, **177** (1995) 128.

9.25. M. Gu and C. J. R. Sheppard, *Zoological Studies*, **34**, suppl. (1995) 96.

Chapter 10

SIGNIFICANCE OF THREE-DIMENSIONAL
TRANSFER FUNCTIONS

So far, we have developed various three-dimensional (3-D) transfer functions for confocal scanning microscopy. These 3-D transfer functions can be classified into three main groups, the coherent transfer function (CTF) for coherent imaging, the optical transfer function (OTF) for incoherent imaging (e. g., fluorescence imaging) and the transmission cross-coefficient (TCC) for partially-coherent imaging. Use of the 3-D transfer functions provides a complete description of the imaging processes through Eqs. (3.2.3), (5.2.2) and (4.3.2). In this chapter, we will discuss, in Sections 10.1-10.5, several particular cases of confocal imaging in terms of the 3-D transfer functions, showing the significance and applicability of the 3-D transfer functions.[10.1, 10.2] As a summary of the book, the main properties of the 3-D transfer functions are listed in Section 10.6.

10. 1   Imaging of Thick Planar Layers

Suppose that the object is a thick structure with no variations in the axial direction. Its object function is a two-dimensional (2-D) function and can be expressed as

$$o(x, y, z) = o'(x, y),$$
(10.1.1)

corresponding to the following Fourier transform:

$$O(m, n, s) = O'(m, n)\delta(s),$$
(10.1.2)

311

where

$$O'(m,n) = \int\int_{-\infty}^{\infty} o'(x,y)\exp[-2\pi i(xm+yn)]dxdy .$$
(10.1.3)

Here for simplicity, we will omit the subscript f in the fluorescent object. Therefore, in terms of Eqs. (3.2.3), (5.2.2) and (4.3.2), the image intensity is, for coherent, incoherent and partially-coherent imaging processes,

$$I(x_s,y_s,z_s) = \left|\int\int_{-\infty}^{\infty} c(m,n,0)O'(m,n)\exp[2\pi i(x_s m + y_s n)]dmdn\right|^2 ,$$
(10.1.4)

$$I(x_s,y_s,z_s) = \int\int_{-\infty}^{\infty} C(m,n,0)O'(m,n)\exp[2\pi i(x_s m + y_s n)]dmdn,$$
(10.1.5)

$$I(x_s,y_s,z_s) = \int\int\int\int_{-\infty}^{\infty} C(m,n,0;m',n',0)O'(m,n)O'^*(m',n')$$
(10.1.6)

$$\exp\{2\pi i[x_s(m-m')+y_s(n-n')]\}dmdndm'dn',$$

respectively. Here x_s, y_s, and z_s are the coordinates of the scan position. The functions $c(m,n,s)$, $C(m,n,s)$, and $C(m,n,s;m',n',s')$ are the 3-D CTF, the 3-D OTF and the 3-D TCC corresponding to three kinds of the confocal imaging process.

In the case of a planar object such that there is no variation in the transverse direction, one has

$$o(x, y, z) = o''(z).$$
(10.1.7)

Therefore the detected intensity in the three cases is given by

$$I(x_s,y_s,z_s) = \left|\int_{-\infty}^{\infty} c(0,0,s)O''(s)\exp(2\pi i z_s s)ds\right|^2 ,$$
(10.1.8)

$$I(x_s,y_s,z_s) = \int_{-\infty}^{\infty} C(0,0,s)O''(s)\exp(2\pi i z_s s)ds ,$$
(10.1.9)

$$I(x_s,y_s,z_s) = \int\int_{-\infty}^{\infty} C(0,0,s;0,0,s')O''(s)O''^*(s')\exp[2\pi i z_s(s-s')]dsds' ,$$
(10.1.10)

where

$$O''(s) = \int_{-\infty}^{\infty} o''(z)\exp(-2\pi izs)dz. \tag{10.1.11}$$

In fact, Eqs. (10.1.4) - (10.1.6) and (10.1.8) - (10.1.10) are independent of the axial and transverse coordinates, respectively. It is noted from Eqs. (10.1.4) - (10.1.6) and (10.1.8) - (10.1.10) that imaging for a thick object structure with no variations in the axial direction is described by the transverse cross-section through the 3-D transfer functions for $s = 0$, while that for planar structures such that there is no variations in the

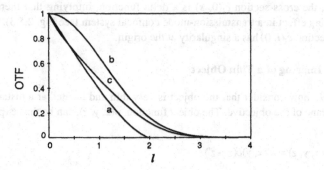

Fig. 10.1   2-D in-focus CTF (a), 2-D in-focus OTF (b), and transverse cross-section of the 3-D OTF (c). The confocal system employs circular lenses and point illumination and collection. Curve a is also the 2-D in-focus OTF for a conventional fluorescence microscope.

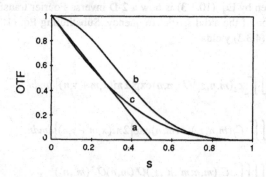

Fig. 10.2   1-D on-axis CTF (a), 1-D on-axis OTF (b), and axial cross-section of the 3-D OTF (c). The confocal system employs circular lenses and point illumination and collection. Curve a is also the 1-D on-axis OTF for a conventional fluorescence microscope.

transverse direction is given by the axial cross-section through the 3-D transfer functions for $m = n = 0$. For a confocal microscope with a point source and a point detector, the two cross-sections of the 3-D optical transfer function for fluorescence imaging are shown as curves labelled with c in Figs. 10.1.1 and 10.1.2.

It is noticed that the transverse cross-section $c_r(l, 0)$ for the reflection case is zero because the corresponding system behaves as a dark-field system (see Fig. 3.2.3 or Fig. 9.1.1 for a high-aperture system), while the cross-section $c_r(0, s)$ is an axially-shifted square function in the range of $0 \le s + s0 \le 1$, which leads to an optical sectioning property in a reflection-mode confocal bright-field system. The axial shift of the 3-D CTF in reflection has been observed in experiments.[10.3, 10.4] For a transmission confocal system, the cross-section $c_t(0, s)$ is a delta function, implying that there is no optical sectioning effect in a transmission-mode confocal system (see Fig. 2.5.3). The transverse cross-section $c_t(l, 0)$ has a singularity at the origin.

10. 2   Imaging of a Thin Object

We now consider that the object is very thin and located at a distance z' from the focal plane of the objective. The object function, $o(x, y, z)$, can be thus expressed as

$$o(x, y, z) = o'(x, y)\delta(z - z').\tag{10.2.1}$$

Here $\delta(z)$ is a delta function. The corresponding Fourier transform is thus

$$O(m, n, s) = \exp(-2\pi i z's)O'(m, n),\tag{10.2.2}$$

where $O'(m, n)$ given by Eq. (10.1.3) is now a 2-D inverse Fourier transform of $o'(x, y)$ and is not a function of the axial spatial frequency. Substituting Eq. (10.2.2) into Eqs. (3.2.3), (5.2.2) and (4.3.2) yields

$$I(x_s, y_s, z_s) = \left| \iint_{-\infty}^{\infty} c_2(m, n, z_s)O'(m, n)\exp[2\pi i(x_s m + y_s n)]dmdn \right|^2,\tag{10.2.3}$$

$$I(x_s, y_s, z_s) = \iint_{-\infty}^{\infty} C_2(m, n, z_s)O'(m, n)\exp[2\pi i(x_s m + y_s n)]dmdn,\tag{10.2.4}$$

$$I(x_s, y_s, z_s) = \iiiint_{-\infty}^{\infty} C_2(m, n; m', n', z_s)O'(m, n)O'^*(m', n')$$

$$\exp\{2\pi i[x_s(m - m') + y_s(n - n')]\}dmdndm'dn',\tag{10.2.5}$$

where $c_2(m, n, z_s)$, $C_2(m, n, z_s)$ and $C_2(m, n; m', n', z_s)$ are called the 2-D defocused CTF, the 2-D defocused OTF and the 2-D defocused TCC, respectively, and given by

$$c_2(m, n, z_s) = \int_{-\infty}^{\infty} c(m, n, s)\exp[2\pi i(z_s - z')s]ds ,$$ (10.2.6)

$$C_2(m, n, z_s) = \int_{-\infty}^{\infty} C(m, n, s)\exp[2\pi i(z_s - z')s]ds,$$ (10.2.7)

$$C_2(m, n; m', n', z_s) = \iint_{-\infty}^{\infty} C(m, n, s; m', n', s')\exp[2\pi i(z_s - z')(s - s')]dsds'.$$

(10.2.8)

It is now clear that the detected image intensity for a thin object is determined by the 2-D defocused transfer functions, which are the 1-D Fourier transforms of the 3-D transfer functions with respect to s. In particular, when the thin object is placed in the focal plane, the 2-D in-focus transfer functions determining the detected in-focus intensity, $c_2(m, n, z_s = z')$, $C_2(m, n, z_s = z')$, and $C_2(m, n; m', n', z_s = z')$, are given simply by the integration of the 3-D transfer functions with respect to the axial spatial frequency component s. In other words, the 2-D in-focus transfer functions are the projections of the 3-D transfer functions in the focal plane. This is a consequence of the projection-slice theorem of the Fourier transform theory.

Note that the 2-D in-focus transfer functions are the projection of the 3-D transfer functions, rather than a cross-section through the 3-D transfer function. A cross-section through the 3-D transfer function at $s = 0$ gives the transfer function for transverse imaging of a thick object, i. e., one which is unchanging in the z direction, rather than a very thin object.

For a transmission confocal system with two equal circular pupils, by integrating the 3-D CTF shown in Fig. 2.5.3 with respect to s, it is easy to derive analytically the 2-D in-focus coherent transfer function,[10.1] $c_2(l)$, given by the auto-convolution of the circular pupil function $P(\rho)$:

$$c_2(l) = \frac{2}{\pi}\left[\cos^{-1}\left(\frac{l}{2}\right) - \frac{l}{2}\sqrt{1 - \left(\frac{l}{2}\right)^2}\right],$$ (10.2.9)

which is shown as curve a in Fig. 10.1.1. It is of course the 2-D in-focus OTF for a conventional fluorescence microscope.

It should be pointed out that an identical result is also obtained for a reflection confocal system by integrating the 3-D CTF shown in Fig. 3.2.3 with respect to s. This is understandable because there is no difference in 2-D in-focus imaging between reflection-mode and transmission-mode systems.

For a similar confocal system with a thin fluorescent sheet as an object, the 2-D in-focus OTF can be derived as

$$C_2(l) = c_2(l) \otimes_2 c_2(l),$$

(10.2.10)

where $\otimes_2$ denotes the 2-D convolution operation. The 2-D in-focus OTF is the 2-D auto-convolution of the 2-D in-focus CTF, which again is also the projection of the 3-D OTF $C(m, n, s)$, rather than a cross-section through the 3-D OTF $C(m, n, s)$. This fact can be used as a check of the accuracy of a calculated 3-D OTF. $C_2(l)$ for confocal fluorescence is shown as curve b in Fig. 10.1.1, and as is already well-known, the normalized cut-off spatial frequency is 4, while the 2-D in-focus CTF $c_2(l)$ for confocal refection or transmission has a normalized cut-off spatial frequency of 2.

10. 3   Imaging of a Line Object

Let us turn to the object that is a line located parallel to the axial axis and varying in strength along its length. In this case, we have, if the object is placed at off-axis distances x' and y',

$$o(x, y, z) = o''(z)\delta(x - x')\delta(y - y'),$$

(10.3.1)

with the corresponding Fourier transform

$$O(m, n, s) = \exp[-2\pi i(x'm + y'n)]O''(s),$$

(10.3.2)

where $O''(s)$ is given by Eq. (10.1.11). Substituting Eq.(10.3.2) into Eqs. (3.2.3), (5.2.2) and (4.3.2) gives

$$I(x_s, y_s, z_s) = \left| \int_{-\infty}^{\infty} c_1(s, x_s, y_s)O''(s)\exp(2\pi i z_s s)ds \right|^2,$$

(10.3.3)

$$I(x_s, y_s, z_s) = \int_{-\infty}^{\infty} C_1(s, x_s, y_s)O''(s)\exp(2\pi i z_s s)ds,$$

(10.3.4)

$$I(x_s, y_s, z_s) = \int\int_{-\infty}^{\infty} C_1(s; s', x_s, y_s)O''(s)O''^*(s')\exp[2\pi i z_s(s - s')]dsds',$$

(10.3.5)

where $c_1(s, x_s, y_s)$, $C_1(s, x_s, y_s)$ and $C_1(s; s', x_s, y_s)$ are the 1-D line CTF, the 1-D line OTF and the 1-D line TCC, respectively, and can be expressed as

$$c_1(s, x_s, y_s) = \iint_{-\infty}^{\infty} c(m, n, s) \exp\left\{2\pi i[(x_s - x')m + (y_s - y')n]\right\} dm \, dn, \qquad (10.3.6)$$

$$C_1(s, x_s, y_s) = \iint_{-\infty}^{\infty} C(m, n, s) \exp\left\{2\pi i[(x_s - x')m + (y_s - y')n]\right\} dm \, dn, \qquad (10.3.7)$$

$$C_1(s; s', x_s, y_s) = \iiiint_{-\infty}^{\infty} C(m, n, s; m', n', s')$$

$$(10.3.8)$$

$$\exp\left\{2\pi i[(x_s - x')(m - m') + (y_s - y')(n - n')]\right\} dm \, dn \, dm' \, dn'.$$

It is noted that the detected image intensity for a line object is thus given by the 1-D line transfer functions which can be obtained by the 2-D Fourier transform of the 3-D transfer functions with respect to m and n. In particular, if the line object is located on the axial axis, the 1-D on-axis transfer functions, $c_1(s, x_s = x', y_s = y')$, $C_1(s, x_s = x', y_s = y')$, and $C_1(s; s', x_s = x', y_s = y')$, are given by the integration of the 3-D transfer functions with respect to m and n. This result implies that the 1-D on-axis transfer functions are the projections of the 3-D transfer functions onto the axial axis.

It should be pointed out that the 1-D on-axis transfer function is the projection of the 3-D transfer function, rather than a cross-section through the 3-D transfer function along the line $m = n = 0$. This cross-section gives the transfer function for axial imaging of very large planar structures, in which the strength varies only in the axial direction.

For the confocal system with two equal circular pupils, the 1-D on-axis CTF for the transmission-mode system can be derived analytically as[10.1]

$$c_1(s) = (1 - 2|s|) \qquad (10.3.9)$$

by integrating the 3-D CTF shown in Fig. 2.5.3 with respect to m and n. Eq. (10.3.9) is shown in Fig. 10.1.2 with a cut-off spatial frequency of 0.5 (see curve a). The 1-D on-axis CTF can alternatively be calculated as the 1-D Fourier transform of the axial variation of the 3-D amplitude point spread function. This 1-D on-axis CTF is different from the cross-section at $l = 0$ through the 3-D CTF, which gives simply a delta-function at the origin. This means that planar structures, which exhibit only axial variations in strength, are not imaged in a transmission confocal microscope. This feature results from the missing cone in the 3-D CTF. Only when there is lateral modulation of the structure is axial imaging obtained. The 1-D on-axis CTF for a reflection-mode system is of the same form as Eq. (10.3.9) but shifted by $(s0 + 1/2)$ along the s axis. This property results from the axial linear phase variation in the corresponding amplitude point spread function.

In the case of confocal fluorescence imaging, the 1-D on-axis OTF for the same system as in Section 10.1 is

$$C_1(s) = c_1(s) \otimes c_1(s),$$ (10.3.10)

which leads to

$$C_1(s) = \begin{cases} 1 - 6s^2 + 6s^3 & , \quad 0 \le s \le 0.5, \\ 2 - 6s + 6s^2 - 2s^3 & , \quad 0.5 \le s \le 1. \end{cases}$$ (10.3.11)

It is depicted as curve b in Fig. 10.1.2, showing a normalized cut-off spatial frequency of unity. The 1-D on-axis OTF for the confocal fluorescence system of a line object can be expressed in a simple algebraic form, even for different exciting and fluorescence wavelengths.[10.1]

10. 4   Imaging of a Point Object

When the object is a single point placed at the positions x', y', and z', then one has

$$o(x, y, z) = \delta(x - x')\delta(y - y')\delta(z - z').$$ (10.4.1)

Therefore the corresponding Fourier transform can be expressed as

$$O(m, n, s) = \exp[-2\pi i(x'm + y'n + z's)],$$ (10.4.2)

implying that a single point includes all spatial frequency components with equal moduli. The detected intensity is thus

$$I(x_s, y_s, z_s) = \left| \iiint_{-\infty}^{\infty} c(m,n,s) \exp\left\{2\pi i\left[(x_s - x')m + (y_s - y')n + (z_s - z')s\right]\right\} dm\,dn\,ds \right|^2,$$ (10.4.3)

$$I(x_s, y_s, z_s) = \iiint_{-\infty}^{\infty} C(m,n,s) \exp\left\{\left[2\pi i(x_s - x')m + (y_s - y')n + (z_s - z')s\right]\right\} dm\,dn\,ds,$$ (10.4.4)

$$I(x_s, y_s, z_s) = \iiint\iiint_{-\infty}^{\infty} C(m,n,s; m',n',s')$$

$$\exp\left\{2\pi i\left[(x_s - x')(m - m') + (y_s - y)(n - n') + (z_s - z')(s - s')\right]\right\} dm\,dn\,ds\,dm'\,dn'\,ds',$$ (10.4.5)

which are the 3-D Fourier transforms of the 3-D transfer functions and therefore represent the 3-D point spread functions. Clearly, if the point object is scanned along the z axis, then $x_s = x'$ and $y_s = y'$. The intensity variations with z are governed by the 1-D Fourier transform of the 1-D transfer functions resulting from the projection of the 3-D transfer functions on the z axis. In general, for any given direction, the intensity variations along this direction are determined by the 1-D Fourier transform of the 1-D transfer functions resulting from the projection of the 3-D transfer functions onto this axis.

10. 5   Imaging with the Extended-Focus Technique

The extended focus method[10.5] is often used in confocal imaging. In this case, the detected intensity of a thick object is given by the integration of the intensity through the axial direction, and can be mathematically expressed as

$$I(x_s,y_s) = \int_{-\infty}^{\infty} I(x_s,y_s,z_s)dz_s, \qquad (10.5.1)$$

which results in the projection of the object in the axial direction. The projections in other directions can be produced by offset of the image data during axial scanning. For coherent, incoherent and partially-coherent cases, Eq. (10.5.1) may be rewritten, according to Eqs. (3.2.3), (5.2.2) and (4.3.2), as

$$I(x_s,y_s) = \int_{-\infty}^{\infty} \left| \int\int_{-\infty}^{\infty} c(m,n,s)O(m,n,s)\exp[2\pi i(x_s m + y_s n)]dmdn \right|^2 ds, \qquad (10.5.2)$$

$$I(x_s,y_s) = \int\int_{-\infty}^{\infty} C(m,n,0)O(m,n,0)\exp[2\pi i(x_s m + y_s n)]dmdn, \qquad (10.5.3)$$

$$I(x_s,y_s) = \int\int\int\int_{-\infty}^{\infty} C(m,n,0;m',n',0)O(m,n,0)O^*(m',n',0) \qquad (10.5.4)$$

$$\exp\{2\pi i[x_s(m-m') + y_s(n-n')]\}dmdndm'dn',$$

respectively. We note that the image intensity for fluorescence and partially-coherent imaging is described by the transverse cross-section of the corresponding 3-D transfer functions through $s = 0$ (or $s' = 0$). The transverse cross-section of the 3-D OTF for a confocal system with a point detector and a point source is shown in Fig. 10.1.1 (curve c). This demonstrates that transverse imaging in confocal fluorescence with the extended-focus method is superior to that for in-focus imaging in conventional fluorescence. It is also seen that strength of high transverse spatial frequencies in confocal fluorescence is the same for either in-focus or extended-focus imaging. Compared with Eqs. (10.1.5) and (10.1.6), Eqs. (10.5.3) and (10.5.4) mean that axial information is suppressed and that only variations with a zero axial spatial frequency can be imaged in incoherent and partially-coherent systems by using the extended-focused technique. However, the image intensity for extended-focus imaging in a coherent system is still described in terms of the 3-D CTF and the image retains axial information. It should be pointed out that imaging of Eq. (10.5.2) is, in fact, nonlinearly dependent on the 3-D CTF and therefore leads to complication for imaging processing.

10. 6   Summary

We have understood that although a practical confocal microscope may be operated in different modes and using different optical arrangements, there are only three imaging processes in terms of the degree of coherence. Under the paraxial

Table 10.6.1 Summary of the 3-D image formation in confocal microscopy.

(reflection and transmission) non-fluorescent samples	fluorescent samples
point detector	

$I(r_s) = \lvert h_1(r_s)h_2(r_s) \otimes o(r_s) \rvert^2$
 $o(r_s)$: *amplitude* reflectivity
       or transmittance
 • coherent imaging
 • coherent transfer function (CTF)
	$I(r_s) = \lvert h_1(r_s)h_2(r_s) \rvert^2 \otimes o_f(r_s)$
	 $o_f(r_s)$: *intensity* reflectivity
	       or transmittance
	 • incoherent imaging
	 • optical transfer function (OTF)

finite-sized detector	
$I(r_s) = \int \lvert [h_1(r)o(r_s - r)] \otimes h_2(r) \rvert^2 D(r)dr$ $D(r)$: detector sensitivity • partially-coherent imaging • transmission cross-coefficient (TCC)	$I(r_s) = \left\{ \lvert h_1(r_s) \rvert^2 \left[\lvert h_2(r_s) \rvert^2 \otimes D(r_s) \right] \right\} \otimes o_f(r_s)$ $D(r)$: detector sensitivity • incoherent imaging • optical transfer function (OTF)

approximation, the 3-D space-invariance holds in the 3-D point spread function for a single lens (Section 2.3). As a result, image formation in confocal systems can be properly described by the 3-D CTF, the 3-D OTF, and the 3-D TCC, respectively. The main properties of the transfer-function description are summarized in Table 10.6.1. In the case of fluorescence imaging, only the single-photon process is considered to be an example. The magnification factor has been assumed to be unity for simplicity.

Table 10.6.2 gives the summary of the image intensity of a single point object in the cases corresponding to Table 10.6.1. The expressions in Table 10.6.2 are sometimes called the 3-D intensity point spread function (IPSF). It is seen that there is no difference of the 3-D IPSF between fluorescence and bright-field imaging. This is a drawback of the description based on the 3-D IPSF. The difficulty can be overcome by using the concept of the 3-D transfer function, which clearly provides the difference between the three imaging processes. A comparison of the imaging properties between conventional and confocal microscopy is given in Figs. 10.6.1 and 10.6.2.

Fig. 10.6.1 includes the passband of the 3-D CTF in conventional and confocal bright-field imaging of high-aperture lenses. As we have learned, the 3-D CTF in conventional imaging is a cap of a sphere, so that the ability of transferring the 3-D information is poor in particular in the axial direction. Use of the confocal geometry actually increases the passband by a factor of two. More importantly, the axial cross-

Table 10.6.2 Comparison of the image intensity of a single point object in the confocal microscopes corresponding to Table 10.6.1.

(reflection and transmission) non-fluorescent samples	fluorescent samples
point detector	
$I(r_s) = \left\| h_1(r_s) h_2(r_s) \right\|^2$	$I(r_s) = \left\| h_1(r_s) h_2(r_s) \right\|^2$
finite-sized detector	
$I(r_s) = \left\| h_1(r_s) \right\|^2 \left[\left\| h_2(r_s) \right\|^2 \otimes D(r_s) \right]$	$I(r_s) = \left\| h_1(r_s) \right\|^2 \left[\left\| h_2(r_s) \right\|^2 \otimes D(r_s) \right]$

section of the 3-D CTF in confocal reflection has a finite bandwidth, which gives the optical sectioning property and thus allows one to perform 3-D imaging. By contrast, there is no such a property in a transmission confocal microscope as the axial cross-section of the 3-D CTF is a delta function. A 4Pi confocal microscope (Section 9.3) virtually comprises the passbands for both reflection and transmission systems and

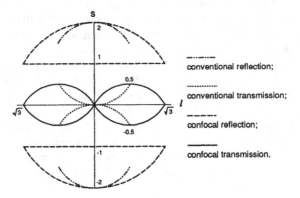

Fig. 10.6.1   Passband of the 3-D coherent transfer functions in conventional and confocal imaging of high-aperture lenses. The numerical aperture of the objective, $\sin\alpha_o$, is 0.866.

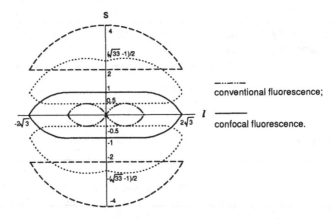

Fig. 10.6.2   Passband of the 3-D optical transfer functions in conventional and confocal imaging of high-aperture lenses. The numerical aperture of the objective, $\sin\alpha_o$, is 0.866. The regions bounded by the dashed and dotted curves are the extra passbands in a type-C 4Pi confocal single-photon fluorescence system.

therefore has a better imaging performance.

The cut-off spatial frequencies of the passband of the 3-D OTF in confocal fluorescence microscopy are twice as large as those in conventional fluorescence, as shown in Fig. 10.6.2. There is no missing cone of spatial frequencies in confocal imaging, while it exists in conventional fluorescence. Use of a 4Pi confocal fluorescence microscope (Section 9.3) extends the passband in the axial direction. This feature may prove to be a main advantage of 4Pi microscopy and provides high axial resolution. Another method, proposed recently,[10.6] for improving axial resolution in confocal fluorescence imaging is called confocal theta microscopy. In this case, the optical axis of the collector lens is placed at an angle θ ($0 < \theta < \pi/2$) with respect to the axis of the objective. As we have learned in Chapters 3, 4 and 5, the 3-D effective PSF for a confocal microscope is the product of two 3-D PSFs for the objective and collector lenses, if a point source and a point detector are used. It can be easily demonstrated that this conclusion also holds in theta microscopy. As a result, the longer axial extension of the 3-D PSF for the objective (Fig. 2.1.3b) can be reduced by the shorter extension of the 3-D PSF for the collector if $0 < \theta < \pi/2$. Of course, the maximum effect occurs when $\theta = \pi/2$. The transfer functions for theta microscopy can be explored in future in terms of the methods described in this book.

References

10.1. C. J. R Sheppard and M. Gu, *J. Microscopy*, **165**, (1992) 377.

10.2. C. J. R Sheppard and M. Gu, 3-D transfer functions in confocal scanning microscopy, in *Visualization in Biomedical Microscopies*, ed. A. Kriete (VCH, Weinheim, 1992), pp. 251-282.

10.3 H. Zhou, M. Gu, and C. J. R. Sheppard, *Optik*, **97** (1994) 94.

10.4 H. Zhou, M. Gu, and C. J. R. Sheppard, *J. Modern Optics*, **42** (1995) 627.

10.5. C. J. R Sheppard, D. K. Hamilton, and I. J. Cox, *Proc. Roy. Soc. Lond. A*, **387** (1983) 171.

10.6 E. H. K Stelzer and S. Lindek, *Opt. Commun.*, **111** (1994) 536.

Appendix 1

HANKEL TRANSFORM

This appendix gives a brief description of the Fourier transform as well as the Hankel transform.

When the three-dimensional (3-D) point spread function for an imaging system has space-invariance, the image of a thick object can be expressed as the 3-D Fourier transform of the corresponding 3-D transfer function multiplied by the 3-D spatial spectrum of the object (Chapter 2). A 3-D Fourier transform can be resolved into a product of a two-dimensional (2-D) Fourier transform in the transverse plane and a one-dimensional (1-D) Fourier transform with respect to the axial direction. The former can be further reduced to a special form called the Hankel transform for a circularly symmetric system. Let us start with the 2-D Fourier transform in Cartesian coordinates, which is given by

$$F(m,n) = \iint_{-\infty}^{\infty} f(x,y) \exp[\pm 2\pi i(xm + yn)] dx dy, \qquad (A.1.1)$$

where $F(m, n)$ is the Fourier transform of the function $f(x, y)$. In this book, we use the following convention for the definition of the direct and inverse Fourier transforms. When the sign in the exponent in Eq. (A.1.1) is positive, Eq. (A.1.1) is called the direct Fourier transform. If the negative sign is chosen, it is called the inverse Fourier transform. In the definition of Eq. (A.1.1), a pre-constant factor has been omitted as it does not affect imaging properties.

A Hankel transform is the 2-D Fourier transform in a polar coordinate. The function $f(x, y)$ can be represented by a function $f(r, \phi)$ if we introduce the following coordinate transformation:

$$x = r\cos\phi,$$

$$y = r\sin\phi,$$
(A.1.2)

and

$$m = l\cos\theta,$$

$$n = l\sin\theta.$$
(A.1.3)

Here r and ϕ are the polar coordinates in the x - y plane, whereas l and θ are the polar coordinates in the m - n plane. Thus Eq. (A.1.1) becomes

$$F(l,\theta) = \int_0^{2\pi}\int_0^{\infty} f(r,\phi)\exp[\pm 2\pi irl\cos(\phi - \theta)]rdrd\phi.$$
(A.1.4)

If $f(r, \phi) = f(r)$, i. e., for a circularly symmetric function, then Eq. (A.1.4) can reduce to

$$F(l) = \int_0^{\infty} f(r)J_0(2\pi rl)2\pi rdr,$$
(A.1.5)

which is called the Hankel transform. Here

$$J_0(x) = \frac{1}{2\pi}\int_0^{2\pi}\exp(ix\cos\phi)d\phi$$
(A.1.6)

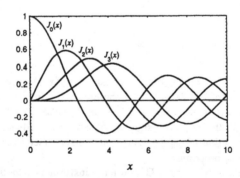

Fig. A.1.1   Bessel functions of the first kind for the first four orders, $J_0(x)$, $J_1(x)$, $J_2(x)$ and $J_3(x)$.

is a Bessel function of the first kind of order zero. If $f(r)$ is circularly uniform within a radius a:

$$f(r) = \begin{cases} 1 & , \quad r \le a, \\ \\ 0 & , \quad r > a, \end{cases}$$

(A.1.7)

thus

$$F(l) = \pi a^2 \left[\frac{2J_1(2\pi al)}{2\pi al} \right],$$

(A.1.8)

where $J_1(x)$ is a Bessel function of the first kind of order unity. Fig. A.1.1 gives the Bessel functions of the first kind for the first four orders. The function $2J_1(x)/x$ is called the Airy function and is shown in Fig. A.1.2.

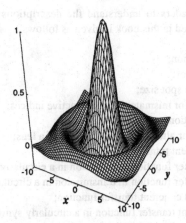

Fig. A.1.2   2-D behaviour of the Airy function $2J_1(x)/x$.

Appendix 2

LIST OF SYMBOLS

In order to allow readers to understand the descriptions more easily, a list of symbols which are often used in this book is given as follows.

a: radius of the pupil function;

a_j: modal amplitude in fibres;

A, A_1, A_2: normalized fibre spot size;

A_r: aberration coefficient for mismatching of refractive indices;

b: radius of the pupil function;

B_t: aberration coefficient for alteration of tube length of a lens;

$c(m, n, s)$, $c(m)$: 3-D coherent transfer function;

$c_r(l, s)$: 3-D coherent transfer function in reflection in a circularly symmetric system;

$c_t(l, s)$: 3-D coherent transfer function in transmission in a circularly symmetric system;

$c_2(m, n, z_s)$: 2-D defocused coherent transfer function;

$c_2(l)$: 2-D in-focus coherent transfer function in a circularly symmetric system;

$c_1(s, x_s, y_s)$: 1-D line coherent transfer function;

$c_1(s)$: 1-D on-axis coherent transfer function;

$C(m, n, s), C(m)$: 3-D optical transfer function;

$C(l, s)$: 3-D optical transfer function in a circularly symmetric system;

$C_2(m, n, z_s)$: 2-D defocused optical transfer function;

$C_2(l)$: 2-D in-focus optical transfer function in a circularly symmetric system;

$C_1(s, x_s, y_s)$: 1-D line optical transfer function;

$C_1(s)$: 1-D on-axis optical transfer function;

$C(m, n, s; m', n', s'), C(m; m')$: 3-D transmission cross-coefficient;

$C(l, s; l', s')$: 3-D transmission cross-coefficient in a circularly symmetric system;

$C_2(m, n; m', n', z_s)$: 2-D defocused transmission cross-coefficient;

$C_2(s; s', x_s, y_s)$: 1-D line transmission cross-coefficient;

$C_w(m, n, s)$, $C_w(m)$: 3-D weak-object transfer function;

$C_w(l, s)$: 3-D weak-object transfer function in a circularly symmetric system;

$C_{w2}(l)$: 2-D in-focus weak-object transfer function in a circularly symmetric system;

$C_{w1}(s)$: 1-D on-axial weak-object transfer function;

$D(r), D(r), D(x, y)$: sensitivity of a detector;

$\overline{D}_0$: thickness of a lens;

$\mathrm{erf}(x)$: error function;

$E(x)$: complete elliptic function;

$E(x, y)$: incomplete elliptic function;

f, f_0: focal length of a lens;

f_1, f_2: mode profiles of single-mode fibres;

F_1, F_2, F_3: 1-D, 2-D and 3-D Fourier transforms;

g: ratio of the fluorescence wavelength to the incident wavelength;

g': ratio of the incident wavelength to the fluorescence wavelength;

$h(x, y, z)$, $h_1(x, y, z)$, $h_2(x, y, z)$, $h_a(x, y, z)$: 3-D amplitude point spread functions;

$h_i(x, y, z)$: 3-D intensity point spread function;

$I(x_s, y_s, z_s)$, $I(r_s)$: detected intensity from a scan point (x_s, y_s, z_s) of a thick object;

$I(x_s, y_s)$: detected intensity for the extended-focus technique or for 2-D in-focus imaging;

j: integer number;

$J_0(x)$: Bessel function of the first kind of order zero;

$J_1(x)$: Bessel function of the first kind of order one;

k: wave vector;

k: modulus of the wave vector;

K, K_r, K_t, K_f: constant of normalization;

L_f: fibre length;

l: radial spatial frequency;

m: spatial frequency vector;

m, n, s: three components of the spatial frequency along x, y, and z directions;

M_1, M_2: magnification matrix of lenses;

$\tilde{n}, \tilde{n}_0$: refractive index of lenses;

N: Fresnel number;

$o(x), o''(z)$: 1-D object function;

$o'(x, y)$: 2-D object function;

$o(x, y, z), o(r)$: 3-D object function;

$O(m, n, s), O(m)$: 3-D Fourier transform of the object function $o(x, y, z)$;

$O'(m, n)$: 2-D Fourier transform of the object function $o'(x, y)$;

$O''(s)$: 1-D Fourier transform of the object function $o''(z)$;

$P(x, y), P_1(x, y), P_2(x, y), P(r)$: pupil functions or apodization functions of a lens;

$\tilde{q}_j$: mode profiles in fibres;

r_s, r: position vector of a scan point, position vector;

r: radial coordinate in a 2-D plane;

r_d: radius for a circular detector;

$r_0:, r_{01}, r_{02}$: spot size for single-mode fibres;

r_s: radius for a circular source;

$\sin\alpha_o$: numerical aperture of the objective in object space;

$\sin\alpha_i$: numerical aperture of the objective in image space;

$\sin\alpha_d$: numerical aperture of the objective in detector space;

s_0: constant spatial frequency offset for a single lens;

$s0$: constant spatial frequency offset for confocal reflection microscopy;

$S(x, y)$: amplitude or intensity of a source;

t: time;

$t(x, y)$: transmittance of a thin lens;

u, v: axial and radial optical coordinates;

$U(x, y), U_0(x, y), U_1(x, y), U_2,(x, y) U_3(x, y)$: fields of light;

$U_0(t)$: temporal field of a light pulse;

v_d: normalized detector radius;

v_s: normalized source radius;

$V_0(\omega)$: amplitude spectrum;

V_f: fibre parameter;

α: maximum angle of convergence of rays of a lens;

β: normalized frequency for a pulsed illumination;

$\tilde{\beta}_j$: propagation constant in fibres;

γ: gradient of the image intensity of a thick layer;

γ', γ'': gradient of the image intensity of thin and thick edge objects;

$\tilde{\gamma}$: degree of coherence;

$\varepsilon, \varepsilon_1, \varepsilon_2$: the normalized radii of the central obstruction of annular lenses

$\delta(x), \delta(x, y), \delta(r)$: 1-D, 2-D and 3-D Dirac delta-functions;

$\Delta\tau$: temporal width of a light pulse (FWHM);

$\Delta\Omega$: spectral width of a light pulse (FWHM);

λ, λ_0: wavelength of the incident light;

λ_f: wavelength of the fluorescence light;

ρ: normalized radius over the lens aperture;

ρ_c: core radius of a single mode fibre;

ω, ω_0: angular frequencies of the light;

$\otimes, \otimes_2, \otimes_3$: 1-D, 2-D and 3-D convolution operations.

Appendix 3

ABBREVIATION

1-D, 2-D, 3-D: one-, two- and three-dimensional;
1-p, 2-p: single- and two-photon;
APSF: amplitude point spread function;
CARS: coherent anti-Stokes Raman spectroscopy;
CTF: coherent transfer function;
CSM: confocal scanning microscope;
CW: continuous wave;
fs: femtosecond (1 fs = 10^{-15} s);
FOCSIM: fibre-optical confocal scanning interference microscope;
FOCSM: fibre-optical confocal scanning microscope;
FWHM: full width at half maximum;
GRIN: graded index;
HWHM: half width at half maximum;
IPSF: intensity point spread function;
N. A.: numerical aperture;
OTF: optical transfer function;
PSF: point spread function;
TCC: transmission cross-coefficient;
UV: ultraviolet;
WOTF: weak-object transfer function.

SUBJECT INDEX